T0271917

# Remediation of Legacy Hazardous and Nuclear Industrial Sites

*Remediation of Legacy Hazardous and Nuclear Industrial Sites* provides an overview of the key elements involved in remediating complex waste sites using the Hanford nuclear site as a case study. Hanford is one of the most complex waste sites in the world and has examples of most, if not all, characteristics of the complex waste sites that exist globally. This book is aimed at a non-technical audience and describes the stages of remediation based on general RCRA/CERCLA processes, from establishing a strategy that includes all stakeholders to site assessment, waste treatment and disposal, and long-term monitoring.

**Features:**

- Informs a non-technical audience of the important elements involved in complex waste site remediation
- Employs the Hanford Site as a case study throughout to explain real-world applications of remediation steps
- Connects the "human" element to the technical aspects through interviews with key current and retired individuals at the Hanford Site
- Includes discussion of stakeholders and the engagement process in remediation
- Demonstrates how all elements of complex waste site remediation from demolition of buildings to groundwater management are interrelated
- Focuses on broader technical and sociopolitical challenges for remediation of a contaminated site

Aimed at a broad audience, this book offers approachable guidance to technical and non-technical readers through a series of real-world examples that cover each important step in the complex waste cleanup process.

**Dr. Stuart T. Arm** is a senior technical advisor who returned to Pacific Northwest National Laboratory (PNNL) in 2019 after 12 years working in industry. Dr. Arm holds a PhD and MEng in chemical engineering from Imperial College, London, and has 30 years' experience in national laboratories and the nuclear industry in the United Kingdom and the United States. He establishes PNNL and DOE strategic plans and objectives for radiochemical process flowsheets and technologies for treating and managing radioactive waste and commercial used nuclear fuel while acknowledging emerging national and international trends. Dr. Arm employs a collaborative and leading approach to technology and flowsheet maturation and integration to establish implementation strategies with multitechnical and organizational elements.

**Dr. Hilary P. Emerson**'s research focuses on the fate and remediation of radionuclides with experimental testing from the laboratory to the field scale. She holds a BS in Environmental Engineering (2009) from the University of Central Florida and a PhD in Environmental Engineering and Earth Science (2014) from Clemson University. Since joining Pacific Northwest National Laboratory in 2019, her work has focused primarily on the development of new remediation technologies for radionuclides and techniques for monitoring remediation technologies.

# Remediation of Legacy Hazardous and Nuclear Industrial Sites

## Perspectives from Hanford

Edited by
Stuart T. Arm and Hilary P. Emerson

CRC Press is an imprint of the
Taylor & Francis Group, an **informa** business

Designed cover image: Andrea Starr, Pacific Northwest National Laboratory

First edition published in 2025
by CRC Press
2385 NW Executive Center Drive, Suite 320, Boca Raton FL 33431

and by CRC Press
4 Park Square, Milton Park, Abingdon, Oxon, OX14 4RN

*CRC Press is an imprint of Taylor & Francis Group, LLC*



ISBN: 978-1-032-35672-3 (hbk)
ISBN: 978-1-032-35889-5 (pbk)
ISBN: 978-1-003-32921-3 (ebk)

DOI: 10.1201/9781003329213

Typeset in Times
by codeMantra

# Contents

## PART III Remedy Screening, Evaluation, and Testing

*Christian D. Johnson, Katherine A. Muller, and Hilary P. Emerson*



## PART IV Remedial Action

*Christian D. Johnson, Katherine A. Muller, and Hilary P. Emerson*

## *PART V Long-Term Stewardship and Future Land Use*
*Vicky L. Freedman and Nicolas J. Huerta*

# Preface

Cleaning up chemical and radioactive contaminants at complex waste sites is a multifaceted, multigenerational challenge. Such sites are defined by the diversity and complexity of the radioactive and hazardous materials contaminating the subsurface environment or stored in a variety of legacy facilities. This book uses the Hanford Site, where World War II and Cold War-era plutonium production has led to one of the largest environmental remediation efforts in history, as a source of case studies examining the strategies and technologies facilitating each step of the remediation process. Here, the steps are divided up by (i) strategy formulation and stakeholder engagement; (ii) site assessment and characterization; (iii) remedy screening, evaluation, and testing; (iv) remedial action; and (v) long-term stewardship. The case studies address subsurface soil and groundwater cleanup; legacy waste remediation; and decommissioning and demolition of legacy facilities. While there are no easy answers, this book offers readers a window into how we historically tackled some of the most urgent and continue to solve some of the most difficult cleanup problems faced by humankind. Potential readers include those interested in the social, regulatory, and engineering dimensions of chemical and nuclear site remediation – including, we hope, the next generation of professionals and students who will take up similar challenges around the globe.

# Acknowledgments

First and foremost, we are indebted to the champions of this book, Tom Brouns and David Peeler, who provided resources and spent countless hours providing feedback to the team. In addition, this book would not have been possible without the careful attention to detail from Andrew Pitman and his dream team of technical editors and graphic designers (Jan Haigh, Melanie Hess-Robinson, Isaiah Steinke, and Megan Bowen) who significantly improved our vision.

The Pacific Northwest National Laboratory is a multi-program national laboratory operated by Battelle Memorial Institute for the Department of Energy under Contract DE-AC05-76RL01830.

# List of Contributors

**Stuart T. Arm**
Pacific Northwest National Laboratory
Richland, Washington

**R. Matthew Asmussen**
Pacific Northwest National Laboratory
Richland, Washington

**Courtney L. H. Bottenus**
Pacific Northwest National Laboratory
Richland, Washington

**Carolyne Burns**
Pacific Northwest National Laboratory
Richland, Washington

**Emily Campbell**
Pacific Northwest National Laboratory
Richland, Washington

**Calvin H. Delegard**
TradeWind Services, LLC
Richland, Washington

**Derek Dixon**
Pacific Northwest National Laboratory
Richland, Washington

**Hilary P. Emerson**
Pacific Northwest National Laboratory
Richland, Washington

**Matthew Fountain**
Pacific Northwest National Laboratory
Richland, Washington

**Vicky L. Freedman**
Sealaska Technical Services
Richland, Washington

**Christopher Grant**
Pacific Northwest National Laboratory
Richland, Washington

**Andrea M. Hopkins**
Washington River Protection Solutions, LLC
Richland, Washington

**Nicolas J. Huerta**
Pacific Northwest National Laboratory
Richland, Washington

**Christian D. Johnson**
Pacific Northwest National Laboratory
Richland, Washington

**Jennifer A. Kadinger**
Washington River Protection Solutions, LLC
Richland, Washington

**Naveen Karri**
Pacific Northwest National Laboratory
Richland, Washington

**Ellen Prendergast Kennedy**
Formerly Pacific Northwest National Laboratory
Richland, Washington

**Amanda R. Lawter**
Pacific Northwest National Laboratory
Richland, Washington

**Adam R. Mangel**
Formerly Pacific Northwest National Laboratory
Richland, Washington

**Jose Marcial**
Pacific Northwest National Laboratory
Richland, Washington

**Katherine A. Muller**
Pacific Northwest National Laboratory
Richland, Washington

**Carolyn I. Pearce**
Pacific Northwest National Laboratory
Richland, Washington

**Andrew T. Pitman**
Pacific Northwest National Laboratory
Richland, Washington

**Nikolla P. Qafoku**
Pacific Northwest National Laboratory
Richland, Washington

**Sarah Saslow**
Pacific Northwest National Laboratory
Richland, Washington

**Michelle M.V. Snyder**
Pacific Northwest National Laboratory
Richland, Washington

**Christopher E. Strickland**
Pacific Northwest National Laboratory
Richland, Washington

**Theodore J. Venetz**
TradeWind Services, LLC
Richland, Washington

**Beric Wells**
Pacific Northwest National Laboratory
Richland, Washington

# Introduction

*Stuart T. Arm, Hilary P. Emerson, and Andrew T. Pitman*

New materials, chemicals, and technologies often interact with the natural environment in ways that are difficult to understand or anticipate. Moreover, historically, the environment was rarely a primary concern of industry or government. During the period of rapid industrial development in the first half of the 20th century, these factors led to large-scale releases of substances that were harmful to people, animals, and the environment. Scientists like Rachel Carson, who wrote *Silent Spring* in 1962, helped shape an environmental movement that changed the way policymakers and the American public think about how humans interact with the environment and shaped our current environmental regulatory system. But the problems of the past still linger: legacy industrial sites are significant liabilities to succeeding generations. This book is meant to introduce people of all professions to the multigenerational technical, management, and social challenges of cleaning up or remediating these sites for beneficial use. Some of the most challenging legacy industrial sites are those associated with nuclear weapons production during World War II and the subsequent Cold War. The focus of this book is on one of the largest and most complex such examples, the United States' Hanford Site, formerly known as the Hanford Engineer Works or the Hanford Nuclear Reservation.

The Hanford Site and the perspectives of the workers and stakeholders who seek to remediate it illustrate the multidimensional character of environmental remediation of complex waste sites. Complex waste sites are those where "remediation progress is uncertain and remediation is not anticipated to achieve closure or even long-term management within the next few decades" (ITRC, 2017). The Hanford Site poses significant technical and non-technical challenges to remediation:

- Technical challenges include cleanup of radioactive and chemical contaminants in soil and groundwater plumes covering >150 $km^2$ (DOE-RL, 2022) and pose a threat to the adjacent Columbia River, and retrieval and treatment of 56 million gallons of solid and liquid mixed chemical and radioactive waste from legacy storage tanks. Notably, regulators may require site-dependent objectives for cleanup of complex sites like this due to the myriad of factors to be considered including future land uses.
- Social challenges require intentional communication with the public. Those leading the cleanup effort must be held accountable to and be transparent with the myriad of stakeholders for the site, including city, state, Tribal, and federal governments and their agencies, as well as the local community and special interest groups.
- Management challenges primarily stem from the scope and timeframe of the project: thousands of workers, hundreds of square miles, decades of project timeline, and multiple Site contractors. For fiscal year 2022, Hanford Site cleanup received $2.6 billion in funding (GAO, 2022). The Site is managed by the U.S. Department of Energy – a federal agency – and operated by four contractors, each responsible for a portion of the mission for a defined contractual period of about a decade. Active cleanup operations are projected to continue through the latter half of this century (DOE-RL, 2021; DOE-ORP, 2020), with long-term monitoring beyond that period.

The purpose of this book is not to present a history or comprehensive technical description of the Hanford Site but to use specific cleanup examples or treatment processes to illustrate different steps of complex site remediation. Nonetheless, some background is warranted given the Site's significance in this book's subject matter. The Hanford Site, established in 1943 during World War II as part of the United States' Manhattan Project, is located in the southeast of Washington State in the

United States. The Site was chosen because it was ideally located, far from major population centers but with rail access to transport workers and materials and the Columbia River to provide cooling water for nuclear reactors. In addition, the relatively flat topography and arid environment were prime for development of an industrial site. The Hanford Site served as the center for production of the radioisotope plutonium-239 during World War II and accounted for approximately 70% of the nation's Cold War plutonium stockpile. Although the production process grew more sophisticated and efficient over time, the essence of the approach stayed the same: nuclear reactors produced plutonium-239 within metallic uranium fuel rods (also fabricated at the site); the fuel rods were dissolved and processed to separate the plutonium from uranium, fission products, and other actinides; and the plutonium finished into oxide, metal, and, from 1949 until 1965, weapon parts. These operations produced large amounts of radioactive and chemical liquid and solid waste; they also contaminated equipment, buildings, soil, and groundwater with various hazards including radioactive elements, heavy metals, organic solvents, and other inorganic salts and compounds. The Site was originally a 640-square-mile area, which was later reduced to its current size of 580 square miles by shedding buffer areas that were considered necessary during the Manhattan Project because of unknown radiological hazards and security concerns.

Nuclear material production at Hanford ceased in the early 1990s with the end of the Cold War. The Site's mission has now transitioned to environmental remediation. Initially, Hanford's remediation mission was largely associated with characterizing the environmental issues and establishing the technical, social, and management frameworks to mitigate them. Today, the mission is focused on implementing remediation within those frameworks with adjustments as needed to respond to new information.

Previous publications have described the history of the Hanford Site and technical operations in detail (Gephart, 2003, 2010). The recognition that much has been accomplished in remediating Hanford since Gephart's seminal work in 2003, as well as maturation of the remediation framework, contributed much to the motivation for this book. Moreover, there are in-depth historical and technical descriptions of specific cleanup activities and treatment processes of the Hanford Site:

- Plutonium processing at the Hanford Site (Delegard et al., 2019; Gerber, 1997)
- Tank Waste Treatment and Stabilization (Peterson et al., 2018; Stewart et al., 1998; Wilmarth et al., 2011)
- Strategies for cleanup of the Hanford Site (Demirkanli and Freedman, 2021; DOE-RL, 2015; Gerber, 2008; Triplett et al., 2010)

While the Hanford Site presents many technical, social, and management challenges, it has also impacted the lives of those professionally associated with it. Families from across the United States moved to the area to establish and develop the Site through the latter half of the last century. Since the remediation mission was initiated, families have also arrived from across the world. A generation of workers has been trained to characterize the environmental issues and establish the remediation strategy at Hanford. A new generation of workers is now largely responsible for implementing remediation and carrying it forward toward completion. With that in mind, most of the Hanford perspectives provided in this book were intentionally prepared by early-career engineers and scientists involved in environmental cleanup, with advice from those in the later parts of their careers. Remediation of complex waste sites is a multigenerational mission, and there is a burden on each generation to nurture and educate the next while positively acknowledging the generational differences in communication styles and methods.

The work at the Hanford Site presents many perspectives on the process of remediating a complex waste site, and this book is organized according to those steps. Each section of this book covers one remediation step and opens with an introduction to the step and its general process before presenting Hanford case studies. The steps to remediating a waste site encompassed by this book include the following:

1. Establishing the remediation approach, including stakeholder engagement approaches
2. Assessing and characterizing the site, including soil, groundwater, infrastructure, and process waste
3. Testing technical approaches to remediation of the site
4. Implementation of the remedial actions, including treatment and disposal of stabilized waste
5. Long-term stewardship of the remediated site

These steps are discussed in the conventional order of implementation, although some may be repeated or revisited at later stages as new information on the waste site is uncovered.

Finally, the co-editors and authors acknowledge the significant contributions of all those who have gone before them in laying the groundwork for remediating the Hanford Site and from whom the baton has been passed.

## REFERENCES


Delegard, C., Emerson, H., Cantrell, K., & Pearce, C. 2019. *Generation and characteristics of plutonium and americium contaminated soils underlying waste sites at Hanford*. PNNL-29203. Pacific Northwest National Laboratory, Richland, WA.

Demirkanli, D. I., and Freedman, V. L. 2021. *Adaptive Site Management Strategies for the Hanford Central Plateau Groundwater. PNNL-32055 Rev. 0*. Pacific Northwest National Laboratory, Richland, WA.

DOE-ORP. 2020. *River Protection Project System Plan. ORP-11242 Rev. 9*. U.S. Department of Energy, Office of River Protection, Richland, WA.

DOE-RL. 2015. *Supplement Analysis of the Hanford Comprehensive Land-Use Plan Environmental Impact Statement*. DOE/EIS-0222-SA-02. U.S. Department of Energy, Office of Richland Operations, Richland, WA.

DOE-RL. 2021. *The 2022 Hanford Lifecycle Scope, Schedule, and Cost Report*. DOE/RL-2021-47 Rev. 0. U.S. Department of Energy, Richland Operations Office, Richland, WA. https://www.hanford.gov/files.cfm/2022_LCR_DOE-RL-2021-47_12-27.pdf

DOE-RL. 2022. *Hanford Site groundwater monitoring report for 2022*. DOE-RL-2022-40 Rev. 0. U.S. Department of Energy, Richland Operations Office, Richland, WA. https://www.hanford.gov/page.cfm/SoilGroundwaterAnnualReports

Gephart, R. E. 2003. *Hanford: A Conversation about Nuclear Waste and Cleanup*. Battelle Press. Columbus, OH.

Gephart, R. E. 2010. A short history of waste managemet at the Hanford Site. *Physics and Chemistry of the Earth*, 35, 298–306. https://doi.org/10.1016/j.pce.2010.03.032

Gerber, M. S. 1997. *History and stabilization of the Plutonium Finishing Plant (PFP) Complex, Hanford Site. HNF-EP-0924*. U.S. Department of Energy, Environmental Management, Washington, D.C.

Gerber, M. S. 2008. *Final Frontier at Hanford: Tackling the Central Plateau. HNF-36881 Rev 0.0*. U.S. Department of Energy. Richland, WA.

GAO. 2022. *Hanford Site Cleanup. GAO-22-105809*. U.S. Government Accountability Office, Washington, D.C. https://www.gao.gov/assets/gao-22-105809.pdf

ITRC. 2017. *Remediation Management of Complex Sites*. RMCS-1, Interstate Technology and Regulatory Council, Washington, D.C. https://rmcs-1.itrcweb.org or https://rmcs-1.itrcweb.org/RMCS-Full-PDF.pdf

Peterson, R. A., Buck, E. C., Chun, J., Daniel, R. C., Herting, D. L., Ilton, E. S., Lumetta, G. J., and Clark, S. B. 2018. Review of the scientific understanding of radioactive waste at the US DOE Hanford Site. *Environmental Science & Technology*, 52(2), 381–396.

Stewart, T. L., Frey, J. A., Geiser, D. W., and Manke, K. L. (1998). Overview of US radioactive tank problem. In *Science and Technology for Disposal of Radioactive Tank Wastes* (pp. 3–13). Springer.

Wilmarth, W. R., Lumetta, G. J., Johnson, M. E., Poirier, M. R., Thompson, M. C., Suggs, P. C., and Machara, N. P. (2011). Waste-pretreatment technologies for remediation of legacy defense nuclear wastes. *Solvent Extraction and Ion Exchange*, 29(1), 1–48.

Triplett, M. B., Freshley, M. D., Truex, M. J., Wellman, D. M., Gerdes, K. D., Charboneau, B. L., Morse, J. G., Lober, R. W., & Chronister, G. B. 2010. *Integrated strategy to address Hanford's deep vadose zone remediation challenges*. ASME 13th International Conference on Environmental Remediation and Radioactive Waste Management, Tsukuba, Japan.

# Part I

# Site Assessment and Characterization

## Stuart T. Arm and Hilary P. Emerson

This section introduces remediation strategy concepts, including the general steps of the environmental remediation process and additional factors to consider with a complex site. The Hanford Site in the United States is used as an example of a complex waste site to illustrate the characteristics and how the environmental remediation process has been implemented. Chapter 1 introduces the general steps of the environmental remediation process and the Hanford Site as a case study for remediation of complex sites. Chapter 2 presents the primary principles of stakeholder engagement and then describes the history and evolution of stakeholder and tribal engagement at the Hanford Site.

### I.1 INTRODUCTION TO ENVIRONMENTAL REMEDIATION

What is environmental remediation? For those not involved in the field, the multiple aspects that define environmental remediation may not be obvious. At its core, environmental remediation is cleanup of pollutants that have been or could be released to the environment (i.e., substances that are no longer in a state of being stored or used for productive purposes), with the purpose of protecting human health and the environment.

Environmental remediation can encompass three broad categories: facilities cleanup (decontamination and decommissioning), waste management, and remediation of contaminated environmental media.

Facilities may require cleaning up if contamination occurred from routine operations performed in the building (e.g., radionuclide contamination of hot cells or mercury contamination of nuclear processing facilities) or if the building itself was made from harmful materials (e.g., asbestos as part of building materials). Waste processing consists of aboveground treatment of waste material that came from historic operations. The waste material is often stored in tanks until it can be treated further. For example, the Hanford Site has many tanks that store previously produced waste from nuclear separation processes related to plutonium production. Another example under the waste processing category would be treatment or disposal of ion exchange resin or granular activated carbon materials as secondary waste from processes that were used to clean up wastewater, groundwater, or vapors.

The broadest category for environmental remediation, and perhaps the first one that comes to mind, is pollutant contamination of environmental media, which includes soil and groundwater,

DOI: 10.1201/9781003329213-1

surface water and sediment, and the atmosphere. Soil and groundwater spans from surface soils through the unsaturated soil and into saturated soil and groundwater aquifers. Surface waters include rivers, lakes, oceans, and the sediments at the floor of those waterbodies. Atmospheric pollution can occur in the air located from ground surface up to higher elevations. All environmental media can interact; groundwater connects to surface water, atmospheric pollution comes down with rain and snow precipitation onto soils and surface waters, and dust can take contamination from the soil into the atmosphere. Thus, depending on the pollutant and the site setting, contamination may transfer between environmental media.

## I.2 CONTAMINANTS

A contaminant (or pollutant) is a physical, biological, chemical, or radiological substance that has been, or potentially could be, released to the environment and is potentially harmful to human health or the environment. While the term "contaminant" can refer to physical or biological substances, that usage is typically outside the scope of environmental remediation because it pertains more to ecological (e.g., invasive species), public health (e.g., communicable diseases), national security (e.g., biological weapons), or food safety (e.g., machine parts or salmonella in food) topics. Contaminant releases may be planned, such as part of disposal or routine operations, or the releases may be unplanned, such as spills or leaks. Contamination may also be encountered when pollutants become incorporated into living organisms, though that is typically addressed as an impact of pollution (i.e., exposure assessment).

In the context of environmental remediation, there are different categories of contaminants including but not limited to metals (e.g., arsenic, mercury, chromium), radionuclides (e.g., uranium, iodine-129), inorganic compounds (e.g., nitrate), and organic compounds (e.g., carbon tetrachloride, benzene). The organic compounds category is often discussed in terms of subcategories of chemicals such as volatiles, chlorinated solvents, hydrocarbons, pesticides, or polychlorinated biphenyls (PCBs). Additionally, there are other substances, such as per- and polyfluoroalkyl substances (PFAS), 1,4-dioxane, pharmaceuticals, and microplastics that are drawing more attention and research of late as potential contaminants. Further, contamination may be present in different forms and phases, including dissolved, non-aqueous phase liquids (NAPL), vapors/gases, solid phases, and particulates.

Remediation of the air, soil, sediment, groundwater, or surface water may be needed to decrease contaminant exposure and bring the risk down to an acceptable level. Risk is the danger contaminants pose for human or other environmental receptors (i.e., ecological receptors such as animals and plant life). An exposure route is the way a chemical enters an organism, which includes the following mechanisms:

- Inhalation (breathing)
- Ingestion (eating or drinking) of water, soils, contaminated food, etc.
- Absorption (skin or eye contact)
- Radiation exposure (ionizing radiation penetrating the body)
- Injection (impact, cut, or puncture to the skin)

Risk occurs when a receptor can be exposed to the contaminant, and an exposure pathway is the route by which a contaminant is transported from a site (contaminated environmental media, stored waste, or a facility) to a receptor. An exposure assessment involves identifying potential receptors and pathways for contamination to reach receptors. There are numerous potential direct or indirect pathways that can connect contaminants to receptors, including but not limited to vapor exposure, contact with contaminated soil, or through the food chain (Figure I.1).

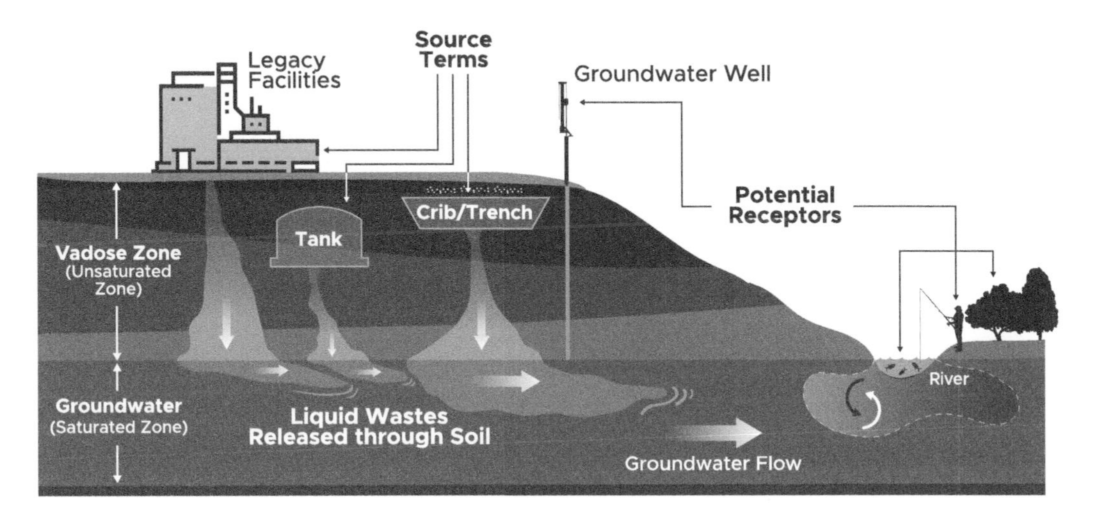


**FIGURE I.1** Potential exposure pathways. Reprinted with courtesy from Melanie Hess-Robinson (ITRC 2017).

The primary objective of environmental remediation is to reduce toxicity, mobility, volume, mass, or concentration of contaminants to decrease the actual or potential risk to human health and the environment and protect groundwater and other environmental resources. These are key elements for determining an environmental remediation strategy.

## I.3 REMEDIATION TECHNOLOGIES

Development of a remediation strategy needs to consider multiple factors and approaches, and often, a multi-pronged solution is adopted. Remediation technologies, engineered controls, and administrative or legal controls can all play a role in the overall strategy. Remediation technologies are used to reduce the toxicity, mobility, volume, mass, or concentration of contaminants directly. Engineered controls (e.g., strategically placed fans or shielding, or an alternative water supply) and administrative controls (e.g., work practices/scheduling, site access controls, building or excavation permits, well drilling prohibitions, easements, and covenants) are implemented to decrease exposure and reduce risk to human health and the environment. An appropriate remedial strategy adopted will depend on the contaminant(s) involved, the site setting, the exposure pathways, receptor locations, and the remedial action objectives (RAOs).

Remediation technologies can be grouped as either ex situ or in situ technologies. Ex situ technologies are aboveground contaminant treatment processes, with the advantage of easy access for construction of well-engineered systems. In situ remediation technologies are applied in the subsurface for treatment of soil/sediment or groundwater, which comes with access constraints for implementation and monitoring. Remediation technologies can also be broadly categorized based on the principle of the treatment: containment, physical, chemical, biological, or thermal processes (Table I.1). Technology types are identified within each category, though many technologies are general and have multiple variations and nuances with respect to the implementation and targeted contaminant(s). Technologies can be deployed as a single technology (e.g., in situ bioremediation), combined technologies (e.g., emulsified vegetable oil with zero-valent iron), or a treatment train of multiple technologies (e.g., a sequence of ion exchange plus biological treatment) to improve treatment effectiveness and/or to address multiple contaminant types, media types, or contaminant media phases. The Federal Remediation Technologies Roundtable (FRTR; https://www.frtr.gov/matrix/default.cfm) and EPA CLU-IN (https://clu-in.org/remediation/) offer resources for additional information on remediation technologies.

**TABLE I.1**
**Non-Exhaustive List of Remediation Technology Examples for both In Situ and Ex Situ Remediation**

| Treatment Category | Technology Class |
|---|---|
| Containment | • Grout wall<br>• Artificial ground freezing<br>• Sheet pile<br>• Hydraulic control<br>• Caps/covers |
| Physical | • Excavation/removal<br>• Encapsulation/solidification<br>• Soil vapor extraction<br>• Pump-and-treat<br>• Dust control |
| Chemical | • Sequestration/adsorption<br>• Ion exchange<br>• Precipitation<br>• Chemical oxidation<br>• Chemical reduction<br>• Hydrolysis<br>• Advanced oxidation processes |
| Biological | • In situ bioremediation<br>• Biosorption<br>• Phytoremediation<br>• Biological water treatment<br>• Soil composting<br>• Bioventing |
| Thermal | • Thermal desorption<br>• In situ vitrification<br>• Melter/glass forming<br>• Electrical resistivity heating<br>• In situ steam injection<br>• In situ smoldering |

## REFERENCE

ITRC. 2017. *Remediation Management of Complex Sites.* Interstate Technology and Regulatory Council, Washington, D.C. https://rmcs-1.itrcweb.org/

# 1 Developing a Remediation Framework

*Katherine A. Muller, Courtney L. H. Bottenus, and Christian D. Johnson*

This chapter introduces remediation strategy concepts, including the general steps of the environmental remediation process and additional factors to consider with a complex site, including waste treatment and disposal, as well as cleanup of subsurface contamination from historical waste release. The Hanford Site in the United States is used as an example of a complex waste site to illustrate the characteristics and how the environmental remediation process has been implemented.

## 1.1 ENVIRONMENTAL REMEDIATION PROCESS

Remediation to address contaminated environmental media, waste processing, and facility cleanup is undertaken following a general process to go from initial characterization to remedy to closure, with iteration and refinement along the way. A variety of factors need to be considered for any site to determine the appropriate remedy to meet remedial action objectives (RAOs) in a reasonable time and a manner acceptable to stakeholders. Complex sites, in particular, have challenges that require suitable technical solutions and sound, but flexible planning. These elements of the environmental remediation process are discussed in the following subsections.

### 1.1.1 General Site Remediation Approach

The regulatory framework provides guidance and requirements for the steps of environmental remediation, including planning, documentation, and approvals. Although the details and terminology differ between regulatory programs, a similar set of steps are followed for environmental remediation. For example, Comprehension Environmental Response, Compensation, and Liability Act (CERCLA) uses the term Feasibility Study (FS), while Resource Conservation and Recovery Act (RCRA) uses the term Corrective Measures Study (CMS), but both activities involve evaluation of potential remedies. The conceptual steps for environmental remediation (Figure 1.1) are described in this section in general terms to avoid program-specific terminology. Depending on the program and whether the remediation is for environmental media, waste treatment, or facility remediation, some steps may be less relevant than others.

Some elements of the environmental remediation process will span all the steps discussed in Section 1.1.1.1. Specifically, engagement and communication with regulatory agencies, the community, and Tribal Nations (if applicable) need to take place throughout the environmental remediation process. Stakeholder perspectives and their engagement are discussed in detail in the next chapter "Stakeholder Perspectives, Environmental Remediation, and the Hanford Site." However, part of that engagement should involve developing a vision for the site upon completion of active remediation (i.e., an end-state vision), which is a key factor guiding the environmental remediation process. For the DOE Office of Environmental Management, an end-state vision means determination of a final risk-informed remedy decision that is:

DOI: 10.1201/9781003329213-2

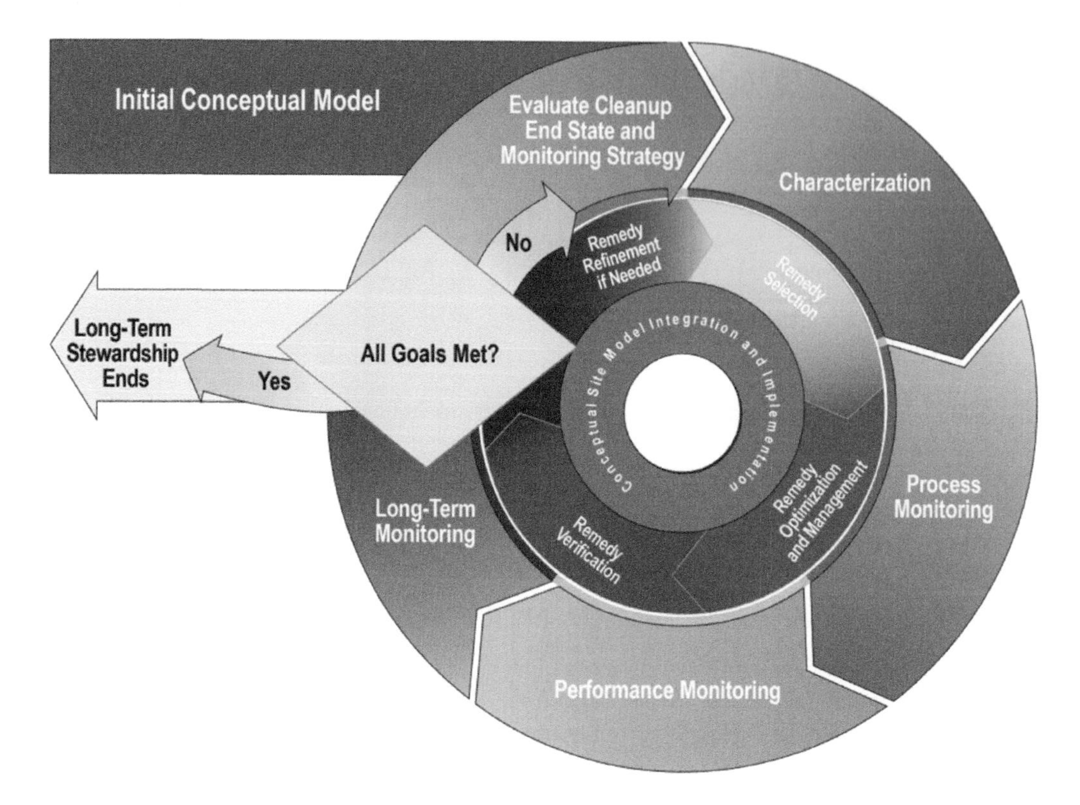


**FIGURE 1.1** Graphical depiction of the generalized remediation approach. Reprinted with permission from Bunn et al. (2012).

1. Protective of human health and the environment for current and future anticipated land use,
2. Inclusive of appropriate regulatory, community, Tribal Nation, and stakeholder acceptance,
3. Respectful of equity, environmental justice, climate resilience, time, and budget factors,
4. Sustainable (minimized operations, maintenance, labor, long-term monitoring, cost, potential future migration, risk, etc.) over the management life cycle, and
5. Ready for transfer to the appropriate entity for long-term stewardship or beneficial use.

The steps of environmental remediation shown in Figure 1.1 are described in Section 1.1.1.1.

### 1.1.1.1 Characterization and Remedy Selection

#### *1.1.1.1.1 Discovery/Preliminary Site Assessment*

Upon discovery, a waste site is initially inspected, and a preliminary assessment is conducted to determine the extent and nature of the pollutant release, likely contaminant migration pathways, and potential receptors. Essentially, this phase is intended to use (typically) limited site information to determine the magnitude and immediacy of potential adverse impacts on human health and/or the environment or, alternatively, conclude that the site does not require further action. This step also involves information gathering about the cause of the contaminant release(s) and those responsible for subsequent remediation.

#### *1.1.1.1.2 Emergency Response*

If a contaminant release is determined to have time-critical potential adverse impacts on human health and/or the environment, then an emergency response is enacted to mitigate the immediate impacts. An example would be the derailment of a train car carrying a hazardous chemical. In this example, release of the chemical may require emergency response such as evacuation of people in the area to avoid inhalation hazards or containment/capture efforts to protect a nearby stream from chemical exposure.

##### *1.1.1.1.3 Planning and Site Characterization*

After discovery and the initial assessment, site characterization is conducted to gather information about the context of the contaminant release, including the pollutants involved, the extent and magnitude of contamination, the nature of the subsurface, the contaminant source (i.e., the nature of stored waste and the storage containers), and/or the nature of facility contamination and building configuration. The characterization information is used to develop a conceptual site model (CSM), which describes the relevant processes, features, and mechanisms that affect contaminant migration and exposure pathways in the context of the site setting (e.g., geology, groundwater flow, surface water interactions, prevailing wind direction). A CSM is a critical decision-making tool used under many regulatory programs. It supports the risk assessments and the development of appropriate cleanup approaches.

Characterization information and the CSM combine to provide a basis to evaluate potential remedies and then down-select and implement a remedy. While characterization is performed early in the process, later steps may identify data gaps that require additional characterization. Characterization is an exploratory information gathering step, distinguished from the process, performance, or compliance monitoring discussed in later stages of the environmental remediation process.

##### *1.1.1.1.4 Remedy Screening*

The remedy screening step builds from the characterization step and CSM to compile a list of remediation technologies that could potentially be used for site remediation. Resources such as the Federal Remediation Technology Roundtable (FRTR) provide a great starting point with technology descriptions, examples, and references. A wide range of technologies have been, and are still being, developed since the 1980s by DOE national laboratories (https://www.energy.gov/national-laboratories), Department of Defense environmental programs (e.g., through Strategic Environmental Research and Development Program [SERDP], Environmental Security Technology Certification Program [ESTCP], Naval Facilities Engineering Systems Command [NAVFAC], Air Force Civil Engineer Center [AFCEC], and U.S. Army Corps of Engineers [USACE]), academia, and the environmental industry. The remedy screening step takes the long list of technologies and initially filters out those technologies clearly not suited for the contaminant types, environmental media, and/or the location of the contaminants within the subsurface. For example, excavation, removal, and disposal are not applicable for treatment of contaminants located very deep in the subsurface. Similarly, soil vapor extraction would not apply to nonvolatile metal contaminants. Technologies remaining on the list after the screening are ranked with respect to suitability for a remedy. At this point, the remaining remedial action alternatives, along with a no-action alternative, undergo a detailed analysis to identify the most effective option that best satisfies the statutory mandates.

##### *1.1.1.1.5 Remedial Alternatives Evaluation*

After potential remediation technologies are identified and ranked in the remedy screening process, a more detailed examination of "remedial alternatives" is conducted for the purpose of comparison and selection of a preferred remedy. The process considers technology effectiveness, implementability, and relative cost. Effectiveness pertains to how well the technology treats the contaminants to meet RAOs given the contamination context (concentrations, distribution, migration, media, etc.). Implementability relates to the maturity of the technology, availability of materials, administrative constraints, and technical/engineering challenges to putting the remedy into action. Multiple technologies may be considered together, either in the form of a treatment train (e.g., to address multiple contaminant types or secondary waste streams) or as different treatments at different locations to address different zones (e.g., the source area vs. dilute groundwater contamination). Preliminary designs are created to better evaluate the achievement of RAOs, short- and long-term effectiveness, implementability, and cost of the remedial alternatives. Throughout this step, any existing data gaps and uncertainties are identified, which may necessitate additional characterization or technology treatability testing. The process should consider risk-informed remedy selection and related decision-making (e.g., National Research Council, 2014).

##### *1.1.1.1.6 Remedy Testing/Demonstration*

Where there are uncertainties about a treatment technology's potential effectiveness, treatability testing or technology demonstrations (in the laboratory and/or at field scale) may be used to gather additional data and experience to reduce those uncertainties. Such work can be important for innovative technologies or for demonstrating the effectiveness of a technology under site-specific conditions. Testing results provide additional performance, implementability, and cost information for a remedial alternative and support remedy selection decisions.

##### *1.1.1.1.7 Interim Remedial Action*

Depending on the site, the extent of contamination, and proximity to potential human or environmental receptors, it may be determined that a remediation technology should be applied as an interim action before a final remedy is selected and approved. The objective of an interim action is to have an expedited response to protect human health and the environment, while characterization, testing, and remedy selection activities may be ongoing. Thus, an interim remedial action may take place during any phase of the environmental remediation process prior to implementation of the final selected remedy. An emergency response is distinguished from an interim remedial action as having a higher degree of urgency in implementation to protect human health and the environment.

##### *1.1.1.1.8 Remedy Selection Decision*

A remedy selection decision is used to document the preferred remedial alternative, including request for public comment, and requiring regulatory agency approval. Documents such as a Statement of Basis (SB), Record of Decision (ROD), or a decision document (DD) are examples of types of remedy selection decision documents used under different regulatory programs. Typically, the remedy selection decision document will define overall cleanup objectives/RAOs, the remediation approach(es), and required outcomes. The document should also include a summary of the nature and extent of contamination, associated risks and/or hazards, the rationale for selecting the remedy, and the criteria for determining a remedy's success. While remedy selection decision documents in the past have been prescriptive, a more prudent practice (to the extent possible) is to include flexibility for remedy assessment and optimization/modification through the application of an adaptive site management approach (discussed in Section 1.1.4).

### 1.1.1.2 Implementation and Process Monitoring

#### *1.1.1.2.1 Remedy Design, Implementation, and Operation*

After the remedy selection decision is approved, a detailed design of the remedy system or implementation work plan is prepared. When the remedial design and work plan is completed and approved, remedy implementation can proceed. The remediation design will include all the elements needed for treating contaminants, including secondary waste disposition, monitoring plans (for process, performance, and compliance monitoring), and institutional controls. After implementation, remedy operation and maintenance (O&M), monitoring, evaluation, and reporting will be ongoing.

### 1.1.1.3 Performance Monitoring

#### *1.1.1.3.1 Performance Assessment*

Remedy performance assessment is an important part of ongoing remedy operations. Strategic performance assessment information from both remediation system monitoring and site monitoring helps verify the expected remedy performance and can provide additional information to update the CSM. As additional information is collected during remedy operations, the updated CSM and performance information may lead to the need for additional characterization or remedy optimization. Guidance on performance assessment (e.g., Truex et al., 2013, 2015, 2017; Johnson et al., 2022) can provide a structured approach for gathering information, updating the CSM, evaluating performance and impacts, and determining appropriate next steps, whether that is continued operation, optimization, transition to another technology, or remedy closure.

##### *1.1.1.3.2 Remedy Optimization*

As remedy operations proceed, additional information is gained, with which the CSM can be updated, and performance assessed. Remedy operations data and/or the outcome from a performance assessment may point to considering remedy optimization to better achieve the objectives and goals, in terms of increasing remedy performance, decreasing the remedy timeframe, and/or reducing costs.

##### *1.1.1.3.3 Remedy Revision*

Revisions to the remedy may be recommended because of a remedy performance assessment and/or remedy process optimization activities. Typically, such revisions need to be documented. For example, CERCLA specifies that significant differences from the remedy defined in the decision document need to be documented in an Explanation of Significant Differences (ESD), whereas a fundamental change to the remedy is documented in a ROD Amendment (EPA, 1999). The adaptive site management approach described in Section 1.1.4 builds in flexibility for a remedy to adapt and evolve as new information is obtained (e.g., from characterization, remedy operations, or performance assessments).

#### 1.1.1.4 Long-term Monitoring

##### *1.1.1.4.1 Long-Term Stewardship*

When a site has met RAOs, but the cleanup goals do not allow for unrestricted use, or when a site has finished the active remediation phase and has entered a passive remediation and monitoring phase, then the site is ready for long-term stewardship (LTS). LTS involves continued monitoring (e.g., Bunn et al., 2012; SRNL, 2023) to confirm or assess contaminant stability, remedy stability, and site maintenance, and to verify the long-term protectiveness of the remedy (documented in periodic reports to the regulatory agency). LTS may also involve continued implementation and management of institutional controls (including land use controls).

##### *1.1.1.4.2 Closure/Site Transition*

Site closure is predicated on meeting the RAOs, which must be demonstrated by compliance monitoring (and perhaps other information, depending on the RAOs), and receiving approval from the regulatory agency. Guidance documents (e.g., EPA, 1992, 2002, 2013, 2014a, 2014b; EPA et al., 2000) describe approaches to demonstrate compliance and provide confidence in the data and lines of evidence supporting remedy completion. If a site is ready for closure, the remedy will need to be demobilized or decommissioned, which may be done under a work plan describing those activities. With the completion of active management and monitoring at a site, the end-state vision will dictate how the site will transition property or facilities to the desired use (e.g., mercantile, industrial, recreational, agricultural, Tribal, or other uses).

### 1.1.2 Regulatory Framework

Environmental remediation is undertaken by the site owner or identified responsible party(ies), with oversight by appropriate regulatory agency(ies), and input from stakeholders. In the United States, multiple federal programs establish regulatory requirements for environmental remediation and the process to follow. Although this text focuses on the regulatory requirements followed within the United States, similar types of regulations are implemented by European, Asian, and international governing bodies/agencies. U.S. federal programs related to environmental remediation are driven by multiple congressional acts, including the following:

- **Comprehensive Environmental Response, Compensation, and Liability Act (CERCLA):** "Superfund" program for cleanup of uncontrolled or abandoned hazardous waste sites, as well as accidents, spills, and other emergency releases of pollutants
- **Resource Conservation and Recovery Act (RCRA):** For control of hazardous waste generation, transportation, treatment, storage, and disposal, and includes oversight of underground tanks
- **Toxic Substances Control Act (TSCA):** Remediation of PCB-contaminated sites
- **Atomic Energy Act (AEA):** Management of source material, special nuclear material, and byproduct material, including radionuclides present in environmental media, wastes, and facilities
- **Uranium Mill Tailings Radiation Control Act (UMTRCA):** For the stabilization, cleanup, and disposal of uranium mill waste

Department of Energy (DOE) orders, state regulatory programs, Tribal Nation programs, consent decrees, and other programs and mechanisms are established under these acts to regulate remediation of waste sites. The regulatory requirements that apply, and thus the agency with regulatory oversight, depend on the nature of the release (e.g., from active operations or legacy disposal), the site history (e.g., as a military base or Manhattan Project site), the waste material or contaminant (e.g., uranium mill tailings or PCBs), and the waste site location (on federal, state, Tribal, or private property). When multiple regulatory agencies and regulations are involved, there is typically an agreement designating the lead agency and the regulatory framework that governs the site's remediation. CERCLA and RCRA programs drive most environmental remediation frameworks. Using Hanford as an example, the historical waste sites from disposal activities resulting from plutonium separation work are regulated under CERCLA, the Hanford tanks are regulated under RCRA, and facility deactivation & decommissioning (D&D) activities are controlled under CERCLA. Furthermore, DOE retains authority under the AEA for management of the radionuclides present in environmental media, waste, or facilities and so implements its own regulations for cleanup of radioactive contamination.

A regulatory framework provides the structure for the environmental remediation process. The framework will guide how RAOs are determined, how remedies are evaluated, and how the performance of a remedy is assessed and potentially optimized over time. As part of the process, management areas are typically defined to facilitate the regulatory process and remedy implementation. A management area may be defined by one or more factors, including spatial location, the impacted material (e.g., groundwater, unsaturated soil, facilities), the type of contaminants, and a specific historical operation or release method. Depending on the size of the site, one or more waste sites (individual locations where a contaminant release occurred) may be grouped into a larger management area, such as an operable unit (OU), particularly if the contaminants are similar or the potential impacts are to the same media (e.g., groundwater). The CERCLA regulatory program defines an OU as a management area that consists of a discrete portion of a remedial response for the purpose of managing contaminant releases or migration or exposure pathways (40 CFR 307.14, 2017). OUs may address geographical portions of a site, specific site problems, and initial phases of a remedial action, or may consist of a set of actions over time or concurrent actions for different parts of a site. A site may be divided into multiple OUs, depending on the site complexity.

### 1.1.3 Complex Site Remediation Considerations

Both technical and nontechnical challenges can potentially impede steps of the site remediation approach. All three categories of cleanup (facility decommissioning and demolition, waste processing, and contaminated environmental media) face similar nontechnical challenges (Table 1.1). However, while facility decommissioning and demolition and waste processing certainly have complexities, such as treatment of widely varying compositions of waste at complex sites, these

**TABLE 1.1**
**Technical and Nontechnical Challenges Characteristic of a Complex Site**

| Technical Challenge | Examples |
|---|---|
| Geological | • Subsurface heterogeneities<br>• Preferential flow paths<br>• Low-permeability geological layers or zones<br>• Fractured bedrock<br>• Karst geology |
| Hydrological | • Variable groundwater velocities<br>• Fluctuating groundwater levels<br>• Groundwater-surface water interactions<br>• Deep groundwater contamination<br>• Perched groundwater |
| Geochemical | • Extreme geochemical conditions (e.g., atypically high or low pH, extreme redox conditions) |
| Contaminant conditions | • Nonaqueous phase liquids (LNAPL or DNAPL)<br>• Recalcitrant or emerging contaminants<br>• High contaminant concentrations<br>• Comingled plumes or multiple contaminants |
| Large-scale or spatial considerations | • Large spatial extent of contamination<br>• Very deep contamination in the subsurface<br>• Number, type, and proximity of receptors |
| **Nontechnical Challenge** | **Examples** |
| Site objectives | • Societal expectations and acceptance<br>• Changing site objectives |
| Managing over long timeframes | • Multiple remediation phases<br>• Future land use<br>• Multiple responsible parties or changes in site management<br>• Loss of institutional knowledge over time<br>• Litigation |
| Funding | • Changes in funding levels over time, including lack of funding<br>• Politics that alter funding/program priorities |
| Land use | • Obtaining site access<br>• Multiple owners<br>• Changing land use or water use |
| Overlapping regulatory responsibilities | • Managing local, state, federal, and Tribal cooperation, and regulations<br>• Changing regulations |
| Institutional controls | • Long-term management and enforcement<br>• Future land use restrictions<br>• Implementation of site access controls |

*Source:* After ITRC (2017).

issues tend to be technically different from subsurface soil and groundwater cleanup. For example, the complex issues associated with waste processing and D&D activities are addressed primarily through aboveground engineering solutions. In contrast, subsurface soil and groundwater cleanup results in a different set of challenges mainly due to limited access to the subsurface and typically the need to remediate large areas. Table 1.1 focuses on some of the specific challenges for remediating subsurface environmental media.

While remediation of any environmental media is frequently not a simple matter, some sites have technical and/or nontechnical challenges that put them in the category of a complex site. As defined by ITRC (2017), a "complex site" is one where the remediation progress is uncertain and achieving closure, or even long-term management, is not expected within a reasonable timeframe. What is a "reasonable" timeframe? Currently, a "reasonable timeframe" has no agreed-upon definition. CERCLA discusses site cleanup within "a time frame that is reasonable given the particular circumstances of the site" (40 CFR 300.430). ITRC (2017) discusses reasonable remediation timeframes and includes some anecdotal information from a survey and potential criteria for assessing a reasonable timeframe. The characteristics of a complex site can span geological heterogeneity, deep contamination, hydraulic variability, large scales, chemical/contaminant conditions (e.g., recalcitrant and/or comingled contaminants), managing remediation elements (active treatment, monitoring, institutional controls, funding, etc.) over long timeframes, and overlapping regulatory responsibilities, to name a few. Table 1.1 provides some examples of potential technical and nontechnical challenges related to soil and groundwater that can make remediation at a site complex.

Complex sites follow the general remediation steps outlined above but may need additional flexibility to incorporate new information into a revised CSM and to adapt the remedy in a timely fashion, versus being locked into a single remedy.

### 1.1.4 Adaptive Site Management

While traditionally the environmental remediation process is described as a linear one (though it can still involve periodic assessment and implementation of remedy optimization or adjustments), decades of experience have resulted in updated thinking and newer approaches that allow the steps in the process to be more dynamically iterative (Figure 1.2). The adaptive site management (ASM) approach is intended to achieve more effective treatment, make better use of available resources, and achieve progress toward site closure and desired end states in a reasonable timeframe without being constrained by a linear and less flexible process. ASM is intended to be an ongoing process of remedy (re)evaluation, optimization, or adjustment.

ASM is being increasingly utilized because it is well suited to help deal with the technical and nontechnical challenges that characterize complex sites, including significant uncertainty in remedy performance predictions (ITRC, 2017). Guidance on ASM for complex sites developed by the Interstate Technology and Regulatory Council (ITRC, 2017) provides the foundation for the ASM approach. This guidance was developed considering earlier work and recommendations by the National Research Council (2003) and EPA (2014a). The steps in ASM (Figure 1.2) include continual refinement of the CSM and site objectives, and deliberate planning, evaluation, and decision metrics for adapting the remedy as needed. The ASM approach overlays on the environmental remediation process, but the remedy decision is not rigidly fixed and can instead be adjusted, if needed, as additional characterization or performance data is collected. That is, the adaptive approach explicitly allows for revisiting the remediation process steps as new information becomes available (e.g., new characterization or remedy performance data) and helps identify and close any data gaps to facilitate understanding. The ASM approach also involves using information to support data-driven modeling for predicting future site behavior.

When cleanup timeframes are expected to be long (e.g., multiple decades), a long-term management plan (LTMP) may also be developed as part of the ASM approach. The LTMP would include consideration of key site complexities and uncertainties, performance metrics and predictions,

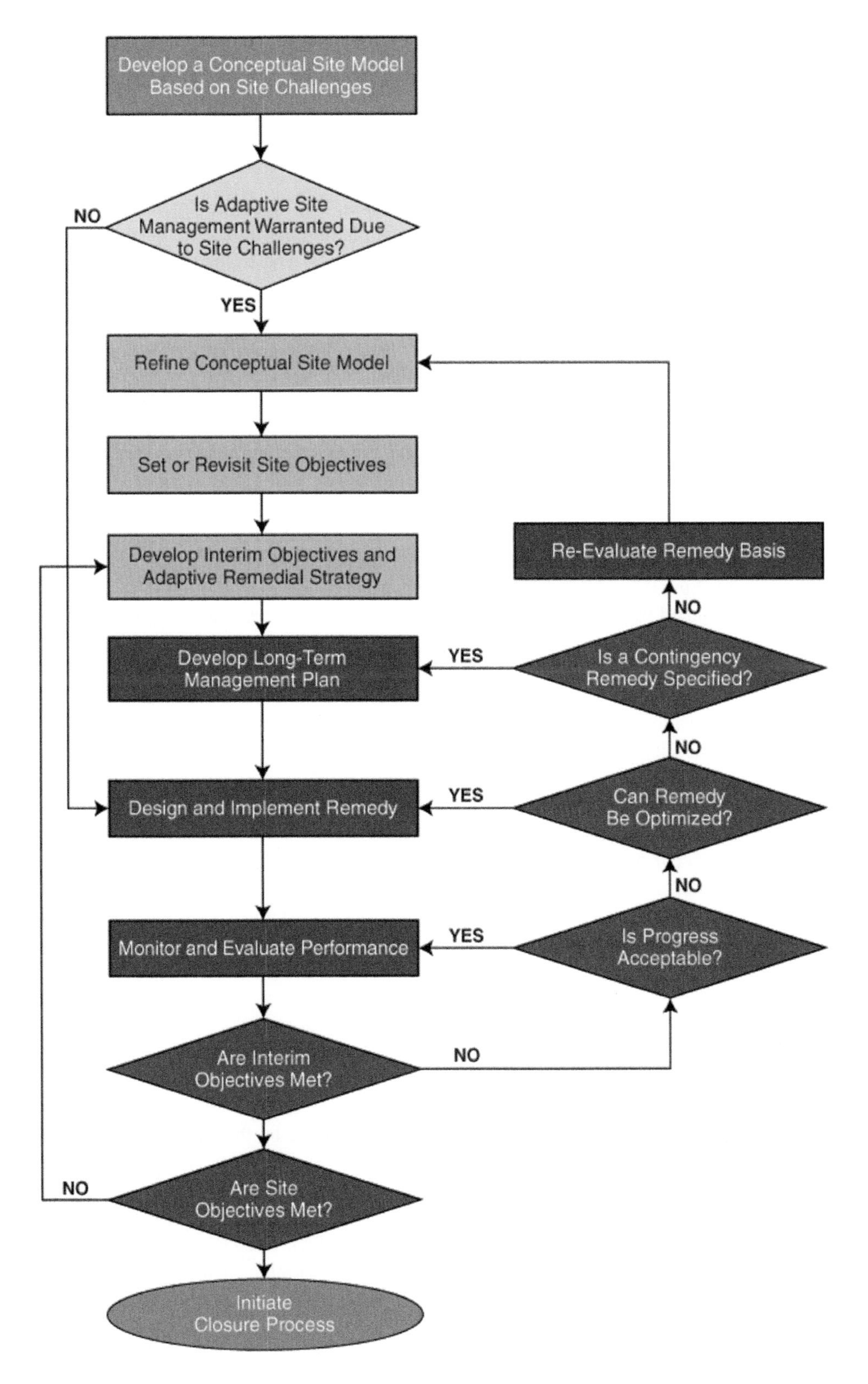


**FIGURE 1.2** Adaptive site management process. Adapted and Modified from ITRC (2017) by Demirkanli and Freedman (2021).

schedule and basis for periodic evaluations, and decision logic for remedy evaluation, optimization, modification, or transition. The LTMP may enable the necessary continuity in decision-making at a complex site (Demirkanli and Freedman, 2021).

## 1.2 CASE STUDY: THE HANFORD SITE

The Hanford Site was established in 1943 during World War II as part of the United States' Manhattan Project. It was the chosen location for initial plutonium production for nuclear weapons and generated nearly 70% of the nation's plutonium stockpile. These operations occurred across nine nuclear reactors located along the adjacent Columbia River for access to water for cooling and five chemical processing canyons (or plants) located in the center of the Site termed the Central Plateau (Figure 1.3). The nuclear reactors produced plutonium-239 within metallic uranium fuel rods (also fabricated at the Site); the fuel rods were dissolved and processed to separate the plutonium from uranium, fission products, and other actinides at the canyons; and the plutonium finished into oxide and metal forms for machining into weapons parts elsewhere. These operations produced large amounts of radioactive and chemical liquid and solid waste; they also contaminated equipment, buildings, soil, and groundwater with various hazards including radioactive elements, heavy metals, organic solvents, and other inorganic salts and compounds.

While many of the cleanup challenges encountered at the Hanford Site are also present at other complex sites, some challenges are more unique to the Hanford Site. For the Hanford Site, even the first step of establishing the regulatory framework was complicated. The Hanford Site contains both legacy wastes regulated by CERCLA and active waste storage and treatment regulated by RCRA. The approach taken at the Hanford Site was to integrate these regulatory programs into a framework through a federal facility compliance agreement. Another major challenge for the Hanford Site is its sheer magnitude in size. To address the diverse nature (e.g., differing geographical areas, contaminant types, contamination level, and media type) of a large site, sites are often divided into smaller sections or OUs. The subdivision into various OUs helps with the distinct planning, regulation, and cleanup activities required on a waste-site–specific level. Still, it is a complex task to determine how to split and define the site into specific OUs. The cleanup of each OU individually progresses based on regulatory decision documents for each OU. Once the OUs have been established, planning also needs to occur on a larger, more holistic scale to determine how efforts at the individual OUs fit together to form the larger cleanup approach. For a site that includes multiple OUs, the overall cleanup strategy sets the OU priorities, all while balancing budget, timeline, and general site goals. This chapter focuses on the complex cleanup approach and structure for the Hanford Site as opposed to individual OUs.

### 1.2.1 The Hanford Site Remedial Approach

#### 1.2.1.1 Regulatory Framework at the Hanford Site

In 1989, the Hanford Site mission transitioned from nuclear material production to waste management and environmental remediation. The evolution was marked by the signing of a comprehensive cleanup and regulatory compliance agreement known as the Tri-Party Agreement (TPA). The *Hanford Federal Facilities Agreement and Consent Order*, also known as the TPA, is an agreement between three government agencies: (i) U.S. DOE, (ii) Washington State Department of Ecology (Ecology), and (iii) U.S. Environmental Protection Agency (EPA). The TPA sets the framework for the Hanford Site cleanup by defining legal responsibilities, prioritizing cleanup actions, and setting enforceable milestones for cleanup compliance.

Further, the TPA also established a process to integrate CERCLA and RCRA requirements to achieve consistent outcomes. Generally, RCRA regulations cover ongoing hazardous and nonhazardous solid and liquid waste storage and remediation, whereas CERCLA handles the cleanup of

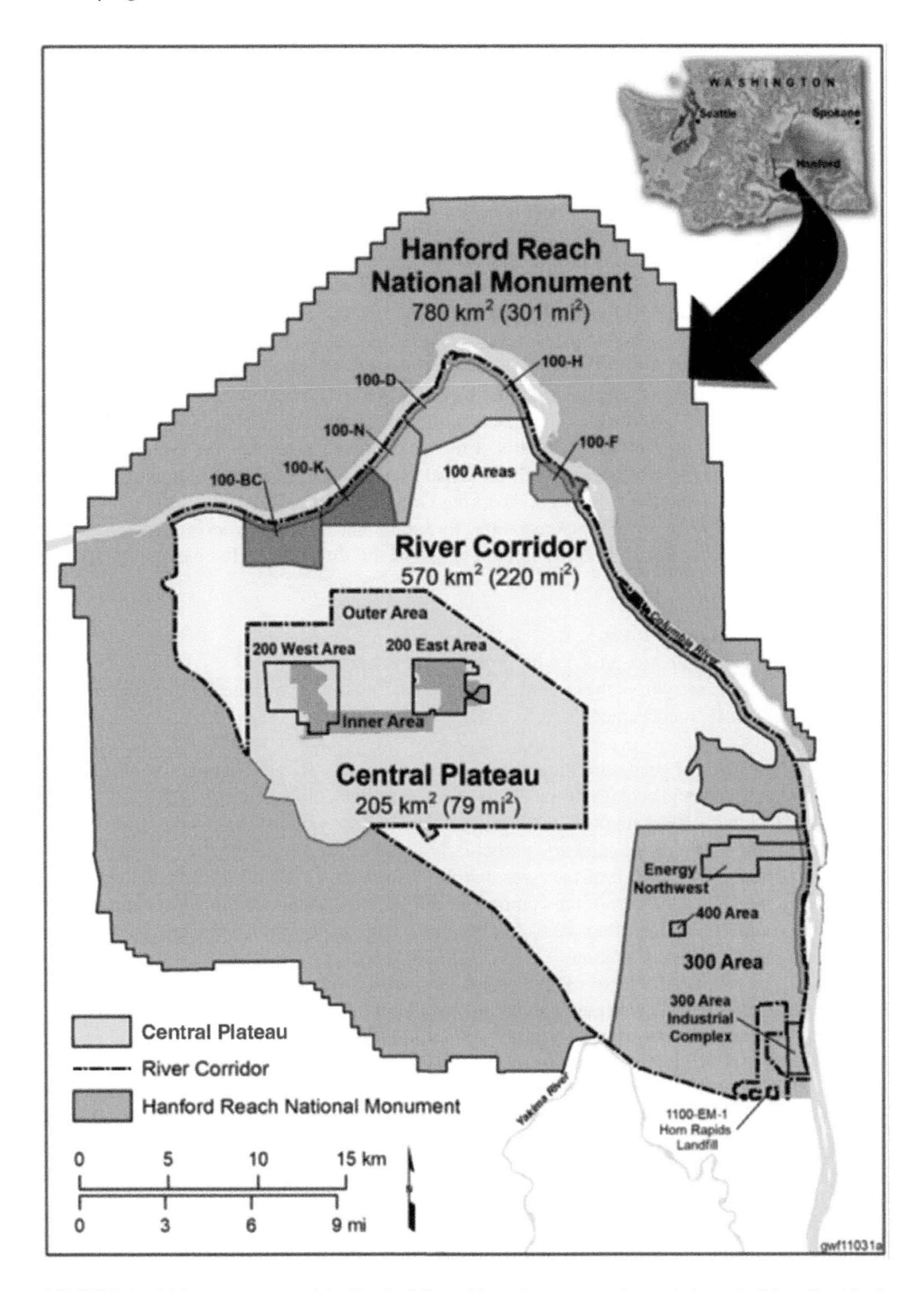


**FIGURE 1.3** Main components of the Hanford Site with nuclear reactors located along the River Corridor in the 100 Areas, chemical processing occurring in the 200 Areas in the Central Plateau, and other research and development occurring in the 300 Areas. Reprinted with permission from DOE/RL (2013).

historical contaminant releases. Because the Hanford Site contains both legacy waste sites and active waste treatment, as well as complex wastes containing both radiological and chemical contaminants, the CERCLA and RCRA programs were integrated into a single framework to implement the environmental regulations that apply to the Hanford Site. In terms of CERCLA requirements, the TPA has split the lead regulatory role for the OUs between EPA and Ecology. The EPA is ultimately responsible for issuing the RODs for remedy decisions.

EPA was initially responsible for implementing RCRA requirements, but Washington State later became a federally authorized hazardous waste program for RCRA corrective actions. The State-authorized corrective action program is carried out under RCW 70.105, "Hazardous Waste Management," known as the *Hazardous Waste Management Act of 1976*, and RCW 70.105D, "Hazardous Waste Cleanup – Model Toxics Control Act."

#### 1.2.1.2 Cleanup Goals at the Hanford Site and Stakeholder Involvement/Engagement

While the initial cleanup framework was set up by the TPA, the Hanford Future Site Uses Working Group (the Working Group) was formed in 1992 to provide input to the Tri-Party Agreement agencies on the potential future for different parts of the Hanford Site (Bergman and Weissberg, 1994). The input was gathered from parties such as Tribal Nations, the State of Washington, State of Oregon, stakeholders, and the public. The Working Group recognized that site cleanup and remediation goals needed to be driven by expectations for future land use. The Working Group provided nine recommendations, three of which framed much of the direction of the cleanup (Bergman, 2011):

1. Protect the Columbia River
2. Deal realistically and forcefully with groundwater contamination
3. Use the plateau away from the river (the center of the Site was termed the "Central Plateau") wisely for waste management

These recommendations recognize that the adjacent Columbia River (Figure 1.3) is a critical resource for the people and ecology of the Pacific Northwest, including the states of Washington and Oregon. The Columbia River supports a multitude of uses that are vital to the economic and environmental well-being of the region, and it is particularly important in sustaining the culture of Native Americans. The 50-mile stretch of the river that flows through the Hanford Site is known as the Hanford Reach and became part of the Hanford Reach National Monument in 2000. Groundwater beneath the Hanford Site flows into the Columbia River, and its protection and remediation are the intent of the second recommendation.

It was also recognized that large volumes of contaminated soil and building debris would be generated during remediation of waste sites and removal of legacy facilities. Consequently, the Working Group called for the center portion of the site, the Central Plateau, to be used for waste management, which led to the 1995 ROD to build the Environmental Restoration Disposal Facility (ERDF), a massive, lined landfill in the center of the Hanford Site. ERDF is regulated under CERCLA and began operating in 1995 and enabled cleanup to begin in earnest for contaminated sites adjacent to the Columbia River. Another waste disposal facility known as the Integrated Disposal Facility (IDF) has been constructed for disposal of low-activity waste primarily from the waste tanks that will be immobilized in glass and regulated under RCRA. Additional discussion on waste disposal is included in Chapter 5.

These recommendations continue to guide cleanup and have been refined into a series of goals for the Hanford Site (Table 1.2), which are articulated in the 2013 Hanford Site Cleanup Completion Framework document (DOE/RL, 2013). The Hanford Site cleanup goals continue to evolve since the first documents were released in 1999, including the "Final Hanford Comprehensive Land-Use Plan Environmental Impact Statement" (DOE/EIS-0222-F, 1999), which is re-visited every five

**TABLE 1.2**
**Hanford Site Cleanup Goals**

| | Description |
|---|---|
| Goal 1 | • To protect the Columbia River |
| Goal 2 | • To restore groundwater to its beneficial use to protect human health, the environment, and the Columbia River |
| Goal 3 | • To clean up the River Corridor waste sites and facilities to:<br>• Protect the groundwater and the Columbia River<br>• Shrink the active cleanup footprint to the Central Plateau<br>• Support anticipated future land uses |
| Goal 4 | • To clean up the Central Plateau waste sites and facilities to:<br>• Protect the groundwater and the Columbia River<br>• Minimize the footprint of areas requiring long-term waste management activities<br>• Support anticipated future land uses |
| Goal 5 | • To safely mitigate and remove the threat of Hanford's tank waste<br>• To safely store tank waste until it is retrieved for treatment<br>• To safely and effectively immobilize tank waste<br>• To close Tank Farms and mitigate the impacts from past releases of tank wastes to the ground |
| Goal 6 | • To safely manage and transfer legacy materials scheduled for off-site disposition including special nuclear material (including plutonium), spent nuclear fuel, transuranic waste, and immobilized high-level waste |
| Goal 7 | • To consolidate waste treatment, storage, and disposal operations on the Central Plateau |
| Goal 8 | • To develop and implement institutional controls and long-term stewardship activities that protect human health, the environment, and Hanford's unique cultural, historical, and ecological resources after cleanup activities are completed |

*Source:* Table recreated from DOE/RL (2013).

years as required by the 1999 ROD. The overarching goals of the Hanford Site cleanup are to protect the Columbia River, consolidate waste, provide long-term stewardship to protect human and environmental health, and protect Hanford's cultural, historical, and ecological resources.

These goals represent principles that guide all aspects of the Hanford Site cleanup while also setting priorities related to the application of resources and the sequence of cleanup efforts to achieve the greatest benefit.

#### 1.2.1.3 Cleanup Approach

To achieve the outlined goals, cleanup actions at the Hanford Site focused on three major components: the River Corridor, the Central Plateau, and tank waste within the Inner Area of the Central Plateau (Figure 1.3).

Geographically, the Central Plateau is often further split into the Inner and the Outer Areas. The Tank Farms, which hold the tank waste, are located within the Inner Area of the Central Plateau. Each of the three major components is further delineated into individual OUs, each with its own unique complexities, challenges, and compliance drivers. Table 1.3 provides a summary of some of the different Hanford Site areas, their primary role during nuclear material production, and some of the associated cleanup challenges.

Due to the multiple site goals, another challenge is how to manage the competing objectives within budget constraints. At the Hanford Site, a risk-based approach was used to identify and quantify the risks at various locations throughout the site (PNL-10651, 1995). Then, given the outlined

**TABLE 1.3**
**Overview of the Hanford Site and Its Contaminated Areas**

| Area | Principal Role in Plutonium Production Mission | Nature of Cleanup Challenges |
|---|---|---|
| | **River Corridor**[a] | |
| 100 Area | • Nine plutonium production reactors, spent fuel storage basins | • Reactor decommissioning (currently maintaining Interim Safe Storage conditions)<br>• Nuclear materials – spent fuel and sludge removal (close to river)<br>• Contaminated waste sites and burial grounds<br>• Contaminated groundwater discharging to the Columbia River (hexavalent chromium and $^{90}$Sr)<br>• Decommissioning and demolition of ancillary facilities |
| 300 Area | • Fuel fabrication facilities, research laboratories | • Contaminated waste sites and burial grounds<br>• Contaminated groundwater discharging to the Columbia River (uranium)<br>• Decommissioning and demolition of surplus facilities<br>• Nuclear materials – high-dose-rate fission product materials |
| | **Central Plateau** | |
| Inner Area | • Five chemical separation facilities<br>• 18 single-shell and double-shell Tank Farms (177 storage tanks)<br>• Liquid waste discharge sites<br>• New and legacy solid waste disposal facilities<br>• Plutonium fabrication facility<br>• Cesium and strontium capsule storage | • 56 million gallons of highly radioactive tank waste<br>• Contaminated waste sites and burial grounds<br>• Contaminated groundwater including tritium, $^{129}$I, $^{99}$Tc, chromium, carbon tetrachloride, and nitrate<br>• Decommissioning and demolition of surplus facilities<br>• Disposition of nuclear materials (e.g., capsules, sludge from spent fuel) |
| Outer Area | • Miscellaneous liquid and legacy solid waste disposal sites | • Contaminated waste sites and burial grounds<br>• Contaminated groundwater including tritium, $^{129}$I, $^{99}$Tc, chromium, carbon tetrachloride, and nitrate |

[a] Most cleanup actions in the River Corridor have been completed with the remaining focus on groundwater remediation and safe storage of decommissioned reactors' pending final disposition.

site goals, an integrated priority list was established to facilitate progress on multiple fronts. Using this risk-based approach, the cleanup order was established and executed. Generally, it was determined the best overall strategy was to address the contamination near the Columbia River (to protect the Columbia River, first and foremost) and move inward to the Central Plateau to shrink the overall active footprint of the site. Still, it is recognized that the Hanford Site will have a "final footprint" that will necessitate long-term management of the remaining waste, which is designated as the Inner Area.

A key strategy employed at the Hanford Site was the use of interim RODs to initiate soil and groundwater cleanup before the final ROD was issued. Interim RODs can allow a site to begin cleanup actions to address immediate threats to human and environmental health and prevent the migration and spread of contaminants further. Additionally, interim RODs allow cleanup to be initiated in one area, while allowing more time to investigate other elements (e.g., extent of groundwater plumes) further before issuing the final ROD and settling on a final cleanup decision. For instance, known source areas can be addressed and remediated, while additional characterization of groundwater contamination is underway. Interim RODs are subject to the same review requirements as final RODs (EPA, 1999); however, compliance with cleanup standards (e.g., Applicable Relevant and Appropriate Requirements, ARARs) may be waived until the final ROD is issued. This can allow a site more time to assess cleanup alternatives, develop new cleanup technologies, or finalize cleanup requirements with regulators and stakeholders.

Due to the scale and extended timeline required for cleanup at the Hanford Site, as the initial cleanup strategy was being developed, it was acknowledged that a level of uncertainty surrounds the cleanup process. Uncertainty exists in the extent and location of contamination, technological developments for remediation and deployment, and regarding what is an acceptable level of risk. Such uncertainties require the cleanup approach to be flexible and to evolve over time and as the cleanup progresses to reveal new information. For instance, initially, the main concerns were over the radioactive contaminants at the Hanford Site; however, the chemical contaminants ultimately also presented a formidable challenge.

#### *1.2.1.3.1 River Corridor*

DOE and stakeholders agree that protecting the Columbia River is a top priority as it is a critical environmental and economic resource for the region. Therefore, waste sites near the Columbia River were given top priority in terms of cleanup actions. The approximately 220-square-mile area adjacent to the Columbia River was designated the River Corridor. The Columbia River played an important role during nuclear material production as it was used as cooling water in the nuclear reactors, and thus, all nine reactors were built near the river. The River Corridor contained reactor operations (known as the 100 Areas), the former fuel fabrication area (known as the 300 Area), nine now retired reactors (B, C, D, DR, F, H, K-East, K-West, and N), and over 760 solid and liquid waste disposal areas. Both the reactor operations and fuel fabrication operations led to the start of the groundwater contamination, primarily with chromium and strontium. Seven of the nine retired reactors have been placed in Interim Safe Storage, or regularly referred to as "cocooned." Reactor cocooning involves putting the reactor in a configuration to safely allow for radiation levels to decay to levels low enough to where the reactor could then be dismantled. The cocooned reactors are checked every five years to verify no contamination is leaking from the sealed core. Additionally, there were over 1,000 structures that needed to be removed, dismantled, and disposed of throughout the River Corridor.

The main risk driver in the River Corridor relates to the fact that the groundwater discharges into the Columbia River. The goals of remedial action throughout the River Corridor are to restore groundwater to drinking water standards (when practical) and to comply with ambient water quality standards for groundwater that discharges directly into the Columbia River. Thus, to reduce risk, the River Corridor cleanup work first focused on removing potential sources of contamination within this area and moving them to a disposal facility (i.e., ERDF) in the Central Plateau. One of the major remedial challenges associated with the River Corridor was the removal, treatment, and disposal of the radioactive sludge that remained in the K Basins as a product of spent fuel storage. The spent fuel has been removed and safely transferred to the Central Plateau (DOE/RL, 2013), while the sludge was likewise removed to the plateau in 2019 (Rush et al., 2020). Even after removal of the waste sources, contaminants in the underlying groundwater still need to be addressed to prevent harm to the Columbia River, achieve the appropriate contaminant groundwater levels, and support future land use goals. Plumes of hexavalent chromium, strontium-90, and uranium are currently being monitored, remediated, and/or hydraulically controlled (e.g., via pump-and-treat systems and permeable reactive barriers) and will be into the future.

The federal government is planning to retain ownership of most of the areas within the River Corridor after cleanup efforts are completed. The 300 Area has been proposed to be designated for industrial use in the future, which is consistent with the future land use vision for the Hanford Site (DOE/EIS-022-SA-02 Rev.0, 2015). To achieve the anticipated future land use, the River Corridor groundwater contamination will be remediated to drinking water standards and ecological standards for water that discharges into the Columbia River.

#### *1.2.1.3.2 Central Plateau*

The Central Plateau is a 79-square-mile area that sits at the center of the Hanford Site and contains former nuclear fuel processing and waste disposal facilities. From a cleanup standpoint, the Central Plateau has been split into two components: the Inner and Outer Areas. The Inner Area contains the major nuclear fuel processing facilities, waste management facilities, and disposal facilities. These facilities include the Plutonium Finishing Plant (PFP) (demolition completed in 2020 of the PFP Main Processing Facility), T Plant, B Plant, Reduction-Oxidation Plant, U Plant, Plutonium Uranium Extraction Plant, ERDF, Tank Farms (includes 149 single-shell tanks [SSTs] and 28 double-shell tanks [DSTs]) containing the waste from historic production operations, and the Waste Treatment and Immobilization Plant. The Outer Area is the portion of the Central Plateau outside the boundary of the Inner Area. Primarily, these two areas are delineated based on the Inner Area containing areas previously and currently used for waste management. It is within the Inner Area, where the most complex and significant contamination exists, and thus, the vision for the Hanford Site is to reduce the footprint down to this area for long-term waste management.

The Central Plateau has housed waste processing and management activities since 1945 (Bergman, 2011). Historically, the irradiated fuel generated from the reactors was taken to the Central Plateau for processing and recovery of plutonium and uranium via chemical separation. The separation processes generated large volumes of mixed radiological and chemical wastes, which included a wide range of contaminants such as various solvents (e.g., carbon tetrachloride), fission products (e.g., cesium-137, strontium-90, and technetium-99), actinides (e.g., americium-241), and chemicals used in the separation processes (e.g., nitric acid). Although the higher activity radiological waste from production was separated and stored in tanks for future processing, cleanup of the Central Plateau is complex because it holds numerous waste disposal sites, active treatment facilities, and areas of deep soil contamination caused by both planned and unplanned waste releases that present an ongoing threat to groundwater. For instance, between 1946 and 1958, several million gallons of tank waste was directly discharged to liquid disposal units on the Central Plateau to accommodate tank storage space for newly generated reprocessing waste (Anderson, 1990). Beyond the planned releases, there were also unplanned releases due to spills, leaks, and overflow events, resulting in uncertainty on the amount, type, and location of contamination throughout the soil and groundwater (Simpson et al., 2006). Since 1996, waste materials and contaminated media from remedial actions throughout the Hanford Site have been brought to the Central Plateau for disposal in ERDF. ERDF is an engineered landfill designed specifically for the disposal of low-level radioactive, hazardous, and mixed wastes generated during cleanup activities. This centralized waste disposal facility provides a huge benefit in terms of streamlining waste disposal efforts and logistics and is regulated under the CERCLA program. Currently, ERDF is comprised of ten large disposal cells with a waste capacity of 21 million tons. Over 20 million tons of cleanup waste has been disposed of in ERDF, and the facility has been designed to accommodate waste into the future (Environmental Restoration Disposal Facility, 2011).

Elements of the Central Plateau cleanup strategy involve addressing both the Inner and Outer Areas, groundwater contamination, deep vadose zone contamination, and working toward overall footprint reduction (e.g., footprint reduction focused on the Outer Area). Since contaminated groundwater on the Central Plateau ultimately flows into the Columbia River, it is essential that soil and groundwater are treated. The Central Plateau groundwater has been split into four OUs (200-PO-1, 200-BP-5, 200-UP-1, and 200-ZP-1) to address the various plumes that contain both radiological and chemical constituents including tritium, $^{129}I$, uranium, technetium, chromium, trichloroethane, and nitrate. Groundwater remedies and treatment systems have been implemented in the Central Plateau through both interim actions and final RODs. Pump-and-treat systems to hydraulically contain and treat contaminant plumes were put into place as interim actions to treat the areas with the highest groundwater concentrations. The goal is to return the groundwater of the Central Plateau to drinking water standards, which is anticipated to be completed through a mixture of active and passive treatments including monitored natural attention. Further investigations conducted through a Remedial Investigation/Feasibility Study will provide the technical basis for the final ROD and remedy selection. Remaining contaminant source areas within the deep vadose zone of the Central Plateau also need to be addressed to protect the underlying groundwater (e.g., BC Cribs and Trenches and U Cribs).

In 2009, DOE updated the overall strategy for the Central Plateau to better achieve cleanup objectives (DOE/RL, 2009) by outlining a more consistent and comprehensive remedial decision-making approach. This new approach included a baseline risk assessment, ecological protection standards, defining soil cleanup levels needed to protect groundwater, creating a master list of contaminants of potential concern, developing standard and appropriate cleanup requirements, and defining a comprehensive decision logic for selecting remedial actions. The process also included anticipated future land use and direct contact exposure pathways associated with that land use when considering remedial decisions.

###### *1.2.1.3.3 Tank Waste*

Geographically, the Tank Farms sit within the Central Plateau; however, tank waste is defined as the third aspect of the Hanford Site cleanup framework because the Tank Farms and tank waste represent a significant, distinct, and complex challenge. Currently, the Hanford Tank Farms consist of 177 underground storage tanks, which hold approximately 56 million gallons of radioactive and hazardous waste. During plutonium production, the reprocessing liquid waste streams were stored in these tanks for future treatment. SSTs had a design life of 20 years based on the anticipated corrosion of the steel liner (DOE/RL, 1998), while the DSTs had design lives of as much as 50 years (Gephart, 2003; PNNL-13605, Rev. 4). Production wastes have been stored in these tanks for longer than the design life span, and some have mechanically failed and leaked waste into the surrounding environment. It is estimated that over one million gallons of waste has leaked from the SSTs (Rodgers, 2023; Agnew, 1997). The risk of further leaks from SSTs has been mitigated by pumping the liquid portion of the waste from them and into sound DSTs. The mechanical integrity of DSTs and the level of waste stored in them, as well as the SSTs, are continuously monitored for potential failure and further leakage.

In 1990, the U.S. Congress passed Public Law 101-510, Section 3137, creating a "Watch List" of concerning tanks at the Hanford Site, also known as *The Wyden Watch List*, in reference to the U.S. Senator Ron Wyden of Oregon who wrote the law. These tanks required specific safety precautions either because of potentially hazardous chemical reactions, uncontrolled heating of the waste from radioactive decay, or the accumulation of flammable gases to concentrations of concern. The waste issues that placed tanks on the lists were as follows: ferrocyanide content (ignitability), high tank temperatures (due to high radiological content), nuclear criticality (plutonium content leading to a nuclear chain reaction), organic materials (if exposed to high temperatures could cause explosive reaction with nitrate and nitrite), and flammable gas (large volumes of flammable gases could be released and ignite in vapor space). Initially, 54 of Hanford's 177 tanks were on the Watch List, and as many as 60 tanks made the list at one time or another. Since the establishment of the Watch List, actions such as waste transfers, waste characterizations, and further studies have closed each of the issues for all tanks (Rodgers, 2023).

Addressing tank waste is highly complex as it requires both management of the remaining waste within the aging tanks and soil and groundwater cleanup efforts to address what has already leaked into the surrounding environment. It is important to recognize, though, that these two efforts are not independent.

Within the tanks, tank waste can exist as supernatant liquid, saltcake solid, and/or sludge solid. The insoluble sludge fraction of the waste consists of metal oxides and hydroxides and contains the bulk of the radionuclides. The saltcake, generated by extensive evaporation of aqueous solutions, consists primarily of dried sodium salts. It is estimated that 24% of the total radioactivity is contained in the supernatant, 20% in the saltcake, and 56% in the sludge (Agnew, 1997).

As mentioned earlier, one step to preventing additional contamination from leaking into the surrounding environment under the tanks was to "interim stabilize" SSTs by removing pumpable liquid and transferring it to DSTs for future treatment. Interim stabilization was conducted from about 1980 to the early 2000s and has lowered the risk of large volumes of waste leaking into the environment. Another measure that DOE has implemented to reduce the environmental threat from past and potential leaks is to install surface infiltration barriers on top of SST farms, including S, SX, T, TX, and TY farms. These barriers mitigate the in-leakage of water that could dissolve waste and potentially leak out to contaminate the surrounding soil. Currently, DOE estimates that 58 SSTs may have leaked in the past and two tanks are currently assumed to be active leakers (Rodgers, 2023). Six of those 58 tanks are in C Farm and have been retrieved to meet TPA requirements. Of the remaining 159 tanks yet to be retrieved (two currently in retrieval operations), 52 SSTs are assumed leakers while two remain known active leakers, whereas one DST (tank 241-AY-102) is considered to have a primary liner leak into the annulus space between the two walls of the tank (Rodgers, 2023).

As retrieval, transfer, treatment, immobilization, and disposition of tank waste proceed, various technical challenges are expected to be encountered. Tank waste is neither uniform nor homogeneous, and as such, tank waste operations involve handling a wide range of tank waste "types" characterized by different physical and chemical properties. For example, slurries and the solids in the slurry have a wide range of particle size, density, and chemical characteristics (Wells et al., 2007). Consequently, depending on the concentration of solids in a waste stream, the slurry may exhibit either a Newtonian (e.g., water-like behavior) or a non-Newtonian behavior. In addition to this nonuniformity of physical and chemical properties of the tank waste, different portions of the waste will be managed, treated, and disposed separately based on their radiological content:

- The liquid (supernate and dissolved saltcake) portion of the wastes will be primarily retrieved and processed into a solid low-activity radioactive waste (LAW) for disposition in a near-surface disposal facility. The LAW would contain about 10% of the total tank waste radioactivity but represent about 90% of the tank waste by volume.
- The solids or sludge portion of the wastes will be primarily retrieved and processed into high-level radioactive waste (HLW) for disposal in a deep geological repository. The HLW would contain most of the radioactivity of the tank waste but constituting less than 10% of the volume.
- A smaller volume of sludge wastes in specific SSTs could potentially be processed into contact-handled transuranic (TRU) waste, representing approximately 1.4 million gallons. The TRU wastes could potentially be transferred to a TRU disposal facility (e.g., Waste Isolation Pilot Plant).

#### *1.2.1.3.4 Active Cleanup Footprint Reduction*

To address the expansive Hanford Site, one key strategy driving the order of cleanup actions was to reduce the active cleanup footprint of the site over time. This strategy helped prioritize cleanup efforts geographically, and thus, some of the first cleanup actions addressed the risks associated with the River Corridor and then moved inward toward the center of the Site. This approach of handling contamination from the "outside-in" is also consistent with the highest priority being to protect the Columbia River. Thus, the cleanup progression involves moving contamination previously located near the shores of the Columbia River inward to the Central Plateau. As the outer portions are addressed, the overall active cleanup footprint is reduced. The long-term objective is to decrease the Hanford Site final footprint (i.e., the Inner Area) to the smallest size practical, and preferably down to only approximately 2% of the original size (Bergman et al. 2011). The envisioned designation of the Inner Area is approximately 10 $mi^2$, although the boundary is subject to change as remediation progresses.

#### *1.2.1.3.5 Long-term Stewardship*

The long-term vision for the Hanford Site was first formally addressed in 1992 by the Working Group. This Working Group included parties at all levels (federal, Tribal, state, and local), who were interested in weighing in on the possible options for the future of the Hanford Site. The long-term scenarios were detailed in "The Future for Hanford: Uses and Cleanup" report. This report, in conjunction with the Federal Register Notice of Intent and the Hanford Remedial Action EIS in 1992, helped to outline future land uses and evaluation of the overarching approaches to site remediation. As the active footprint of the site decreases, various land areas will be transitioned over for new land uses. The proposed future land uses of the areas near and within the Hanford Site (Hanford Comprehensive Land-Use Plan; DOE, 1999) and adjacent areas that are part of the Hanford Reach National Monument are shown in Figure 1.4.

Even upon completion of active cleanup efforts within the River Corridor, the federal government will likely retain ownership of most of the land to support the long-term stewardship of the area for protection of both human and environmental health long into the future (DOE/RL, 2013). While the majority of the Hanford Site is set for conservation and preservation land use, the land nearest the city of Richland is anticipated to be released for industrial use once cleanup efforts are completed.

## 1.3 REMEDIAL STRATEGY EVOLUTION AT THE HANFORD SITE

An integrated cleanup strategy at complex sites should evolve over time. Evolution is often driven by new information, whether it be from new characterization efforts, remedy performance, development of new remedies, or long-term monitoring. This new information should then be incorporated into the cleanup strategy both on the individual management area level and more broadly across the site. At the Hanford Site, there are many examples of how the cleanup strategy adapted throughout the process.

As remediation progressed at the Hanford Site, the cleanup strategy has also evolved to a more comprehensive approach. There has been an attempt to shift strategy from a more piecemeal approach to one that is more holistic (DOE/RL, 2013). This is captured in the fact that OUs at Hanford have been reorganized several times as waste sites have moved through the cleanup process. This allowed for a more efficient use of cleanup resources and balanced responsibilities between regulatory agencies (e.g., EPA and Ecology).

Another example of purposeful strategy evolution is the strategic use of interim action RODs. Interim action RODs enabled remediation work to begin prior to the full characterization and investigation phases. These quick remedial actions paired with characterization and monitoring provided updated knowledge of the nature, extent, and the future threat of remaining contaminants. This improved site knowledge was then incorporated into the final ROD. In some instances, the final ROD did not change from the interim ROD; however, in other cases the newly gathered information did alter the ROD. Use of interim RODs allowed the Hanford Site to adapt the cleanup over time.

Similarly, cleanup efforts are also monitored through time. The CERCLA five-year review process assesses the protectiveness of remedies to determine whether remedies are performing as expected or if changes may be required. At the Hanford Site, some cleanup remedies have evolved as a direct outcome of this review process. For example, the initial ROD for the 300-Area groundwater plume was based on an expectation that the uranium plume would dissipate below levels of concern without active treatment. However, monitoring indicated natural attenuation processes were not sufficient to be protective, and thus, active remediation was pursued and deployed. In a similar vein, the lack of information, in the form of uncertainties, also requires any cleanup strategy

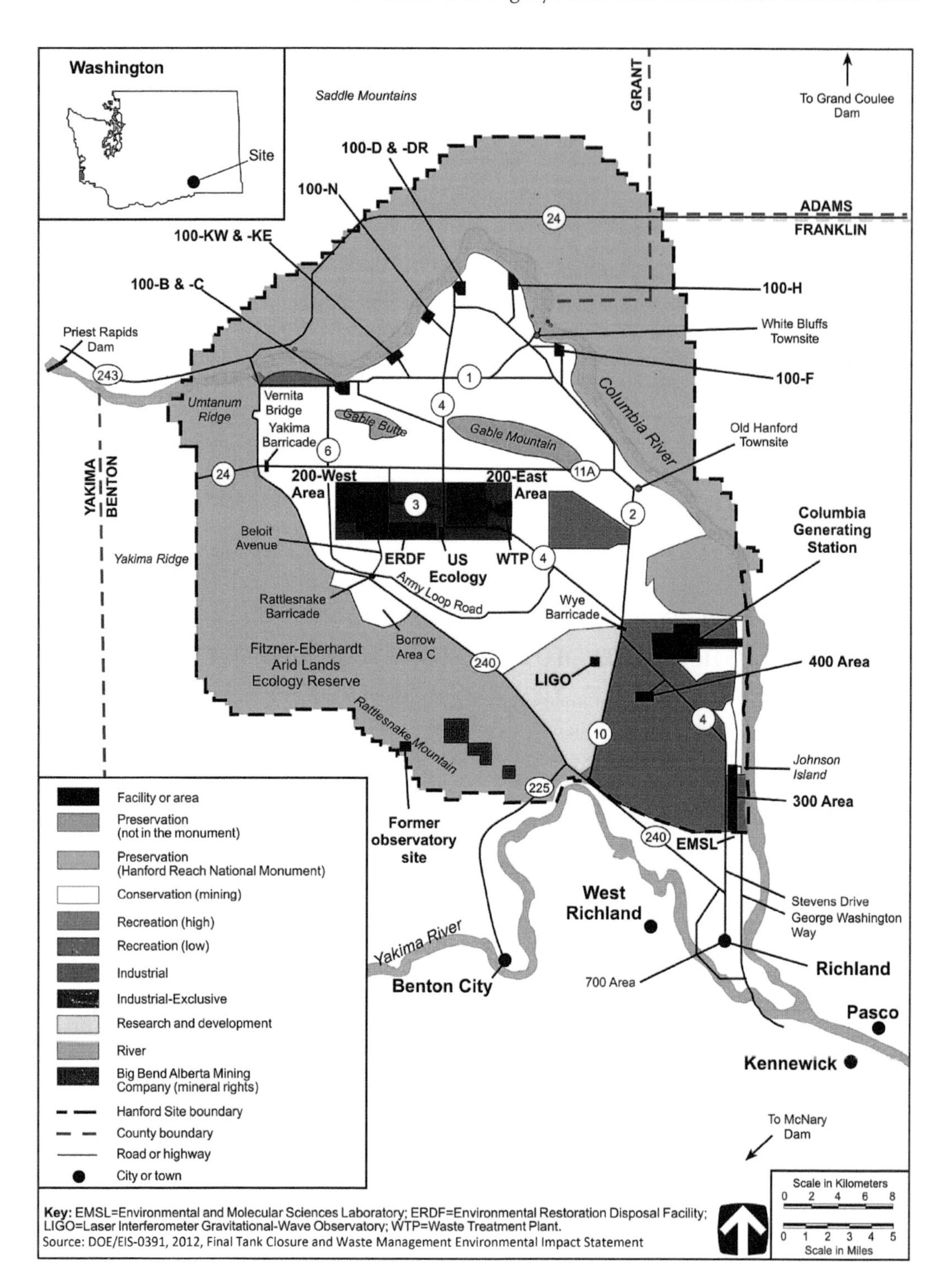


**FIGURE 1.4** Proposed future land use for the Hanford Site. Note: Most of the area in white is reserved for the management and protection of archeological, cultural, ecological, and natural resources and related uses that require the protection of human health and ecological receptors. Limited and managed mining (e.g., quarrying for sand, gravel, basalt, and topsoil for governmental purposes only) could also occur. Reprinted with permission from DOE/RL (2015).

to be flexible and to evolve over time and as the cleanup progresses. The acknowledgment of uncertainty shows up directly at the Hanford Site when considering the possible future outcomes for different geographical areas throughout the Hanford Site. This message was clearly the backdrop for "The Future for Hanford: Uses and Cleanup" (1992) document, which states that uncertainty exists, and it may alter the future land use recommendations and projections.

## 1.4 GENERAL NEED FOR REMEDIAL STRATEGY EVOLUTION AT COMPLEX SITES

In general, cleanup strategies at complex cleanup sites will likely continue to evolve. Some of the common reasons include the following: (i) uncertainties, (ii) new site information, (iii) shifting priorities or limited resources, or (iv) improved scientific understanding or new technological developments.

A cleanup strategy must pivot and adapt to address such complexities and shifts. There are far too many unknowns and uncertainties to expect that all possible outcomes and strategies can be identified and understood a priori. Recognizing not only the level of uncertainty, but also when, and if, the uncertainties can be resolved or mitigated, will keep sites from needlessly spending time and resources. Balancing the risk associated with uncertainty with the impact on the overall objectives and success is key. For instance, both the exact location of contaminants and their concentration and speciation within the subsurface will always have a level of uncertainty associated with them. This uncertainty will play a role in evaluating cleanup alternatives and achievable cleanup objectives. Likewise, intrinsic spatial heterogeneity will only increase these levels of uncertainty. This results in the inherent trade-off between uncertainty and budget. There is often this interplay between how much uncertainty can exist while still allowing for successful design and execution of the targeted remedial actions and the need for additional characterization efforts. In the same way, characterization efforts will need to balance the need for additional information with the budgetary and time constraints.

As cleanup efforts progress, new information is often obtained. Additional information can come in the form of newly discovered waste sites, supplementary characterization information, or remedy performance information. Such data can be incorporated into site conceptual models to improve overall understanding and/or lead to strategic cleanup adjustments. Long-term monitoring information can also provide awareness about the required cleanup actions or available future land uses.

With any long-term remediation effort, there is also the need to adjust to external circumstances. The strategy may need to adjust to changing budgets or shifting remedial priorities, which can speed up, slow down, or shift the cleanup efforts. Technical understanding is also dynamic. For instance, maximum allowable contaminant concentrations, required for protecting human and environmental health, may be updated through time or new remedial technologies may become available, or improved, allowing for contaminants to be cleaned up to lower levels than originally possible.

## REFERENCES


40 CFR 307.14. 2017. "Comprehensive Environmental Response, Compensation, and Liability Act (CERCLA) Claims Procedures, Subpart A, Definitions" U.S. Code of Federal Regulations.

Agnew, S.F. 1997. *Hanford Tank Chemical and Radionuclide Inventories: HDW Model Rev.4*. LA-UR-96-3860, Los Alamos National Laboratory, Los Alamos, NM.

Anderson, J.D. 1990. *A History of the 200 Area Tank Farms*. WHC-MR-0132, Westinghouse Hanford Company, Richland, WA.

Association of State and Territorial Solid Waste Management Officials (ASTSWMO) Remedial Action Focus Group. 2017. "Interim Record of Decisions."

Bergman, T.B. 2011. *Hanford Site Central Plateau Cleanup Completion Strategy*, CHPRC-01187-FP Rev.0. U.S. Department of Energy, Richland Operations Office, Richland, WA. https://www.osti.gov/servlets/purl/1004619

Bergman, T.B. and D.M. Weissberg. 1994. "The Hanford Future Site Uses Working Group." United States. Available at: https://www.osti.gov/biblio/75874.

Bunn, A.L., D.M. Wellman, R.A. Deeb, E.L. Hawley, M.J. Truex, M. Peterson, M.D. Freshley, E.M. Pierce, J. McCord, M.H. Young, T.J. Gilmore, R. Miller, A.L. Miracle, D. Kaback, C. Eddy-Dilek, J. Rossabi, M.H. Lee, R.P. Bush, P. Beam, G.M. Chamberlain, J. Marble, L. Whitehurst, K.D. Gerdes, and Y. Collazo. 2012. *Scientific Opportunities for Monitoring at Environmental Remediation Sites (SOMERS): Integrated Systems-Based Approaches to Monitoring*. DOE/PNNL-21379, Prepared for Office of Soil and Groundwater Remediation, Office of Environmental Management, U.S. Department of Energy, Washington, DC, by Pacific Northwest National Laboratory, Richland, WA.

Demirkanli, I., and V.L. Freedman. 2021. *Adaptive Site Management Strategies for the Hanford Central Plateau Groundwater*. PNNL-32055, Pacific Northwest National Laboratory, Richland, WA. Available at: https://www.pnnl.gov/main/publications/external/technical_reports/PNNL-32055.pdf.

DOE. 2022. *PFAS Strategic Roadmap: DOE Commitments to Action 2022-2025*. U.S. Department of Energy, Washington, D.C. https://www.energy.gov/pfas/articles/pfas-strategic-roadmap-doe-commitments-action-2022-2025#:~:text=The%20%E2%80%9CPFAS%20Strategic%20Roadmap%3A%20DOE,or%20potential%20releases%20of%20PFAS

DOE. 2005. *Hanford Soil Inventory Model, Rev.0.*

DOE (U.S. *Department of Energy*). 1998. *Groundwater/Vadose Integration Project Specification*. Draft A, DOE/RL-89-48. Department of Energy, Richland Operations Office, Richland, WA. https://pdw.hanford.gov/document/D198200850

DOE /EIS-022-SA-02 Rev.0. 2015. *Supplement Analysis of the Hanford Comprehensive Land-Use Plan Environmental Impact Statement*. U.S. Department of Energy, Richland, WA.

DOE/RL. 1998. *Groundwater and Vadose Zone Integration Project Specification*. DOE/RL-98-48, Draft C, U.S. Department of Energy, Richland Operations Office, Richland, WA.

DOE. 1999. *Final Hanford Comprehensive Land-Use Plan Environmental Impact Statement (HCP EIS)*, Hanford Site, Richland, Washington. DOE/EIS-0222-F. https://www.energy.gov/nepa/articles/eis-0222-final-environmental-impact-statement

DOE/RL. 2009. *Central Plateau Cleanup Completion Strategy*. DOE/RL-2009-81, U.S. Department of Energy, Richland Operations Office, Richland, WA.

DOE/RL. 2013. *Hanford Site Cleanup Completion Framework*. DOE/RL-2009-10, Rev.1, U.S. Department of Energy, Richland Operations Office, Richland, WA.

DOE/RL. 2015. *Supplement Analysis of the Hanford Comprehensive Land-Use Plan Environmental Impact Statement*. DOE/EIS-0222-SA-02, Rev 0, U.S. Department of Energy, Richland Operations Office, Richland, Washington.

Environmental Restoration Disposal Facility. 2011. "The Hanford Site Fact Sheet." Available at: https://www.hanford.gov/files.cfm/%20ERDF_Fact_Sheet_May2021.pdf.

EPA. 1992. *Methods for Evaluating the Attainment of Cleanup Standards, Vol. 2: Groundwater*. EPA 230/R-92/014, U.S. Environmental Protection Agency, Office of Policy, Planning, and Evaluation, Washington, DC.

EPA. 1999. *A Guide to Preparing Superfund Proposed Plans, Records of Decision, and Other Remedy Selection Decision Documents*. EPA/540/R-98/031, U.S. Environmental Protection Agency, Office of Solid Waste and Emergency Response, Washington, DC.

EPA. 2002. *Guidance on Environmental Data Verification and Data Validation (QA/G-8)*. EPA/240/R-02/004, U.S. Environmental Protection Agency, Office of Environmental Information, Washington, DC. Available at: https://www.epa.gov/sites/default/files/2015-06/documents/g8-final.pdf.

EPA. 2013. *Guidance for Evaluating Completion of Groundwater Restoration Remedial Actions*. OSWER 9355.0-129, U.S. Environmental Protection Agency, Office of Solid Waste and Emergency Response, Washington, DC.

EPA. 2014a. *Groundwater Remedy Completion Strategy*. OSWER 9200.2-144, U.S. Environmental Protection Agency, Office of Solid Waste and Emergency Response, Washington, DC.

EPA. 2014b. *Recommended Approach for Evaluating Completion of Groundwater Restoration Remedial Actions at a Monitoring Well*. OSWER 9283.1-44, U.S. Environmental Protection Agency, Office of Solid Waste and Emergency Response, Washington, DC.

EPA, DOE, NRC, and DOD. 2000. *Multi-Agency Radiation Survey and Site Investigation Manual (MARSSIM)*. EPA 402/R-97/016, U.S. Environmental Protection Agency, Washington, DC. Available at: https://www.epa.gov/radiation/download-marssim-manual-and-resources.

Future for Hanford: Uses and Cleanup. 1992. *Summary of the Final Report of the Hanford Future Site Uses Working Group*. Department of Energy, Richland Operations Office, Richland, WA. https://pdw.hanford.gov/document/D196123428

Gephart, R.E. 2002. A short history of Hanford waste generation, storage, and release. PNNL-13605 Rev. 4, Pacific Northwest National Laboratory, Richland, WA. https://www.pnnl.gov/main/publications/external/technical_reports/pnnl-13605rev3.pdf

Gephart, R.E. 2003. *Hanford: A Conversation about Nuclear Waste and Cleanup*. PNNL-SA-37974. Battelle Press, Columbus, OH. https://www.pnnl.gov/publications/hanford-conversation-about-nuclear-waste-and-cleanup

ITRC. 2017. *Remediation Management of Complex Sites*. RMCS-1, Interstate Technology and Regulatory Council, Washington, DC. Available at: https://rmcs-1.itrcweb.org or https://rmcs-1.itrcweb.org/RMCS-Full-PDF.pdf.

Johnson, C.D., K.A. Muller, M.J. Truex, G.D. Tartakovsky, D. Becker, C.M. Harms, and J. Popovic. 2022. "A Rapid Decision Support Tool for Estimating Impacts of a Vadose Zone Volatile Organic Compound Source on Groundwater and Soil Gas." *Groundwater Monitoring and Remediation*, 42(1):81–87. https://doi.org/10.1111/gwmr.12468.

National Research Council. 2003. *Environmental Cleanup at Navy Facilities: Adaptive Site Management*. National Academies Press, Washington, DC.

National Research Council. 2014. *Best Practices for Risk-Informed Decision Making Regarding Contaminated Sites: Summary of a Workshop Series*. The National Academies Press, Washington, DC. Available at: https://doi.org/10.17226/18747.

PNL-10651. 1995. *Development of a Risk-Based Approach to Hanford Site Cleanup*. Pacific Northwest Laboratory, Richland, WA.

Rodgers, M.J. 2023. *Waste Tank Summary Report for Month Ending* August 31, *2023*. HNF-EP-0182 Rev.428. Washington River Protection Solutions LLC, Richland, WA.

Rush, J., T. Dillsi, and M. St. Germaine. 2020. "Sludge Removal: Success and Partnership with T Plant." *WM2020 Conference*, March 8–12, 2020, Phoenix, AZ.

Simpson, B.C., Corbin, R.A., Anderson, M.J., and Kincaid, C.J. 2006. Hanford Soil Inventory Model (SIM) Rev. 1 Users Guide. PNNL-16099. Pacific Northwest National Laboratory, Richland, WA. https://www.osti.gov/biblio/895180

SRNL. 2023. *ALTEMIS: Advanced Long-Term Environmental Monitoring Systems*. Savannah River National Laboratory, Aiken, SC. Available at: https://srnl.doe.gov/factsheets_2023/altemis-fact-sheet-print.pdf.

Truex, M.J., D.J. Becker, M.A. Simon, M. Oostrom, A.K. Rice, and C.D. Johnson. 2013. *Soil Vapor Extraction System Optimization, Transition, and Closure Guidance*. PNNL-21843, Pacific Northwest National Laboratory, Richland, WA.

Truex, M.J., C.D. Johnson, D. Becker, M.H. Lee, and M.J. Nimmons. 2015. *Performance Assessment for Pump-and-Treat Closure or Transition*. PNNL-24696, Pacific Northwest National Laboratory, Richland, WA.

Truex, M.J., C.D. Johnson, T. Macbeth, D.J. Becker, K. Lynch, D. Giaudrone, A. Frantz, and H. Lee. 2017. "Performance Assessment of Pump-and-Treat Systems." *Groundwater Monitoring and Remediation*, 37(3):28–44. https://doi.org/10.1111/gwmr.12218.

Wells, B.E., M.A. Knight, E.C. Buck, S.K. Cooley, R.C. Daniel, L.A. Mahoney, P.A. Meyer, A.P. Poloski, J.M. Tingey, W.S. Callaway, G.A. Cooke, M.E. Johnson, M.G. Thien, D.J. Washenfelder, J.J. Davis, M.N. Hall, G.L. Smith, S.L. Thomson, and Y. Onishi. 2007. *Estimate of Hanford Waste Insoluble Solid Particle Size and Density Distribution*. PNWD-3824, Pacific Northwest National Laboratory, Richland, WA.

# 2 Stakeholder Perspectives and Environmental Remediation

*Ellen Prendergast-Kennedy*

The major focus of this chapter is on the processes that have evolved over time to advance stakeholder, Tribal, and community engagement and participation in the remediation process for the Hanford Site, which has had a profound positive impact on awareness and remediation strategy development. As such, some information presented in Chapter 1 is reiterated here in the context of stakeholder engagement to help the reader's comprehension of this important topic.

## 2.1 STAKEHOLDERS

The three main recommendations of the Hanford Future Site Uses Working Group described in Section 1.2.1.2 are being addressed by the two primary missions of the Department of Energy (DOE): (i) waste management and (ii) remediation of environmental media. These are two complex and long-term undertakings that require open and informed communication and stakeholder engagement among Hanford stakeholders, the public, and decision-makers (DOE et al. 2017). Engaging consistently and transparently with stakeholders is critical to the successful implementation of DOE's missions. The term stakeholders can generally be understood as any individual, groups of individuals, or organizations affected by or who can affect a project (Freeman 1984), as well as those groups of individuals who maintain a stake or claim in the operations and decisions of an organization or project and may include those that are internal or external to a project or organization (Carroll 1991). Figure 2.1 depicts the wide variety of stakeholders that participate in environmental remediation decision-making activities at the Hanford Site. They include regulators, federal agencies, state governments, Tribal Nations, contractors, local business leaders, members of academic and scientific institutions, and members of the community. All have varying roles, levels of influence, goals, interests, and concerns related to environmental remediation projects, processes, and associated outcomes that DOE is responsible for making at the Hanford Site.

The key stakeholders are described below in terms of their roles and responsibilities.

- **U.S. DOE:** The DOE Richland Operations Office (RL) and the Office of River Protection (ORP) manage and operate the Hanford Site. Both field offices have integrated responsibility for cleanup of the Hanford Site: RL is responsible for cleanup of facilities, soil and groundwater, and waste disposal, and ORP manages the storage, retrieval, and treatment of tank waste. Both field offices report to the DOE Headquarters Office of Environmental Management in Washington, DC. The U.S. DOE is one of the three agencies that signed the Hanford Federal Facility Agreement and Consent Order (the "Tri-Party Agreement", TPA). Overall, the U.S. DOE is responsible for assuring that Hanford Site cleanup is compliant with the regulatory requirements and with the TPA.
- **U.S. Environmental Protection Agency (EPA):** The EPA has a lead regulatory oversight of DOE cleanup activities occurring on the Hanford Site that are being conducted under the Comprehensive Environmental Response, Compensation, and Liability Act (CERCLA). EPA is one of the three TPA agencies.

 DOI: 10.1201/9781003329213-3

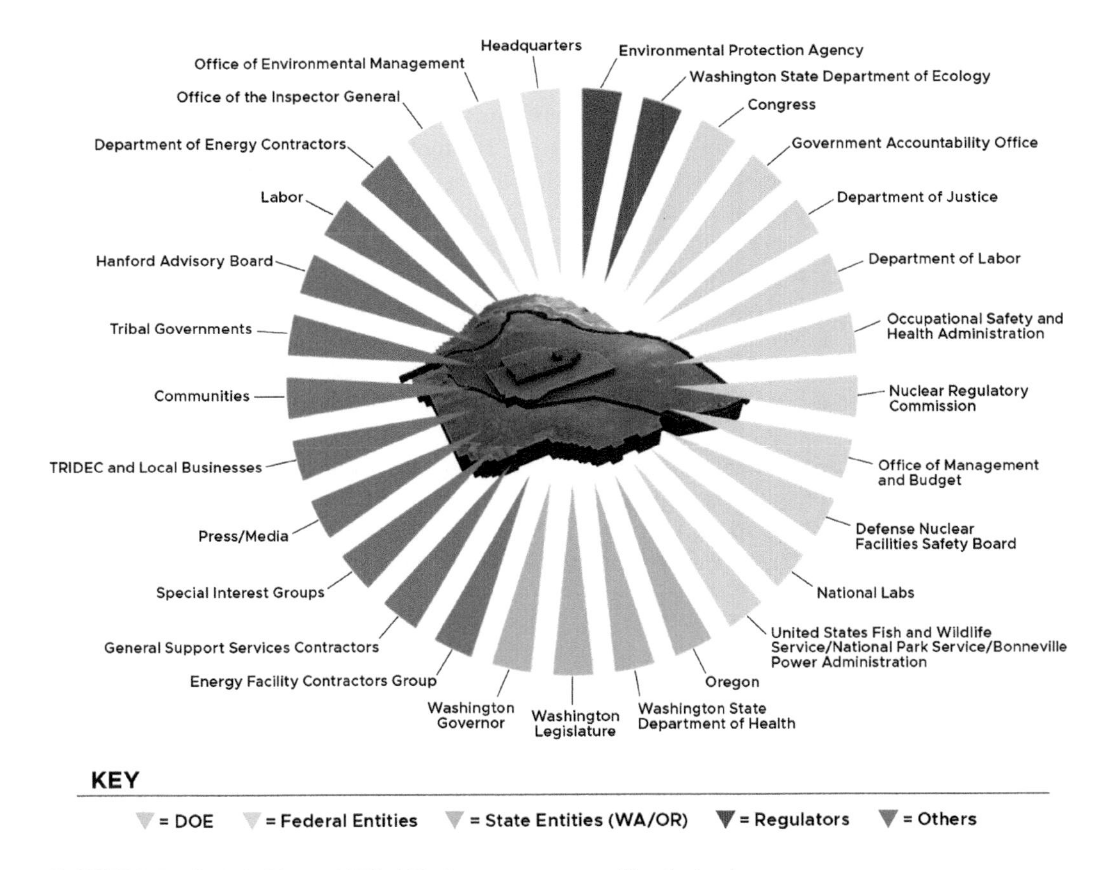


**FIGURE 2.1** Stakeholder and Tribal Nation engagement at Hanford today.

- **Washington State Department of Ecology (Ecology):** Ecology provides a lead regulatory oversight of DOE activities being conducted under Washington's Hazardous Waste Management Act. In that role, Ecology is "responsible for oversight of tank waste treatment and storage, waste management activities" (DOE et al. 2017, p. 3). Ecology is one of the three TPA agencies.
- **Tribal Nations:** Federally recognized Tribal Nations have a unique status as sovereign nations and a special trust relationship[1] with the U.S. federal government, defined by U.S. policy, treaties, and statutes. Due to this unique and important legal status, Tribes with ancestral and historical ties to the Hanford Site are not considered to be, nor are they treated as, a stakeholder group. In recognition of this status, the DOE consults with Tribal Nations in a government-to-government manner regarding policies and decision-making occurring at the Hanford Site.

The Hanford Site is located on lands that were extensively used by Tribal Nations for more than 10,000 years. American Indian descendants of the area's original inhabitants include members of the Yakama Nation, Confederated Tribes of the Umatilla Indian Reservation, and the Nez Perce Tribe, as well as the Wanapum Tribe of Priest Rapids have strong cultural, ancestral, and historical ties to the Hanford Site and the Columbia River. The Yakama Nation, Confederated Tribes of the Umatilla Indian Reservation, and the Nez Perce Tribe who are federally recognized Tribes ceded large amounts of land, including lands that overlap with portions of the Hanford Site, to the U.S. government in three separate treaties.[2] Through these treaties, the Tribes retained and continue to retain rights at the Hanford Site, including rights to fish and hunt at all "usual and accustomed places," and the privilege of hunting, gathering roots and berries, and pasturing horses and cattle on open and unclaimed land.[3] The Wanapum Tribe of Priest Rapids, who lived year-round on the lands where the Hanford Site is located and chose not to participate in the treaty process, also relied upon natural resources (e.g., plants, roots, salmon, deer, and elk) that were seasonally available and gathered culturally important foods and medicinal plants at sacred and ceremonial areas. In 1943, Tribes and settlers and farmers who resided in towns and farms established on the Hanford Site were forced to leave when the lands were seized by the federal government as part of the top-secret Manhattan Project war effort for purposes of plutonium production (DOE 2005). Despite the terms of the treaties, Tribal members were unable to access important resources or conduct cultural practices associated with these important resources. Representatives of the Yakama Nation, Confederated Tribes of the Umatilla Indian Reservation, and the Nez Perce Tribe also participate in the Hanford Advisory Board (HAB), workshops, and roundtables described below, but formal decision-making occurs within the context of government-to-government consultation. Additional information regarding participation of Tribal Nations in remediation decision-making is described in Section 2.3.

- **Oregon State Department of Energy:** It keeps track of ongoing cleanup activities at Hanford and potential downstream impacts on the Columbia River.
- **Hanford Advisory Board (HAB):** The HAB was created by DOE in 1994 to advise the TPA agencies regarding cleanup policies at the Hanford Site. It is discussed in more detail in Section 2.3.
- **Washington State Department of Health:** The Washington State Department of Health's Division of Radiation Protection regulates Hanford radioactive air emissions.
- **Washington Department of Fish and Wildlife:** The Washington Department of Fish and Wildlife is responsible for monitoring programs at the Hanford Site to prevent injury to fish, wildlife, and their habitats (DOE et al. 2017, p. 17).
- **National Park Service:** The National Park Service manages the Manhattan Historical National Park and B-Reactor, a key part of the Park.
- **U.S. Fish and Wildlife Service:** The U.S. Fish and Wildlife Service manages the Hanford Reach National Monument/Saddle Mountain National Wildlife Refuge, which surrounds the Hanford Site, under a Permit and Memorandum of Understanding with DOE (DOE et al. 2017, p. 18).
- **Hanford Natural Resource Trustee Council:** The Hanford Natural Resource Trustee Council is a collaborative Working Group that advocates for natural resource planning and restoration as part of the decision-making process in the remediation of the Hanford Site.
- **Local Organizations:** Public and private organizations that serve the local area at the city and county levels interact with TPA agencies on Hanford cleanup issues. These organizations include, but are not limited to, the Tri-Cities Industrial Development Council; Central Washington Building Trades Council; Hanford Atomic Metal Trades Council; Hanford Communities; Benton, Franklin, and Grant County governments; and the city governments of Richland, West Richland, Pasco, and Kennewick (DOE et al. 2017, p. 17).
- **Public:** TPA agencies involve the public in several ways, including via public meetings, hearings, workshops, and public comment periods.

## 2.2 REGULATORY FRAMEWORK

To understand the roles of stakeholders at the Hanford Site, a reminder of the regulatory framework and agreement documents that govern DOE's environmental remediation decision-making from Chapter 1 is helpful in the context of the public participation processes used at Hanford. The TPA was signed by Ecology, DOE, and the EPA in 1989 (DOE et al. 2017). The TPA is a comprehensive agreement created to guide compliance with hazardous waste environmental laws, specifically the CERCLA of 1980, Resource Conservation and Recovery Act (RCRA) of 1976, and Washington's Hazardous Waste Management Act (HWMA) of 1973. It also provides a legal framework for DOE's decision-making process for cleanup. The TPA prioritizes regulatory commitments, establishes enforceable milestones, and provides a basis for budget requests.[4] The TPA is a legally binding agreement/action plan that includes deadlines and milestones for specific compliance and cleanup actions so that the DOE is compliant with RCRA and CERCLA (DOE et al. 2017, p. 3). CERCLA requirements include the need for the creation of a Community Relations Plan that describes how the public will be informed and involved in the cleanup process (DOE et al. 2017, p. 4). The TPA references the Community Relations Plan, first issued in 1990 (DOE et al. 2017, p. 4), and provides a general description of public involvement activities. The most recent version of this plan is the 2017 Public Involvement Plan (DOE et al. 2017), which is the sixth revision of the document.

DOE's Public Involvement Plan outlines and provides guidance on the variety of public participation processes at Hanford, which vary depending upon the regulatory requirements and scale and scope of impacts that might result from various decisions (DOE et al. 2017). The stakeholders described above are involved to varying degrees depending on the public engagement process. The 2017 Public Involvement Plan outlines the different mechanisms for public and stakeholder engagement, which are described in more detail below.

## 2.3 ENGAGEMENT ACTIVITIES AT THE HANFORD SITE

Stakeholder engagement can range from limited to expansive efforts influencing the ability of a stakeholder to have an impact on decision-making. The International Association for Public Participation's model Spectrum of Public Participation outlines this range (see Figure 2.2). On the one end of the spectrum, more limited stakeholder engagement may focus on informing the public about key project decisions with minimal opportunity for stakeholder input. On the other end of the spectrum, in a more encompassing stakeholder engagement strategy, stakeholders are involved early in the decision-making process – or may even be involved in initial planning – allowing stakeholder values, needs, issues, and concerns to be identified, addressed, or even incorporated into the project.

Stakeholder and Tribal engagement at the Hanford Site has evolved, moving from an era of secrecy and limited public engagement to one of openness and active engagement (INFORM) where the public and Tribal governments and representatives play an oversight and decision-making role (COLLABORATE and EMPOWER). The need for secrecy ended following the end of the Cold War

INCREASING IMPACT ON THE DECISION

| | INFORM | CONSULT | INVOLVE | COLLABORATE | EMPOWER |
|---|---|---|---|---|---|
| PUBLIC PARTICIPATION GOAL | To provide the public with balanced and objective information to assist them in understanding the problem, alternatives, opportunities and/or solutions. | To obtain public feedback on analysis, alternatives and/or decisions. | To work directly with the public throughout the process to ensure that public concerns and aspirations are consistently understood and considered. | To partner with the public in each aspect of the decision including the development of alternatives and the identification of the preferred solution. | To place final decision making in the hands of the public. |
| PROMISE TO THE PUBLIC | We will keep you informed. | We will keep you informed, listen to and acknowledge concerns and aspirations, and provide feedback on how public input influenced the decision. | We will work with you to ensure that your concerns and aspirations are directly reflected in the alternatives developed and provide feedback on how public input influenced the decision. | We will look to you for advice and innovation in formulating solutions and incorporate your advice and recommendations into the decisions to the maximum extent possible. | We will implement what you decide. |



**FIGURE 2.2** Spectrum of Public Participation. Reprinted with permission from International Association for Public Participation (2018).

with the dissolution of the Soviet Union in 1991, and there was a shift from a focus on weapons production to a focus on safety and environmental protection and cleanup (DOE-EM 2021c). In addition, during the late 1980s and early 1990s, DOE released thousands of pages of reports and documentation making the public aware of waste discharges, the presence of radioactive contaminants, and leaking underground storage tanks at the Hanford Site (Clarke 1999). Key legislation also influenced this trend, including the Nuclear Waste Policy Act (NWPA) of 1982 and CERCLA, which emphasized inclusion of the public and Tribal Nations[5] in federal decision-making involving cleanup of hazardous and nuclear materials. Various mechanisms have been employed to conduct stakeholder engagement at the Hanford Site and are described in more detail below. These include workshops that were led by TPA agencies, such as the Future Site Uses Workshops conducted in 1992, an intensive stakeholder engagement strategy piloted by staff at Pacific Northwest Laboratory (precursor to Pacific Northwest National Laboratory) beginning in the early to mid-1990s, creation of the HAB in 1994, and a variety of other public engagement processes including public meetings and various public notification processes. DOE has also engaged separately with Tribal Nations via formal government-to-government consultation and staff-to-staff meetings.

As described in Chapter 1 and reiterated here in the context of stakeholder engagement, the "Future Site Uses Working Group" meeting series in 1992 were held to provide substantive input to the TPA agencies regarding remediation efforts at the Hanford Site (Hanford Future Site Uses Working Group [HFSUWG] 1992). The Working Group was comprised of federal, Tribal, state, and local government entities, as well as representatives from constituencies concerned about future uses and cleanup, including labor, environmental, academic, agricultural, economic development, and public interest groups. The Working Group was charged with selecting a range of future land use options, as well as "appropriate cleanup scenarios to make these future land uses possible" (HFSUWG 1992, p. 3). The Working Group developed nine recommendations that influenced the direction of cleanup to focus on the Columbia River and groundwater remediation while also protecting human health and safety, as well as important cultural and environmental resources. Recommendations provided are listed below (HFSUWG 1992):

1. Protect the Columbia River
2. Deal Realistically and Forcefully with Groundwater Contamination
3. Use the Central Plateau Wisely for Waste Management
4. Do No Harm During Cleanup or with New Development
5. Cleanup of Areas of High Future Use Value Is Important
6. Clean Up to the Level Necessary to Enable the Future Use Option to Occur
7. Transport Waste Safely and Be Prepared
8. Capture Economic Development Opportunities Locally
9. Involve the Public in Future Decisions about Hanford

Several future site use options proposed by the Working Group include agriculture, wildlife, Native American uses, industry, waste management, research/office, recreational, and commercial. These meetings and other future land use decisions culminated in the Comprehensive Land Use Plan Environmental Impact Statement. Tribal Nations were consulted on the EIS, as well as other members of the public, and helped shape the final document.[6]

> People participated [in the Working Group meetings] but they were wary, especially DOE – Tri-Party Agreement was scary to agencies, but thinking back, the working relationship was amazing.
>
> —*Pat Serie (previously affiliated with EnviroIssues, Inc.)*

Other efforts to engage Tribal Nations and stakeholders regarding input on specific environmental remediation strategies include a multi-step, in-depth engagement process initiated by Pacific

Northwest Laboratory staff (precursor to PNNL) to complement the Working Group meetings (personal conversations with Pat Serie on July 20, 2021, and Gretchen Hund on June 8, 2021) beginning in the early to mid-1990s. The overall purpose was to find ways to enable stakeholders to better understand environmental remediation technologies and make informed decisions about the application and acceptability of deploying these strategies. The strategy established an iterative method to involve and engage with interested stakeholders, which was subsequently used at other DOE sites (i.e., the VOCs in Arid Soils Integrated Demonstration Project).[7,8] The strategy was considered ground-breaking at the Hanford Site because of its aim to listen to and learn from interested parties, find common ground, and provide opportunities for meaningful involvement in environmental remediation decisions. This method proved extremely valuable at a time when public trust in DOE was low, because it focused on making stakeholders feel heard; it can be categorized as COLLABORATE and EMPOWER stakeholder engagement approach from the International Association for Public Participation's "Spectrum of Public Participation" (see Figure 2.1).

In addition, specific visual aids were developed to foster discussion regarding criteria for evaluating stakeholder acceptance of environmental remediation technologies (e-mail dated June 14, 2021; Figure 2.3). A software known as ProTech, an early-computer–based tool, was used to compare innovative technologies being considered and tested with the current technologies in use (Hund 2021 | e-mail dated September 30, 2021). Side-by-side comparisons were used based on these criteria, to foster stakeholder discussion on these technologies and make them more understandable (Hund 2021 | e-mail dated September 30, 2021).

> Instead of going to "convince" people, it was much more of "we can learn something from talking to people" and maybe design a test we want to do with the technology in a much better way, having that info in hand. I ended up becoming in charge of stakeholder involvement.
>
> —*Gretchen Hund (previously affiliated with Battelle and PNNL)*

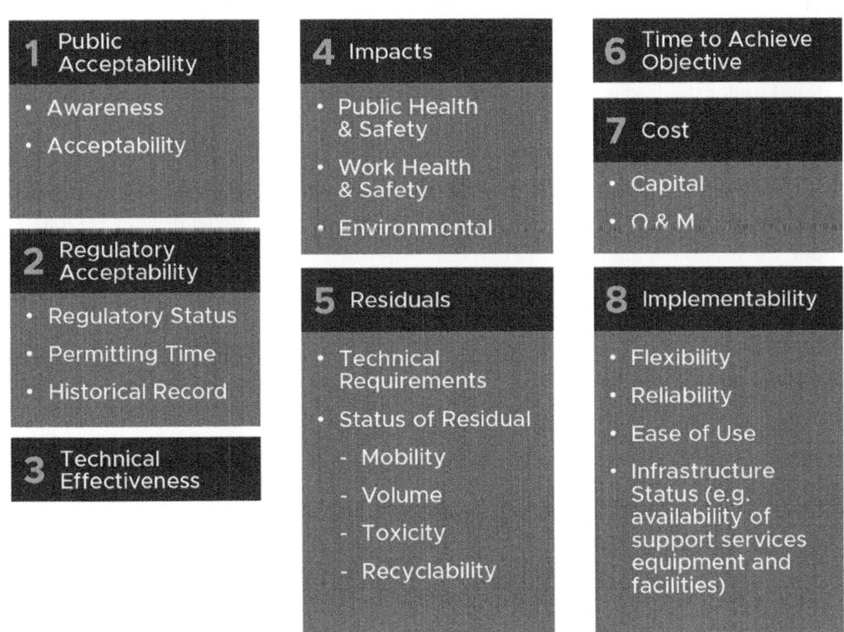


**FIGURE 2.3** Example criteria for evaluating stakeholder acceptance of environmental remediation technologies.

This method involved an intensive effort to work at the community level to assure that community members that may have a stake were identified. Best stakeholder identification practices were implemented and included partnering with knowledgeable individuals in the community to identify key community stakeholders, as well as a snowball sampling or chain referral approach.[9] Stakeholders identified to participate in the workshops for environmental remediation technologies included DOE, EPA, Washington State Department of Ecology, Tribes, academics, Hanford Site contractors, local leaders, and environmental and health groups.

The steps in the process include the following:

1. Conduct initial one-on-one interviews with identified stakeholders to identify and explore their views about the specific environmental remediation technology.
2. Facilitate roundtable group meetings with similar stakeholder groups.
3. Send out relevant reports ahead of time, which not only allowed stakeholders to see that their views were being heard but also provided opportunities to understand other stakeholder perspectives and concerns.
4. Facilitate integrated workshop where the focus is on finding common ground between various stakeholders.

Additional best practices included making sure all stakeholder's interests and concerns were heard and the reliance on easy-to-read visuals.

> We sat down with people in the area ...and they told us all these things we had not thought about. That's an example of if you don't go into these communities, talk to people, live/eat/breathe in that community, you will miss something.
>
> —*Gretchen Hund*

The creation of the HAB was a key step in the evolution of stakeholder engagement at the Hanford Site. Created in 1994, the HAB has been in existence for almost 30 years and is the primary method for engaging multiple stakeholders and making environmental remediation decisions. The HAB was established by DOE to provide consensus advice to the TPA agencies on cleanup policy decisions. It was chartered under the Federal Advisory Committee Act as a DOE-Environmental Management Site Specific Advisory Board (EM SSAB) and continues to serve in that capacity. It is one of the 8 DOE-Environmental Management Site Specific Advisory Boards (EM SSAB) created to provide the "assistant secretary for environmental management and designees with advice, information, and recommendations on issues affecting the program at various DOE sites" (DOE-EM 2022). Others include the Idaho Cleanup Project Citizens Advisory Board, Northern New Mexico Citizens' Advisory Board, Nevada SSAB, Oak Ridge SSAB, Savannah River Site Citizens Advisory Board, Portsmouth SSAB, and Paducah Citizens Advisory Board (DOE-EM 2022). Specific issues of focus for the HAB include environmental restoration, and cleanup management of waste and nuclear materials including disposition of waste, long-term stewardship, and future land use (DOE 2022).

> [The HAB] had a huge impact – as it opened up, it became clear how complicated it [Hanford Site cleanup] was. Stakeholders became VERY knowledgeable about it. The decisions DOE made over the years, not just cleanup decisions, were impacted [by the] HAB. The general public would show up for meetings and listen and were aided by the HAB. But the HAB had an awful lot of very detailed, concrete effects on DOE.
>
> —*Pat Serie*

There are 32 members and their alternates representing a range of stakeholders affected by Hanford Site cleanup activities. Board membership "is carefully considered to reflect a full diversity of viewpoints in the affected communities and region (DOE 2022)." In addition to the parties that signed the TPA, additional HAB participants include the following: Washington State Department of Health, representatives of the Nez Perce Tribe, Yakama Nation, the Confederated Tribes of the Umatilla Indian Reservation, the State of Oregon, and individuals representing local government interests, local business interests, Hanford workforce, local environmental interests, regional citizen, environmental and public interest organizations, local and regional and public health, universities, and the "public at large."[9]

> It's a shared decision-making that works. Not just input.... Shared decision-making [is a way] that will meet the range of needs of stakeholders as is humanly possible.
>
> —*Pat Serie*

The Board is intended to be an integral component for some Hanford Tribal and general public involvement activities, but it is not intended to be the sole conduit (DOE-EM 2021d). Public involvement is informed by the Hanford Federal Facility and Agreement and Consent Order Hanford Public Involvement Plan, which was last updated in 2017. TPA agencies conduct several other forms of formal public involvement such as public meetings, hearings, and workshops, which are held on an as-needed basis.

> The TPA agencies assess public interest and areas of public concern regarding specific actions based on consultations with Tribal Nations, and discussions with State of Oregon representatives, the Hanford Advisory Board (HAB), stakeholders, and interested members of the public. Based upon those interactions and regulatory requirements, the TPA agencies determine the need for a public meeting or other outreach forum.
>
> *(TPA 2017, p. 7)*

The TPA agencies also engage in informal methods of outreach, including "workshops, focus groups, classroom visits, open houses, community events, and meetings with professional organizations, local governments and civic organizations" (TPA 2017, p. 8). TPA agencies rely on public notifications as a means for communicating opportunities for the public to provide comments via multiple methods including e-mail lists, posting in social media, newspapers, public service announcements, and notifications in the federal register (DOE et al. 2017, p. 6).

Other Working Groups exist to engage stakeholders in DOE decision-making. The Hanford Natural Resources Trustee Council, "a collaborative working group chartered to address natural resources impacted by Hanford Site releases of hazardous substances", is an example of one such Working Group format.

Representatives of the Yakama Nation, Confederated Tribes of the Umatilla Indian Reservation, and the Nez Perce Tribe also participate in the HAB and in the workshops and roundtables described above, but formal decision-making occurs within the context of government-to-government consultation.

The NWPA assigned responsibility to DOE "to provide for the development of repositories for the disposal of high-level radioactive waste and spent nuclear fuel, to establish a program of research, development, and demonstration regarding the disposal of high-level radioactive waste and spent nuclear fuel, and for other purposes" (DOE 2004). It also gave authority to Tribal Nations to participate in the siting decisions and required the DOE to provide financial support to Tribal governments to accomplish this (DOE 2004). The Yakama Nation, the Nez Perce Tribe, and Confederated Tribes of the Umatilla Indian Reservation were identified as affected Tribes and became eligible to receive federal funding and participate in the DOE's decision-making process at Hanford (Clarke 1999). DOE formally consults with the three federally recognized Tribal Nations in a government-to-government manner in recognition of their sovereign status and in accordance with the NWPA and other cultural resource protection mandates. DOE also engages with Tribal Nations on a staff-to-staff level on multiple activities occurring at the Hanford Site in addition to cleanup decision-making activities. Under cooperative agreements with the DOE,

> Tribal staff and consultants of the Yakama, Nez Perce, and CTUIR are engaged on a daily basis with DOE and its contractors. The principal activities by tribes include reviewing and commenting on plans and documents, participating in meetings at the request of DOE, monitoring cultural resource sites, participating in site surveys, and identifying issues that will require additional consultation with elected officials on a government-to-government level.
>
> *(DOE-EM 2021a)*

While not federally recognized, since the 1940s, the Wanapum Tribe of Priest Rapids who have strong historical and ancestral ties to the Hanford Site worked hard to establish and maintain relationships with predecessor agencies first involved in the Manhattan Project and later Cold War activities, including the U.S. Army Corps of Engineers and later the Atomic Energy Commission (DOE 2021). As a result of those early and ongoing efforts by Wanapum leaders to build these relationships, today, DOE continues to recognize the Wanapum's strong cultural and historic ties to the Hanford Site and consults with the Wanapum Tribe of Priest Rapids in accordance with various federal cultural resource protection mandates and applicable DOE cultural resource and Tribal Nation policies.

Having multiple stakeholder engagement structures including the HAB that builds on trust and transparency has enabled remediation to continue in a manner that meets the goals and concerns of multiple stakeholders, as well as consulting Tribal Nations. With a look ahead to the future, it is important to continue to evaluate the effectiveness of stakeholder engagement at the Hanford Site and keep an open mind toward creative and expanded engagement such that the momentum of success continues.

## NOTES

1. DOE's *Working with Indian Tribal Nations, A Guide for DOE Employees* issued in 2000, "all federal agencies, including DOE, have a permanent legal obligation to exercise statutory and other legal authorities to protect tribal land, assets, resources, and treaty rights." This "trust responsibility" is a legally enforceable obligation, a duty, on the part of the U.S. Government to protect the rights of federally recognized Indian tribes. The language in Supreme Court cases suggests that it entails legal duties, moral obligations, and the fulfillment of understandings and expectations that have arisen over the entire course of dealings between the U.S. government and Indian tribes. See https://www.energy.gov/sites/prod/files/DOE%20Guide%20to%20Working%20with%20Tribal%20Nations.pdf.
2. The three treaties are available at https://www.yakama.com/about/treaty/, https://www.fs.usda.gov/Internet/FSE_DOCUMENTS/stelprdb5108216.pdf, and https://ctuir.org/departments/office-of-legal-counsel/codes-statutes-laws/treaty-of-1855/
3. Ibid.
4. www.hanford.gov/page.cfm/TriParty
5. Other key legislations specific to expanding the involvement of Tribal Nations in federal decision-making during this timeframe include the passage of the Native American Graves and Repatriation Act in 1990 (NPS 2022) and the 1992 updates to the National Historic Preservation Act's implementing regulations at 36 CFR 800, which expanded Federally recognized Tribes' role in consultation and participation in federal decision-making process regarding projects or activities that might impact significant cultural resources.
6. See https://www.energy.gov/nepa/articles/hanford-comprehensive-land-use-plan-eis-helps-doe-preserve-unique-resources
7. Several technical publications were generated as a result of these efforts. See McCabe et al. (1995) at https://www.osti.gov/servlets/purl/86938 and Peterson et al. (1994).

8 A book was later published by Gretchen Hund, Jill-Engel-Cox (former PNNL staff member), and Kim Fowler (current PNNL staff member), by Battelle Press entitled, *A Communications Guide for Sustainable Development, How Interested Parties Become Partners.*
9 See HAB (2021) for a complete list of specific organizations and interests.

## REFERENCES

Carroll, A.B. 1991. The Pyramid of Corporate Social Responsibility: Toward the Moral Management of Organizational Stakeholders. *Business Horizons,* 34:4, 39–48.

Clarke, K.V. 1999. Environmental justice and Native Americans at the Department of Energy Hanford Site, *Fordham Environmental Law Review*, 10:3. Accessed September 15, 2021 at https://ir.lawnet.fordham.edu/cgi/viewcontent.cgi?article=1518&context=elr.

DOE. 2000. Working with Indian Tribal Nations, A Guide for DOE Employees. U.S. Department of Energy, Washington, D.C. https://www.energy.gov/sites/prod/files/DOE%20Guide%20to%20Working%20with%20Tribal%20Nations.pdf

Freeman, R.E. 1984. *Strategic Management: A Stakeholder Approach.* Boston: Pitman.

HAB. 2021. Hanford Advisory Board Membership. https://www.hanford.gov/files.cfm/2021_Membership_List_v3.pdf

Hanford Future Site Uses Working Group (HFSUWG). 1992. *The Future for Hanford: Uses and Cleanup the Final Report of the Hanford Future Site Uses Working Group.* U.S. Department of Energy, Richland Operations Office. Richland, WA. https://pdw.hanford.gov/document/D196123428

International Association for Public Participation. 2018. "Thought Exchange." https://thoughtexchange.com/blog/enhance-public-participation-iap2/

National Park Service. 2022. *Native American Graves Protection and Repatriation Act, Getting Started.* Accessed April 11, 2022 at https://www.nps.gov/subjects/nagpra/getting-started.htm.

Peterson, T.S., G.H. McCabe, P.J. Serie, and K.A. Niesen. 1994. *Phase II Stakeholder Participation in Evaluating Innovative Technologies: VOC-Arid Integrated Demonstration, Groundwater Remediation System*, prepared for Thomas Brouns, Pacific Northwest Laboratory, Seattle, WA: Battelle HARC.

United States Department of Energy (DOE). 2004. *Nuclear Waste Policy Act, as Amended with Appropriations Acts Appended.* Accessed September 29, 2021 at https://www.energy.gov/sites/default/files/edg/media/nwpa_2004.pdf.

United States Department of Energy (DOE). 2005. *Hanford Cultural Resources Management Plan (DOE-RL-98-10, Rev. 0).* Richland, WA. Accessed September 22, 2021 at https://www.hanford.gov/files.cfm/han_cult_res_mngmt_plan_full_doc.pdf.

United States Department of Energy (DOE), Washington State Department of Ecology, and United States Department of Environmental Protection. 2017. *Hanford Federal Facility Agreement and Consent Order Hanford Public Involvement Plan.* Accessed September 15, 2021 at https://www.hanford.gov/files.cfm/FacAgreementand-Consent-Order_FINAL.pdf.

United States Department of Energy (DOE). 2021. *Department of Energy's Tribal Program.* Accessed September 28, 2021 at https://www.hanford.gov/page.cfm/INP.

United States Department of Energy (DOE). 2022. *Hanford Advisory Board.* Accessed November 4, 2022 at https://www.hanford.gov/page.cfm/hab.

United States Department of Energy, Office of Environmental Management (DOE-EM). 2021a. *Site Programs and Cooperative Agreements.* Accessed September 29, 2021 at https://www.energy.gov/em/site-programs-and-cooperative-agreements.

United States Department of Energy, Office of Environmental Management (DOE-EM). 2021c. *Early Environmental Management History (1989-1992) Timeline.* Accessed September 29, 2021 at https://www.energy.gov/em/early-environmental-management-history-1989-1992-timeline.

United States Department of Energy, Office of Environmental Management (DOE-EM). 2021d. *Early Environmental Management History (1989-1992) Timeline.* Accessed September 29, 2021 at https://www.energy.gov/em/secretary-hazel-r-oleary-administration-january-1993-january-1997-timeline.

United States Department of Energy, Office of Environmental Management (DOE-EM). 2022. *EM Site Specific Advisory Board.* Accessed November 7, 2022 at https://www.energy.gov/em/em-site-specific-advisory-board.

# Part II

# Site Assessment and Characterization

Christian D. Johnson, Katherine A. Muller, and Hilary P. Emerson

When the release of a contaminant into the environment is discovered, the first steps toward remediation are to assess the nature and extent of the contamination. This part of the book focuses on these initial steps. Depending on the nature of the contaminant release and the characteristics of the release site, initial steps may also include triage to identify any immediate potential adverse impacts on human health and the environment and carry out emergency response actions. After any necessary emergency response, a preliminary site assessment is conducted to provide the initial information to support further planning and site characterization. Then, a remedy screening may be conducted (Part III). (Refer to Section 1.2 of Chapter 1, "Remediation Strategy for Complex Waste Sites," for context on how these steps fit into the broader remediation process.)

## II.1 PLANNING AND SITE CHARACTERIZATION

Complex site assessment begins with gathering information about the site. Field data must be collected to determine the current conditions including the nature and extent of the contamination. If available, historical information may also be gathered. Knowing which contaminants were involved, and when and where they were used, generated, handled, stored, or discharged, can provide critical insight into both past releases and potential sources of current and future contamination. Historical records include notes by previous site owners and operators, purchasing records, permits, regulatory documents, maps, aerial photographs, or accounts from staff.

These preliminary data gathering efforts feed into a conceptual site model (CSM), a descriptive framework that describes the relevant processes, features, and mechanisms that affect contaminant migration and exposure pathways in the context of site geology, groundwater flow, surface water interactions, prevailing wind direction, waste sources, contaminant distribution, and other factors. A CSM includes both a graphical and written summary of site-specific information that is relevant for decision-making. Any data gaps identified in a CSM at this stage guide additional investigation

DOI: 10.1201/9781003329213-4

to ensure that future activities based on the model adequately protect human health and the environment. A quantitative CSM can be used to identify risk levels, potential exposure pathways, and regulatory requirements, and to develop site-specific risk levels based on future land use. Importantly, a CSM is a *dynamic tool* – it is updated and refined as new data become available throughout the remediation process (ITRC 2017). A CSM can then be used to develop a predictive model tool that can be used in later steps for decision-making.

Both field data and historical data gathering present challenges at complex waste sites, and availability of historical information may be uneven. Records may be classified or proprietary, depending on the activities formerly occurring at the site, and important individuals or documents may no longer be available. For example, because work at the Hanford Site's Plutonium Finishing Plant (Chapter 4) pertained to nuclear weapons development, some historical materials about it remain classified. And even when data are available, they may be challenging to interpret; the sheer number of industrial and chemical processes tested at the Hanford Site's 300 Area (Chapter 3) has made it difficult to extrapolate from historical records to present-day contamination. The interpretation of historical data may also be complicated by changes in the owners or contractors who operated the facilities or collected data over a site's operational lifetime, varying approaches used to estimate properties, or shifts in nomenclature.

Characterizing and modeling contaminant fate and transport in the subsurface requires measuring a number of physical and hydraulic properties of soil and sediments. There can be tremendous spatial variability and geologic heterogeneity across a site. For example, within a span of 10 ft of depth within the Hanford Site's perched water zone near the B Tank Farm, the saturated hydraulic conductivity varies over five orders of magnitude, significantly impacting the movement of water through these layers of the vadose zone (Rockhold et al. 2018a, 2018b, 2020). These variabilities are due, in large part, to the geologic history of the site. Pleistocene-age cataclysmic floods deposited material consisting of coarse gravel to boulder sediments within stratigraphic units, with deposition and erosion processes from succeeding floods resulting in lenses of sand and silt surrounded by sand and gravel (Martin 2011).

Additional site characterization may include sampling and analysis of buildings, surfaces, water sources, air, subsurface sediments, and groundwater. The characteristic sources, types, and pathways of sediment and groundwater contamination at complex waste sites, and their environmental impact, are conceptually illustrated in Figure II.1. The illustration depicts the sources of contamination from facilities, tanks, and near-surface disposal structures (e.g., cribs) that could release liquid waste streams through the soil to groundwater. However, it should be noted that there are

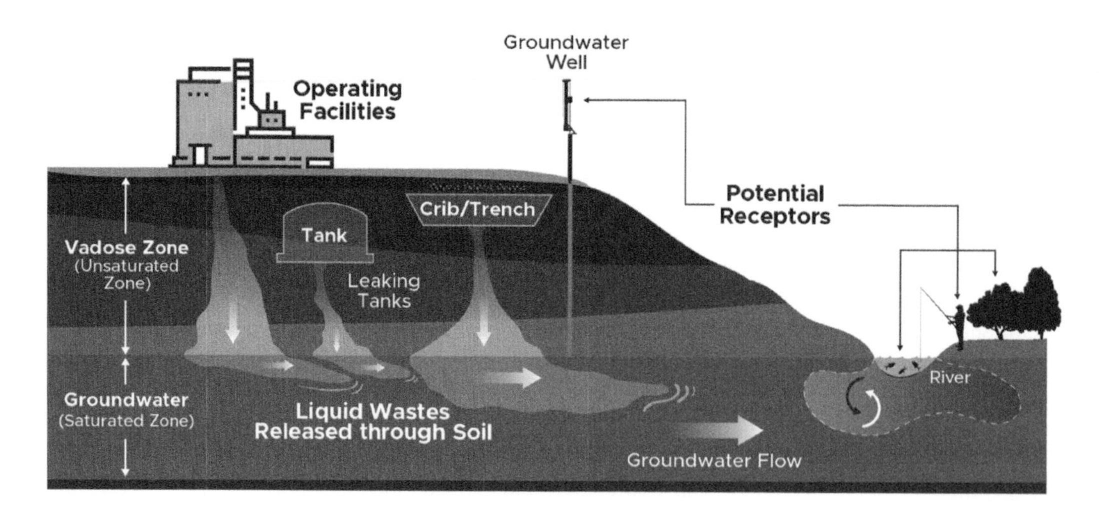


**FIGURE II.1** Sources and types of contamination at complex waste sites.

other contaminant exposure pathways, including air, surface water, animals, or vegetation. Even if thorough historical recordkeeping with respect to waste characteristics and contaminant releases is available, subsurface characterization will likely still be required – if not already conducted – to understand the potential movement of contaminants within the subsurface. Potential sources of future subsurface contamination from stored wastes and contaminated facilities must be characterized for long-term stability in terms of the physical containment, as well as the waste chemistry.

One challenge in characterizing stored wastes at these sites is that their chemical and physical properties may change over time. The radioactive and hazardous waste from approximately 40 years of plutonium production at the Hanford Site is currently stored in underground storage tanks. The characteristics of the waste are slowly changing as a result of complex chemistry and physics including relatively high levels of alkalinity, ionic strength, radiation emissions, radiolytic heat generation, evaporative loss, condensation, and water intrusion. These slow changes become important in the context of the multi-decade timeline for waste remediation at complex sites. For example, at the Hanford Site, organic chelators in the waste are an important consideration for developing treatment approaches because they raise the solubility of some radionuclides, such as strontium-90 and transuranic metals. But at relevant time scales, these chelators are slowly breaking down to non-chelating organic acids as a result of radiolysis due to interactions with gamma irradiation (Toste et al. 2013). Previously complexed radionuclides precipitate from solution as the chelators break down to simpler non-chelating organic compounds, and thus necessitate a modified treatment approach over time.

Future remedial decisions are made based on sound data, and data quality objectives (DQOs), which are also determined during initial planning and characterization, are developed to assure the soundness of these data. DQOs are developed to ensure a systematic, objectives-based guide for the site characterization process to allow for effective future decision-making – they guide the collection of *the right data at the right time with no data gaps and no surplus data* (ITRC 2017). DQOs help guide the process of formulating a problem, identifying the decisions to be made, specifying quality requirements for the decisions, and developing a defensible sampling and analysis plan (PNNL 2024). DQOs may include the following parameters: acceptance criteria like allowable uncertainty levels and resolution requirements, data collection objectives, and a data collection and analysis plan. DQOs may also be based on applicable or relevant and appropriate requirements (ARARs), additional federal, state, or local requirements, and targeted risk levels, which must also be identified as part of the remediation process. Applicable requirements are any substantive requirement that pertains directly to laws that specifically address a hazardous substance, remedial action, location, or other circumstances. Moreover, relevant and appropriate requirements include any additional standards that, while not applicable to the hazardous substance, remedial action, location, or other circumstance, address problems or situations sufficiently similar to those encountered at the site. For example, there may be specific regulations for soil conservation and drainage control for soil used to cover landfills if the soil is taken from a surface mining site.

Once ARARs for the site are identified and preliminary remedial action objectives (RAOs) are formulated, depending on the contaminants, site, and ARARs, the next step is to rank hazards. An example ranking system by the EPA is described in their Hazard Ranking System Guidance Manual (EPA 1992). These types of risk assessments can be helpful in determining the timelines and prioritizing specific contaminants, sources, or regions of a site for cleanup. At complex sites, site-specific objectives for contaminant source removal, containment, and exposure prevention are generally defined based on regulatory requirements, regardless of the technical ability to meet them. It may be beneficial to segment a complex site into small units, depending on the timeline (e.g., need for early actions to reduce site size), budget (e.g., need for a phased approach to cleanup), future land use, dissimilarity of contaminants, area, or subsurface unit (e.g., groundwater versus vadose zones or differing geochemical environments).

Because of the uncertainty of cleaning up a complex site, alternative cleanup objectives (e.g., other future land uses) or different remediation approaches may need to be developed. Alternate site objectives would likely be considered at a later step in the process (i.e., corrective action decision and record of decision phases based on regulatory decision documents and timelines) based on technical or economic practicability to restore affected areas to beneficial uses based on local, state, federal, or other regulations. However, interim objectives may be implemented at any step of the process to yield measurable, incremental progress toward site goals and help to develop a step-by-step approach for the overall cleanup strategy. This site management approach is known as adaptive site management (ASM) and is a relatively new concept described in more detail in Chapter 1, Section 1.4.

Additional non-technical challenges should also be considered in these initial assessment steps, including the need for accountability and transparency in communicating remedial decisions to stakeholders and the general public, as described in Chapter 2, "Stakeholder Perspectives, Environmental Remediation, and the Hanford Site." Overall, this step begins to develop the technical approach for cleanup. Presuming the preliminary findings from the initial site planning and characterization step identifies risks to be mitigated, and then, remedial efforts will continue to remedy screening, evaluation, and testing as described in Part III of this book. After promising remediation technologies are identified, treatability testing can begin.

## II.2 HANFORD SITE ASSESSMENT AND CHARACTERIZATION

This part of the book includes site assessment and characterization case studies from remediating different areas and historical processes across the Hanford Site, including Chapter 3 "Hanford 300 Area Uranium Plume," Chapter 4 "Hanford Plutonium Finishing Plant and Subsurface Characterization," Chapter 5 "Hanford Tank Waste Characterization," and Chapter 6 "Hanford Tank Integrity Characterization."

Chapter 3 describes the history of uranium processing operations in the 300 Area, including waste production, releases to the subsurface, and the subsequent characterization of the subsurface. Historical data gathering and subsurface characterization were vital for developing the CSM, which was used to make remediation decisions with respect to the uranium plume generated from these historical activities (Chapter 11). Developing the CSM for a site having waste releases from several different processes and a subsurface groundwater source that is impacted by seasonal variation of the adjacent river water stages was challenging. Characterization occurred over several years and was conducted via multiple types of sampling, including collection of soil or sediment, groundwater, and river water.

The history of the PFP and the subsequent characterization, including the buildings, waste disposal structures, and the impacted sediments, are the focus of Chapter 4. Historical data were important for understanding the current and future hazards within the structures and the subsurface, as well as for developing the CSM for the PFP. Additional characterization was required to understand the extent of contamination due to the complexity of the chemical processes and wastes and the release of contamination within the subsurface Characterization was vital to understand the potential movement of plutonium beneath the ground surface. However, the final remedial action for the cleanup of plutonium in the subsurface has not yet been completed and may require additional characterization depending on the results from the next steps of the process.

Physical and chemical characterization of Hanford tank waste is discussed in Chapter 5. Hanford tank waste characterization by direct chemical and physical analysis is more valuable than by analyzing the records of the processes that initially generated it because of their complexity, the incomplete nature of records, and changes over time during storage (e.g., effects of high temperatures from radioactive decay). Nonetheless, process records inform the characterization, especially where tank waste is difficult to sample due to access and safety concerns, as well as heterogeneity across the tanks from different physiochemical processes. The analytical methods and pathways employed to characterize the waste for subsequent processing and the types of facilities and ancillary capabilities needed are described. Finally, the means by which characterization results are recorded, disseminated, and presented to the Hanford community and its stakeholders are described.

Chapter 6 describes the characterization of the structural integrity of the underground tanks storing historical process waste. This is especially important as decisions are required if the tanks are to be modified to enable retrieval of the wastes, or used in a role for which they were not originally intended. At the Hanford Site, many of the tanks are several decades beyond their design life, sometimes with incomplete records of the source waste streams, and subjected to a variety of extreme process conditions such as high heat levels. Finite element analyses of thermal and operating loads are summarized, which included the soil, tank components, and the contact conditions between each of the structural features. Such analyses have shown the structural margin of the analyzed tanks is wider than originally conceived and enables greater flexibility in their continued use.

## REFERENCES


EPA. 1992. *Hazard Ranking System Guidance Manual*. United States Environmental Protection Agency, Washington, DC.

ITRC. 2017. *Remediation Management of Complex Sites*. Interstate Technology and Regulatory Council, Washington, DC. https://rmcs-1.itrcweb.org/.

Martin CJ. 2011. *Overview of Hanford Hydrogeology and Geochemistry*. DOE, Richland, WA.

PNNL. 2024. *Data Quality Objectives in Visual Sample Plan*. Pacific Northwest National Laboratory, Richland, WA. https://www.pnnl.gov/projects/visual-sample-plan.

Rockhold ML, JL Robinson, K Parajuli, X Song, Z Zhang, and TC Johnson. 2020. "Groundwater characterization and monitoring at a complex industrial waste site using electrical resistivity imaging." *Hydrogeology Journal* 28 (PNNL-SA-147892): 2115–2127. https://link.springer.com/article/10.1007/s10040-020-02167-1

Rockhold ML, X Song, JD Tagestad, PD Thorne, GD Tartakovsky, and X Chen. 2018a. *Sensitivity Analysis of Contaminant Transport from Vadose Zone Sources to Groundwater*. Pacific Northwest National Lab. (PNNL), Richland, WA.

Rockhold ML, FA Spane, TW Wietsma, DR Newcomer, CR Clayton, DI Demirkanli, MJ Truex, MMV Snyder, and CJ Thompson. 2018b. *Physical and Hydraulic Properties of Sediments from the 200-DV-1 Operable Unit*. Pacific Northwest National Laboratory, Richland, WA.

Toste, AP, TJ Lechner-Fish, and RD Scheele. 2013. "Organics in a Hanford mixed waste revisited: myriad organics and chelator fragments unmasked." *Journal of Radioanalytical and Nuclear Chemistry* 296: 523–530.

# 3 Characterization of a Uranium Groundwater Plume along the Columbia River

*Amanda R. Lawter and Michelle M.V. Snyder*

The remediation process begins with site assessment and characterization. As discussed in Chapter 1, this essential initial step is used to inform decisions about cleanup and move the remediation process forward. In this chapter, the historical waste releases that occurred within the 300 Area of the Hanford Site, and specifically the resulting uranium plume, will be presented as a case study.

## 3.1 BACKGROUND

The 300 Area is adjacent to the Columbia River on the southeastern end of Hanford, approximately one mile north of the city of Richland (Figure 3.1). Uranium fuel fabrication took place in the 300 Area from 1943 to 1988 (Gerber 1993). In addition to producing uranium fuel rods, the 300 Area was home to research laboratories where scientists worked to improve the efficiency of transforming uranium into plutonium. Although the 300 Area was the closest processing facility to a residential

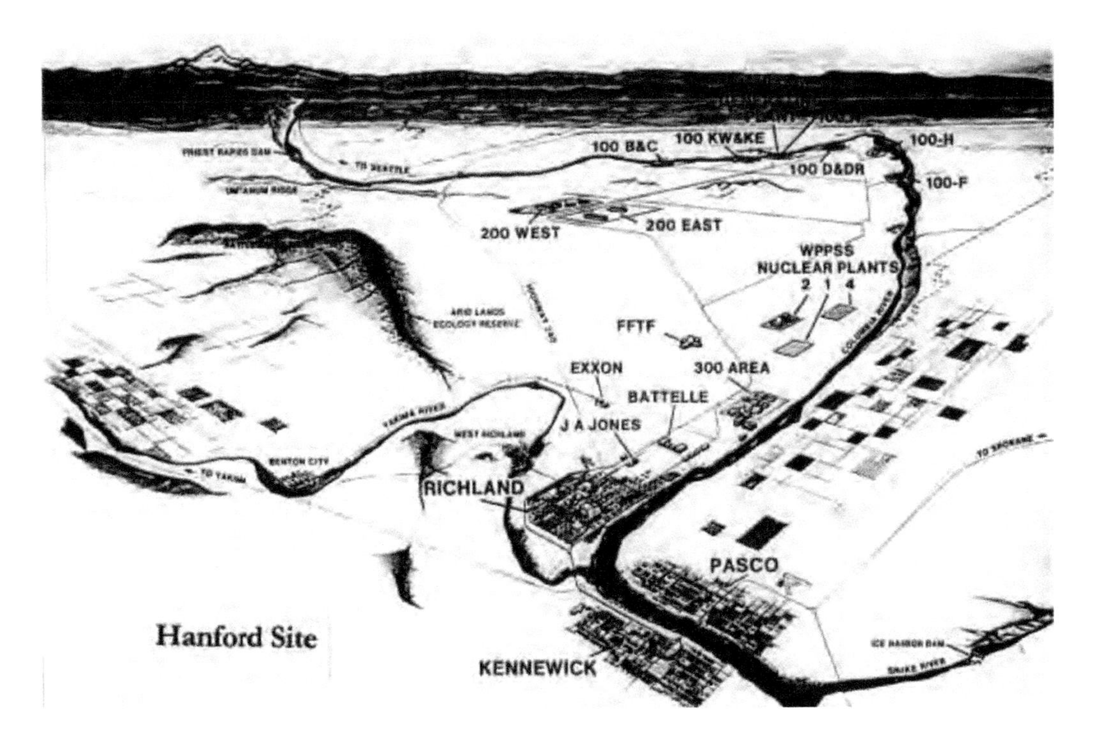


**FIGURE 3.1** Map of the Hanford Site showing the proximity of the 300 Area to the city of Richland. Reprinted with permission from Briggs (2001).

DOI: 10.1201/9781003329213-5

area, the amounts of radioactivity contained in uranium fuel prior to irradiation and the manufacturing process were considered to pose a low risk to human health. Figure 3.2 shows common radiation doses compared to those expected to occur from operation of nuclear facilities.

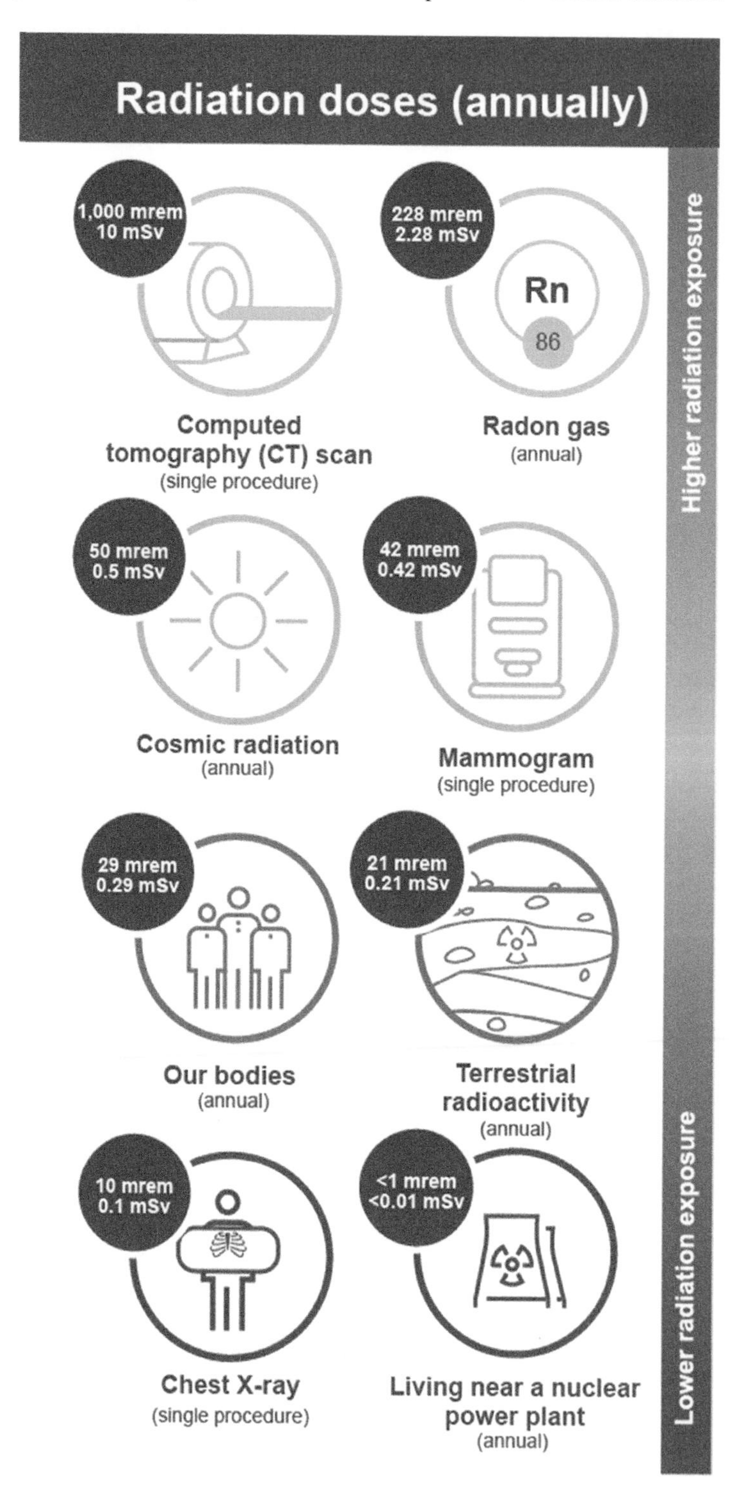


**FIGURE 3.2** Relative doses from various radiation sources. Reprinted with permission from NCRP (2009).

During the fabrication process, uranium metal was received at the 300 Area fuel fabrication facilities as billets, or metallic blocks. The billets were first heated and then extruded into uranium fuel rods. During this process, a mixture of water and oil was used to keep fine particles that were produced during cutting from sparking. The uranium fuel rods, or slugs, were then dipped into different molten metals to create a protective coating in a process called "canning" or "jacketing" (Gerber 1993). After a final dip into an aluminum-silicon alloy, the slugs were put into thin aluminum tubes that were held inside steel jackets. The canning process protected the uranium from corrosion while also helping to contain the radionuclides that were created later when the uranium was exposed to neutrons inside the reactors. Finally, the canned uranium was cut to the required length, welded shut, cleaned, and then shipped to an on-site reactor by railcar.

Fuel production in the 300 Area continued to support the N reactor until it shut down in 1987 (Gerber 1992a,b). At the N reactor, the exterior of the uranium fuel rods had to be more durable due to an improved cooling process that required the fuel rods to withstand increased pressure (Harvey 2000). To accomplish this, a coextrusion process was used, where the uranium fuel rods were jacketed with multiple metals and extruded simultaneously, reducing the occurrence of air bubbles and other flaws. Instead of using aluminum, this process introduced new metals, including copper (Gerber 1992a). Due to the decreased demand for fuel production around the 1970s, several buildings in the 300 Area transitioned to other missions (Gerber 1992b).

## 3.2 WASTE RELEASES AND CONTAMINATION

The fuel fabrication and research activities in the 300 Area produced large amounts of waste, including uranyl nitrate hexahydrate, ammonium nitrate, hexone, and neutralized acid waste that was discharged to the ground (Zachara 2007).

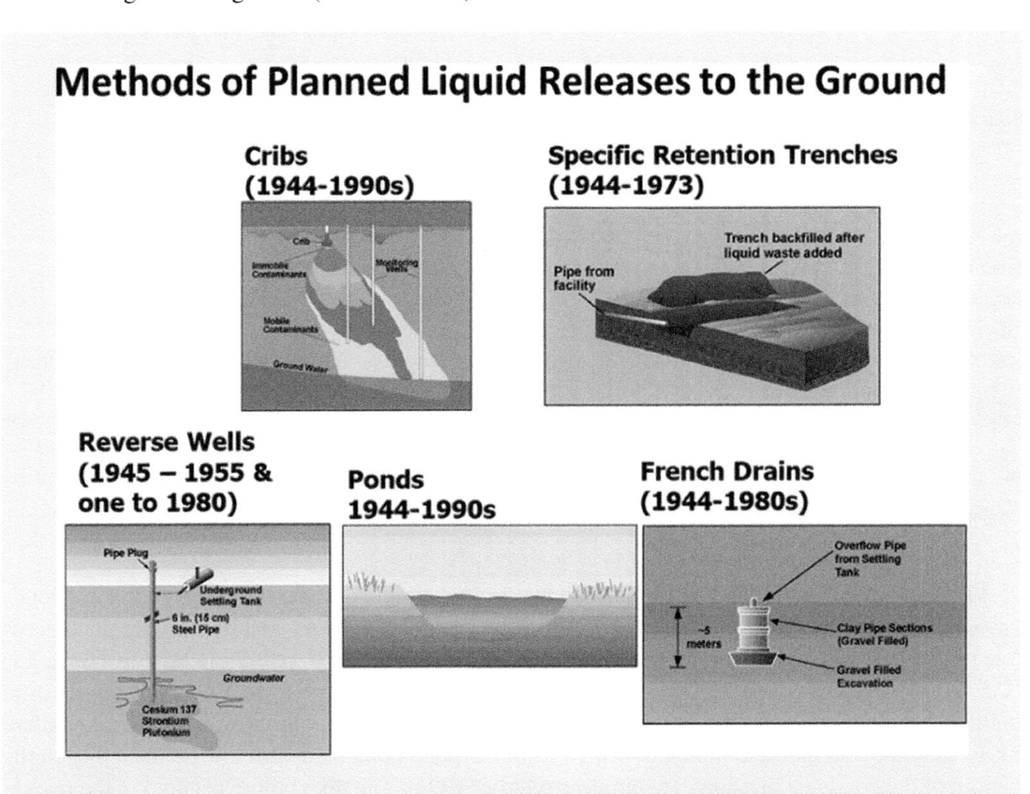


**FIGURE 3.3** Methods of liquid waste releases to the ground that were historically utilized on the Hanford Site. Reprinted with permission from Gephart 2003 (PNNL-13605, Rev 4).

- **Cribs:** Underground boxlike structures
- **Trenches:** Open trenches backfilled with gravel
- **Reverse wells:** Liquid waste pumped down wells into the subsurface or sometimes the underlying aquifer
- **Ponds:** Surface sites filled with waste
- **French drains:** Vertical buried concrete pipes
- **Unplanned release:** Including spills and leaks

**TABLE 3.1**
**300-FF-1 Waste Sites (ROD 1996;DOE 2023)**

| Facility Description/Designation | Years of Service | Waste |
|---|---|---|
| South Process Pond (316-1) | 1943–1975 | • Process wastes<br>• Water treatment filter backwash |
| North Process Pond (316-2) | 1948–1975 | • Process wastewater<br>• Slurry coal fly ash |
| North Process Pond Scraping Disposal Area (618-12) | 1948–1964 | • Sludge from North Process Pond<br>• Coal fly ash |
| Process Trenches (316-5) | 1975–1994 | • Process wastewater |
| Process Trench Spoil Area | 1991 | • Disposal location for sediments excavated from the active portions of the east and west trenches |
| Process Sewer System (within 300-FF-1) | 1943–1994 | • Process wastewater (cooling water and low-level radioactive liquid wastes from fuel fabrication)<br>• Laboratory wastes<br>• Chemical spills |
| Sanitary Sewer System (Sanitary Trenches) | Post-1954 to 1996 | • Sanitary sewage<br>• Septic tank overflow<br>• Cooling water<br>• Small quantities of photographic chemicals |
| Ash Pits | 1943–1997 | • Slurry coal fly ash |
| Filter Backwash Pond | 1987–1998 | • Water treatment filter backwash |
| Retired Filter Backwash Pond (Infiltration Basin within South Process Pond) | 1975–1987 | • Water treatment filter backwash |
| Landfills (1a-1c) | Unknown | • Various landfills used for burning debris. Waste from 1c was removed during Remedial Investigation |
| Landfill 1d | 1962–1974 | • Located north of the west end of the sanitary trenches. Used as a burn pit |
| Burial Ground No. 4 (618-4) | 1955–1961 | • Miscellaneous uranium-contaminated materials |

300 Area waste sites included 31 miles of underground piping, the South and North Process Ponds (also called 316-1 and 316-2), various process sewers, trenches, ash pits, and backwash pond areas (see Figure 3.3), as well as the 300–3 Aluminum Hydroxide site, and Landfills 1a thru 1d (Table 3.1) (U.S. DOE 1996). The South Process Pond was in operation from 1943 to 1975 and was the first unlined disposal facility for liquid waste in the 300 Area. In 1948, a South Process Pond dike failed, likely caused by an increase in liquid levels within the pond combined with low permeability in the bottom of the pond due to a clay-like build-up (Gerber 1993). The dike failure resulted in the release of 14.5 million gallons of waste (containing 5.4–28 kgs of uranium) to the Columbia River, and the North Process Pond was put into service at this time (Gerber 1993). The South Process Pond was

dredged and repaired before being put back into service, after which periodic dredging occurred regularly at both ponds with the dredged material added to the dikes to further increase strength (Gerber 1993). Both ponds received 1.5–11.4 million liters of waste per day; the total mass of primary contaminants in the hazardous wastes included 33,600–59,000 kgs uranium, 241,000 kgs copper, 117,000 kgs fluoride, 113,000 kgs aluminum, and 2,060,000 kgs nitrate, along with unspecified amounts of nitric acid and sodium hydroxide (EPA 2013). When the process ponds were phased out in 1974 and 1975, liquid waste was discharged to the 300 Area Process Trenches (also known as the 316–5 waste site) until 1994. The last known liquid uranium–contaminated effluent was disposed of in the 300 Area at the process trenches in 1985 (DOE 2013). From 1985 until 1994, only nonhazardous effluent was disposed of in these trenches (DOE 2013).

There are three different operable units (OU) that were established in the 300 Area for cleanup purposes: the 300-FF-1 OU, 300-FF-2 OU, and the 300-FF-5 groundwater OU (see Figure 3.4). The 300-FF-1 OU is approximately 117 acres, and contains contaminated soils, structures, debris, and solid waste burial grounds. The 300-FF-1 OU received most of the liquid waste generated during fuel fabrication and is composed of two categories of waste: the process waste sites, which primarily received the liquid waste, and the burial ground, which primarily received solid waste (Table 3.1). The 300-FF-2 OU contains contaminated soils, debris, burial grounds, and groundwater within the 300 Area that is not included in 300-FF-1, the 400 Area, or the 600 Area. The 300-FF-5 OU includes the groundwater beneath the 300-FF-1 and 300-FF-2 OU (ROD 1996). In 1989, the 300-FF-1, 300-FF-2, and 300-FF-5 OU were added to the National Priorities List for cleanup when the Tri-Party Agreement was signed (Chapters 1 and 2).

### 3.2.1 The 300-Area Uranium Plume

The uranium groundwater plume, likely originating from the North and South Process Ponds (and a lesser amount from some of the process trenches), has an area of 0.24 $km^2$ with a maximum concentration of 1.44 mg/L in 2020 (Cline 2020). An estimated 70,000 kg of uranium has been removed from the 300 Area through excavation of soil, with approximately 4,000 kg remaining in the sediment and another 62 kg in the groundwater (PNNL-17034). The amount of uranium reaching the Columbia River was estimated to be 200–430 kg/yr (DOE-RL 1994a; Fitz and Arntzen 2007). DOE-RL (1994a) predicted that the changing concentration of uranium in groundwater would result in a decrease in uranium discharged to the Columbia River; their model predicted 15 kg/yr would reach the river by 2018. The concentration of the uranium plume varies seasonally, with higher concentrations generally found near source areas (e.g., below former waste sites) and decreased concentrations (due to dilution by infiltrating river water) moving east toward the Columbia River during high water table conditions. High water table conditions are caused by seasonal fluctuations in the Columbia River depth (or stage), which flows into the site boundary and elevates the groundwater table along the river as shown in Figure 3.5. During low water table conditions, uranium remobilized during the high water table period migrates toward the river, causing a higher concentration along the shoreline.

Most of the contaminated waste disposal sites, including the surrounding soil, were excavated from 1991 to 2004. Although 95% of the 300 Area waste sites have been cleaned up, as much as 4,000 kg of residual uranium persists in the vadose zone (the unsaturated zone above the groundwater table) (DOE 2019; PNNL-17034). During excavation and prior to backfilling in the South and North Process Ponds, sediment samples were analyzed to determine uranium content in the remaining soil; results of this analysis confirmed the persistence of uranium (Peterson et al. 2008), but the continued source of uranium to the groundwater in the 300 Area has not been well defined. Characterization has shown that periodic rewetting (called the periodically rewetted zone; see Figure 3.5) by the river water causes additional uranium from the vadose zone to mobilize. However, due to the lower alkalinity and specific conductance in river water compared to groundwater, uranium is less mobile in river water, causing some of the uranium to be redeposited on the sediments (Peterson 2008).

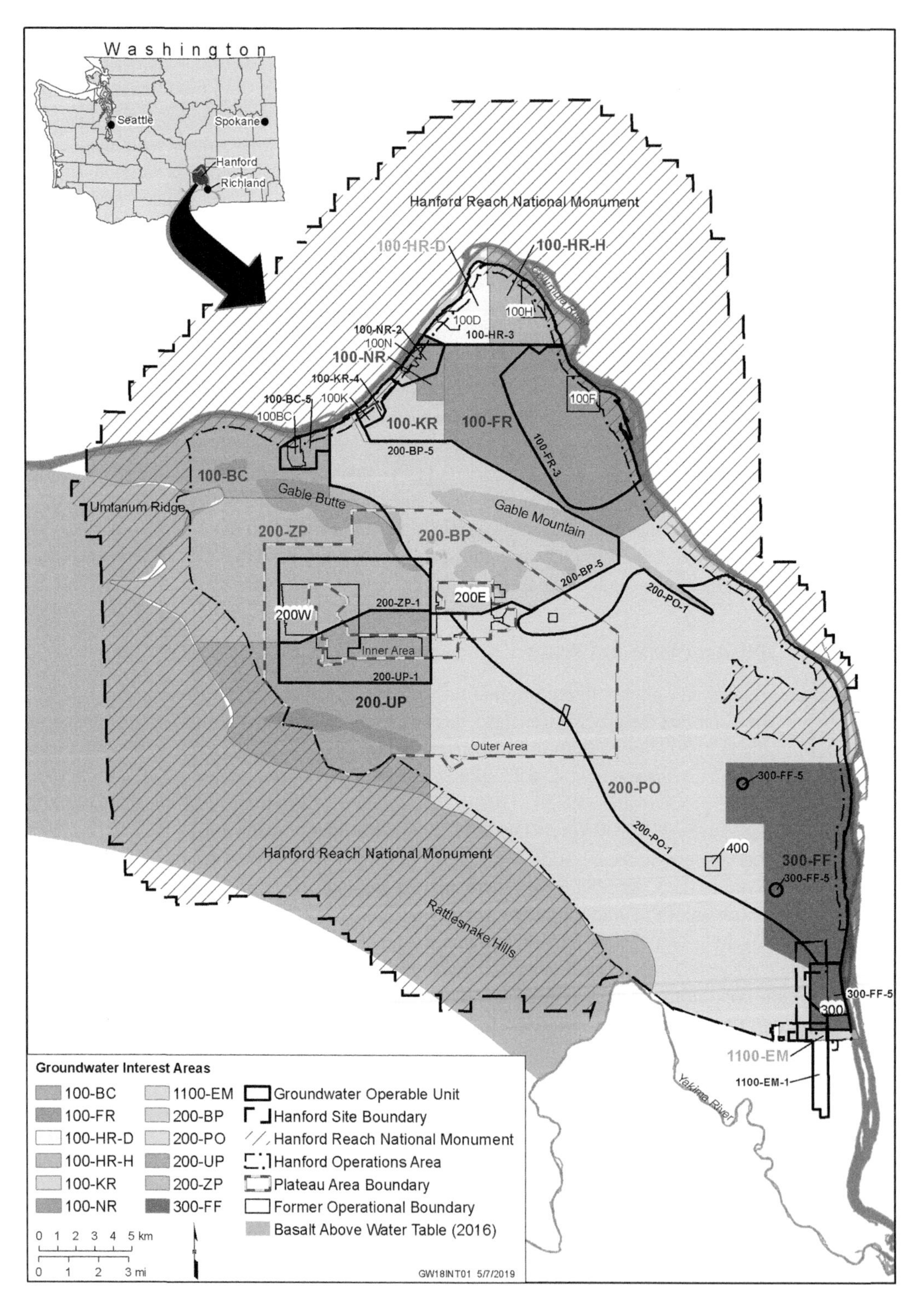


**FIGURE 3.4** Map of Hanford Site operable units, 300-FF sites are labeled in dark gray box in the bottom left. Reprinted with permission from DOE (2019).

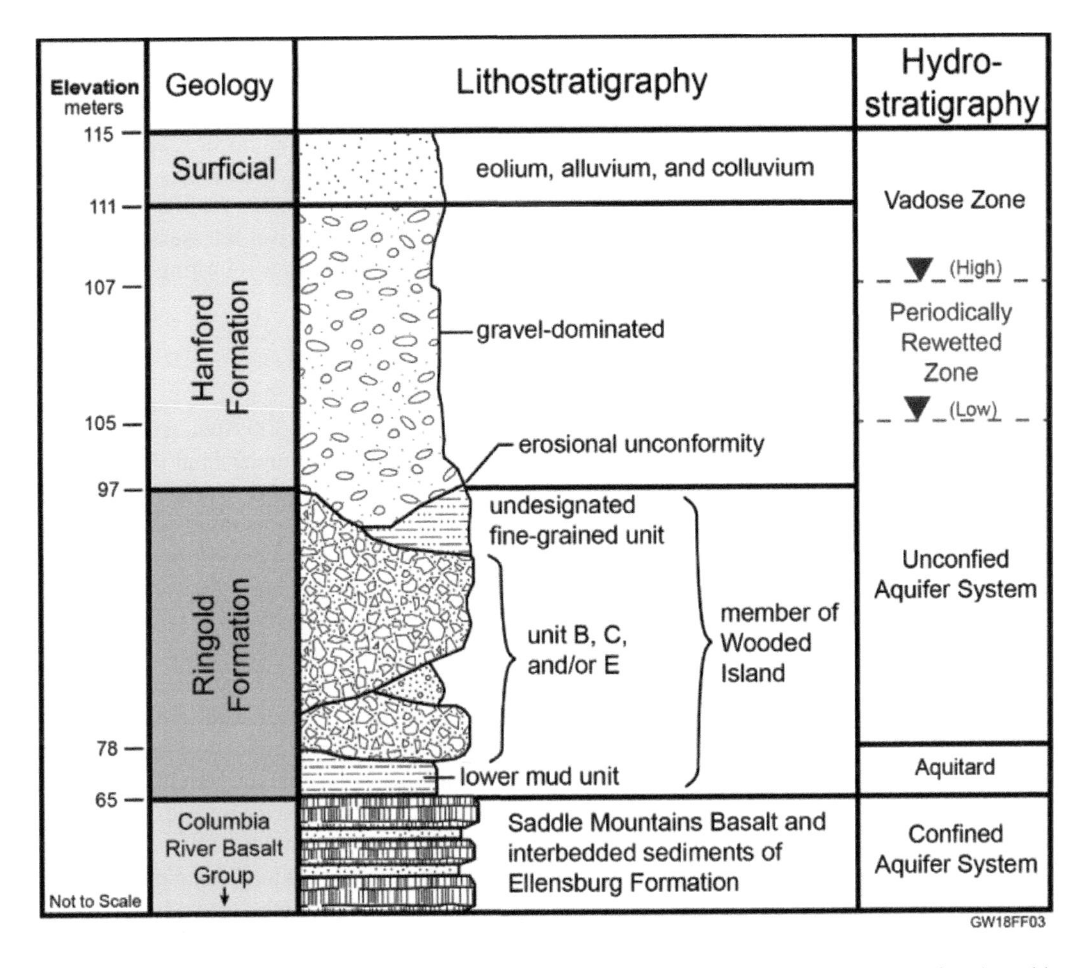


**FIGURE 3.5** Geology of the 300-FF OU, showing the periodically rewetted zone, where uranium is mobilized during seasonal periods of high river stage. Reprinted with permission from DOE (2019).

### 3.2.2 The Mighty Columbia

The Columbia River is an important local, regional, and national resource. The river is important in Native American culture, as well as one of the greatest hydropower producers in the world, while also providing irrigation, recreation, and other ecological and economical values to the area. Reducing pollution to the Columbia River has been a priority for Hanford Site cleanup, which is why cleanup started in the River Corridor, including the 300 Area. Characterization and monitoring of uranium in the 300 Area have shown that the largest plume is found along the shoreline and is migrating toward the river (Zachara 2012). In a study in 1956, when the 300-Area disposal ponds were still actively used, dilution between the ponds to the river was calculated to be more than 500 times as the waste mixed with groundwater (Haney 1957). Discharge to the ponds was measured at $6.4 \times 10^{-7}$ microcurie per milliliter (μc/mL), with only 1/500th of that reaching the Columbia River for a final concentration in the river of $7 \times 10^{-7}$ μc/mL in 1954 (Haney 1957). The report indicated no concerns, as the uranium concentration in the river was relatively close to background levels and suggested that additional measurements during different seasons (e.g., during various river stages, as this report was conducted at low river stage) and evaluations prior to adding proposed dams to the river were recommended. In 1964, the estimated exposure to nearby residents in Richland, Washington, was 0.0078 rem (or 0.16 rem for "maximally exposed" individuals who ate 220 kg of aquatic foods and spent more than 1,000 hours on or near the river per year) (Walters et al. 1992), while the annual exposure limit for individuals recommended by the Federal Radiation Council in 1960 was 0.5 rem per year (Jones 2005). Along with the slow-moving, consistent releases to the river through the process ponds, large spills or discharges have occurred; as mentioned earlier, the South Process Pond dike failure in 1948 resulted in the release of 14.5 million gallons of waste (containing 5.4–28 kgs of uranium) to the Columbia River (Gerber 1993). An estimated 430 kg of uranium is released into the Columbia River per year from the 300-FF-5 operable unit in the 300 Area of the Hanford Site; for comparison, irrigation canals on the opposite side of the river release an estimated 1,750 kg of uranium per year from agricultural fertilizers in irrigation water returning to the river (Yokel and Priddy 2010).

## 3.3 CHARACTERIZING THE SUBSURFACE

To determine the location of contaminant plumes within the Hanford Site's 300 Area, several puzzle pieces must be put together using multiple resources for stakeholders to understand the extent of contamination and begin to move into the Remedial Investigation or Feasibility Study phase. The first piece of the puzzle is determining the location and chemistry of waste disposed of in the area. Records beginning in the 1940s can help piece this information together, and the information shown in Table 3.1 that identifies waste sites, their contents, and the timeframe each site was used is an example of that process knowledge.

Knowing the history of the waste is helpful, but due to the movement of the contaminants within the subsurface (both those released in the 300 Area or potentially migrating from other areas into the 300 Area) and potential unknown or undocumented releases, additional information is needed. Characterization of the subsurface and groundwater is important to determine where the contaminants are now, and what form they are in (e.g., is the contaminant in a form that is easily mobilized or is it in an immobile form?). This information can be used to design a remediation strategy and to inform modeling efforts to predict where the contaminants will move to next. Early characterization of the 300 Area included geophysical techniques using metal detectors, magnetometers, acoustics, ground-penetrating radar, and infrared imaging studies to locate buried wastes, as well as sediment characterization including geological descriptions and temperature and moisture depth profiles (Phillips and Raymond 1975). After the Hanford Site was listed on the National Priorities List and the Tri-Party Agreement was signed in 1989 (Chapters 1 and 2), characterization work continued with surface radiological surveys and manually collected sediment samples analyzed for uranium and other contaminants such as copper and strontium (Teel and Olson 1990). Since then, various characterization activities have continued, with advances in technology and instrument sensitivity allowing for increasingly detailed information about the subsurface (breakout 2). For example, samples collected prior to 1957 were analyzed for gross beta concentrations, but after 1957, samples were able to be analyzed for specific radionuclides (Walters et al. 1992), giving a much clearer picture of contamination and contaminant transport. Another example is the use of advanced solid-phase characterization techniques; a study by Stubbs et al. (2009) combined the use of electron microprobe analysis to locate uranium within sediment samples with a focused-ion beam to collect and prepare the located uranium for analysis by a transmission electron microscope. This combination of microscopy techniques resulted in the identification of six different mineralogical hosts for the uranium, providing information that was not previously available and highlighting the complexity of uranium in the 300 Area (Stubbs et al. 2009).

When characterization and monitoring data are combined with historic knowledge of waste chemistries and locations, contaminant plume maps can be created, like the plume map shown in Figure 3.6. As new information is gained, the maps are updated to continuously refine and improve the available data. Groundwater and subsurface characterizations have revealed three main plumes of contaminated groundwater in the 300 Area: the uranium plume, tritium plume, and 110-Area plume (refer to Figure 3.6). In addition to uranium, the contaminants in the primary uranium plume

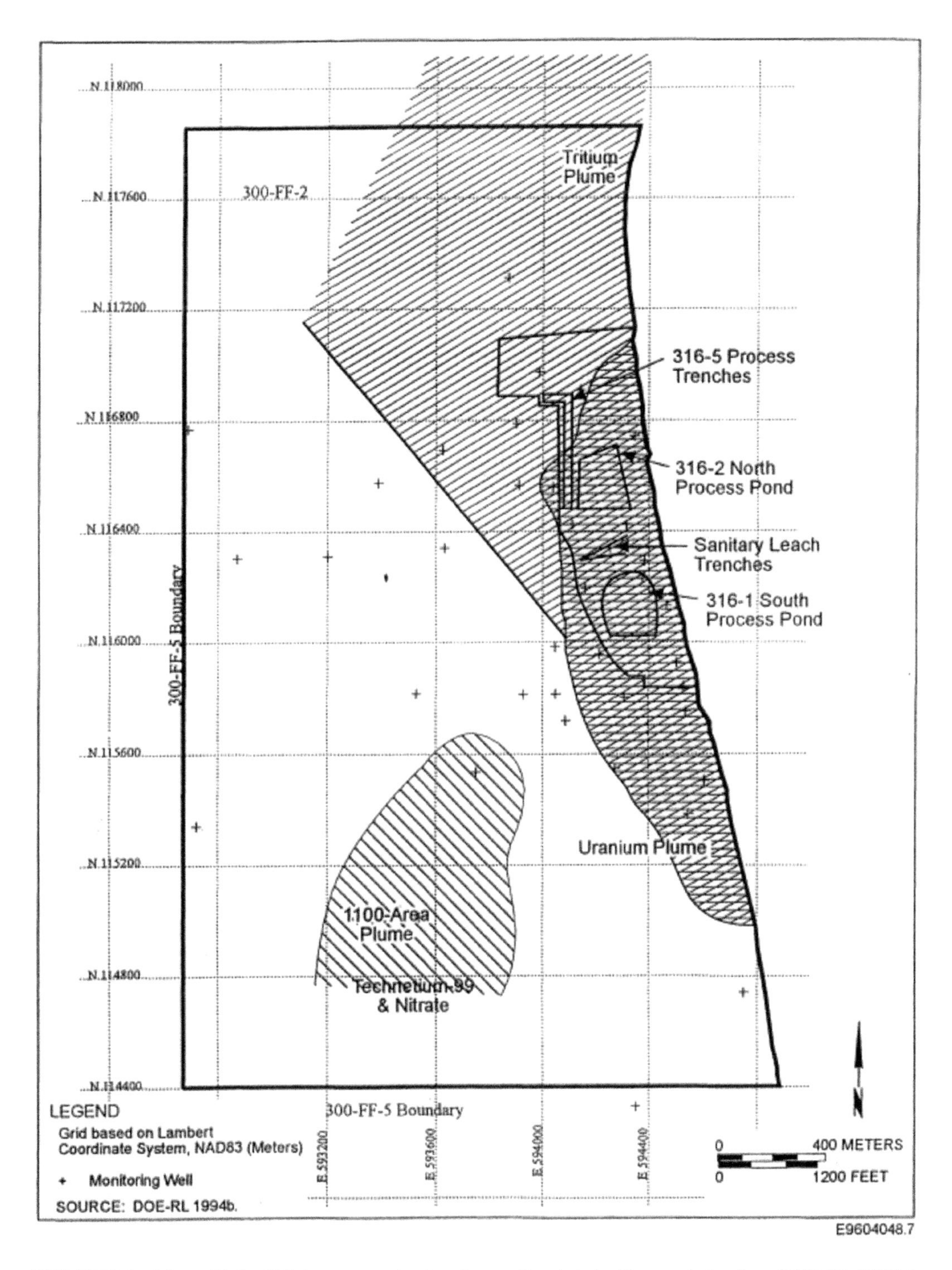


**FIGURE 3.6** Map with the 300 Area groundwater plumes. Reprinted with permission from DOE-RL (1994a).

include total coliform, bacteria, chloroform, dichloroethane, trichloroethane, nickel, copper, and $^{90}Sr$ (DOE-RL 1994a). This is the only plume of the three that is a result of uranium fuel fabrication activities in the 300 Area. A second plume (tritium) is present through the northern and eastern portions of the 300-FF-5 OU and consists of tritium originating from the release of tritium gas from irradiated debris in one of the burial grounds. The third plume (1100 Area) is in the southwestern portion of the 300-FF-5 OU and consists of $^{99}Tc$ and nitrate from sources to the southwest of the 300 Area (DOE-RL 1994).

Multiple combinations of characterization techniques have been used throughout the Hanford Site to create reliable conceptual site models that include contaminant locations and movement within the subsurface (Last and Horton 2000; Truex et al. 2017). These techniques have included geophysical, geochemical, and microbial characterization methods.

### 3.3.1 Geophysical Characterization Techniques

- Ground-penetrating radar widely used for detection and mapping of underground structures (e.g., buried pipelines) and debris
- Electromagnetics, including metal detectors and frequency-domain electromagnetics used to detect and map underground debris or structures made of magnetic or conductive material buried relatively shallow (<30 ft)
- Electrical resistivity tomography, another electromagnetic tool, that can also be used for early leak detection
- Seismic reflection used to survey subsurface hydrogeology
- Borehole logging where sensors are used inside existing boreholes to monitor contaminants (using gross gamma-ray, spectral, and/or neutron moisture logging), temperature, or lithologic characterization (Figure 3.7)

### 3.3.2 Geochemical Characterization Techniques

- Physical characterization can include (but are not limited to) geological descriptions, particle size analysis, specific surface area measurements, bulk density, permeability, porosity, and moisture content
- Direct analysis measurements of contaminant concentrations along with other (noncontaminant) chemical constituents of sediment or groundwater samples
- Sediment extractions include various techniques, including water, acid, or sequential extractions, and can be used to characterize the concentration, mobility, and mineral associations of select contaminants

These and other characterization methods can be applied in a tiered approach, used by PNNL for Hanford Site sediment characterization (Brown and Serne 2008) to minimize cost and risk while maximizing scientific information gained (Figure 3.8):

- Tier I: Determine contaminants present and basic chemical properties
- Tier II: Use Tier I data to identify sediment samples to further characterize using specialized extractions and other analytical techniques to determine the type and extent of contamination in the sample
- Tier III: Using data from Tiers I and II, select samples are further characterized to identify detailed information on the contaminant(s) present, including oxidation, chemical, and/or chemical state of the contaminant(s)

### 3.3.3 Microbial Characterization Techniques

- **Microbial Density Estimation:** It is used to measure the total number of microorganisms present in an environmental sample. Values are represented as colony-forming units on laboratory growth media or as cell equivalent determinations from molecular-based assays.

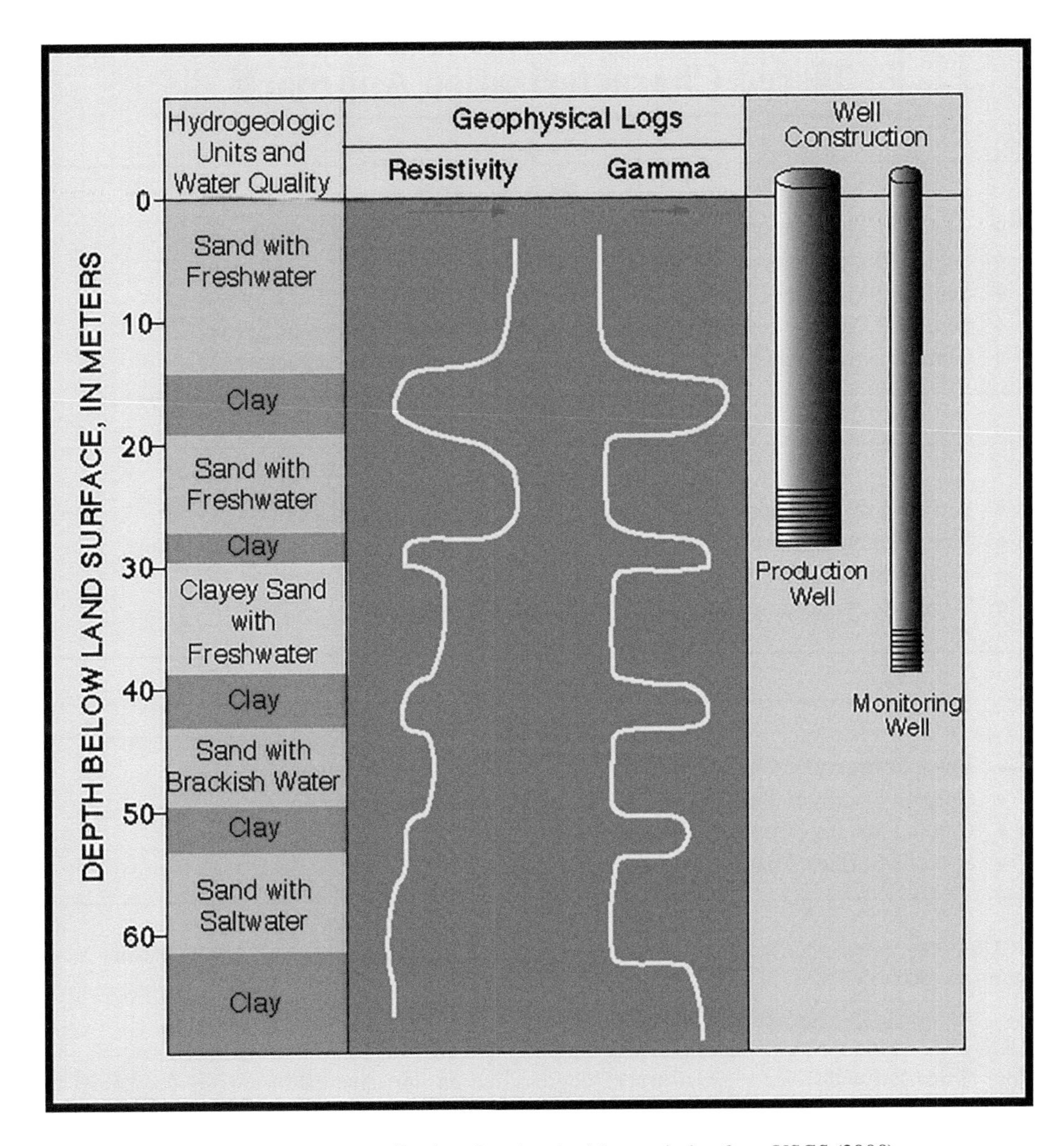


**FIGURE 3.7** Spectral gamma log collection. Reprinted with permission from USGS (2000).

- **Microbial Diversity:** It is used to provide a comprehensive inventory of the different types of bacteria present in an environmental sample. Inferences can be made about physiological potential regarding environmental processes and predominant environmental conditions.

## 3.4 GROUNDWATER MONITORING

After characterization is complete and selected remedies have been implemented, it is important to continue monitoring to ensure remedial actions are performing as expected. At the Hanford Site, environmental monitoring is conducted "...to demonstrate that discharges are at safe planned levels, identify trends and anomalies, and provide detection of unplanned releases to the environment"

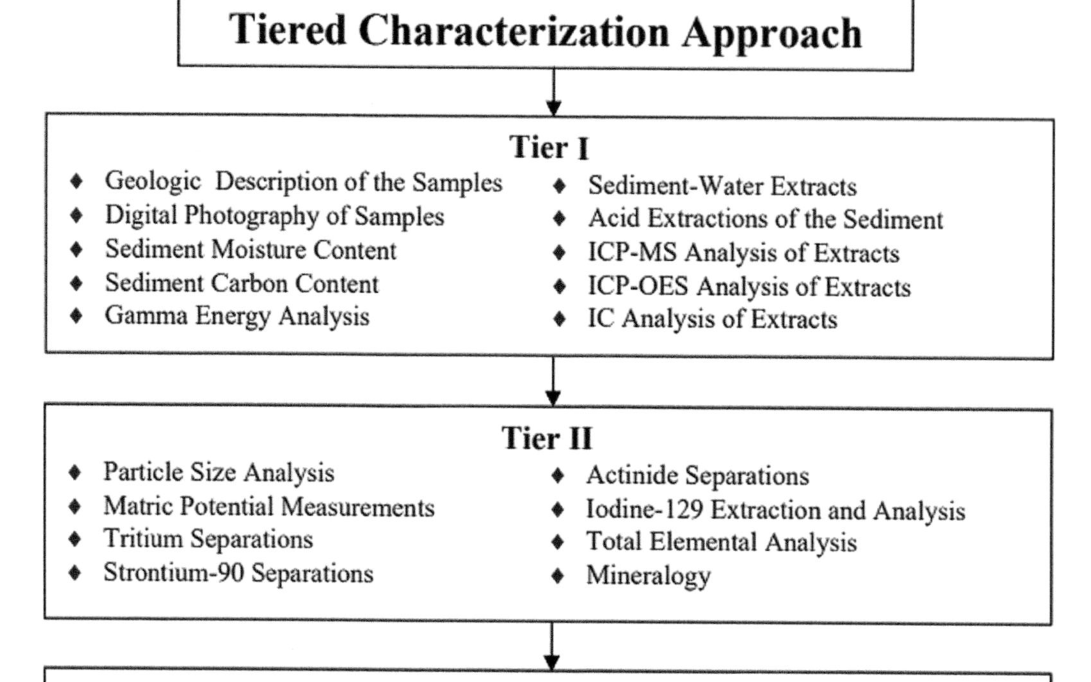


**FIGURE 3.8** Tiered approach to sample analysis and characterization. Reprinted with permission from Brown and Serne (2008).

(DOE 2022). Monitoring includes planned sampling and analysis of groundwater, river water, and biota associated with river seeps to track progress and ensure remediation targets (explained in Chapter 5) are being met. For over 40 years, the groundwater monitoring network has continued to track uranium concentrations in the groundwater (Zachara et al. 2012; DOE 2021). A combination of historical records and annual monitoring of the groundwater has shown that residual uranium sources in the vadose zone, periodically rewetted during the high river stage, provide a persistent source of uranium to the groundwater in the 300-FF-5 OU.

Monitoring of the groundwater is conducted under DOE's Groundwater Remediation Project (DOE-RL 2002). Because of fluctuations to the water table beneath the 300 Area, trends in discharge and contaminant concentrations fluctuate seasonally (PNNL-17034). The groundwater is monitored via sampling from three downgradient wells and one upgradient well for four months in a row during each biannual sampling period (Figure 3.9). The wells used to monitor groundwater in the 300 Area cover three hydrologic regions, including the upper unconfined aquifer, the lower part of the unconfined aquifer, and the uppermost confined aquifer. Samples are analyzed for volatile organic compounds and uranium. Groundwater sampling and analysis, including scheduling, is conducted by the site contractor following the Hanford Site Environmental Monitoring Plan (DOE-RL 2018), which is based on requirements laid out in two DOE orders (DOE O

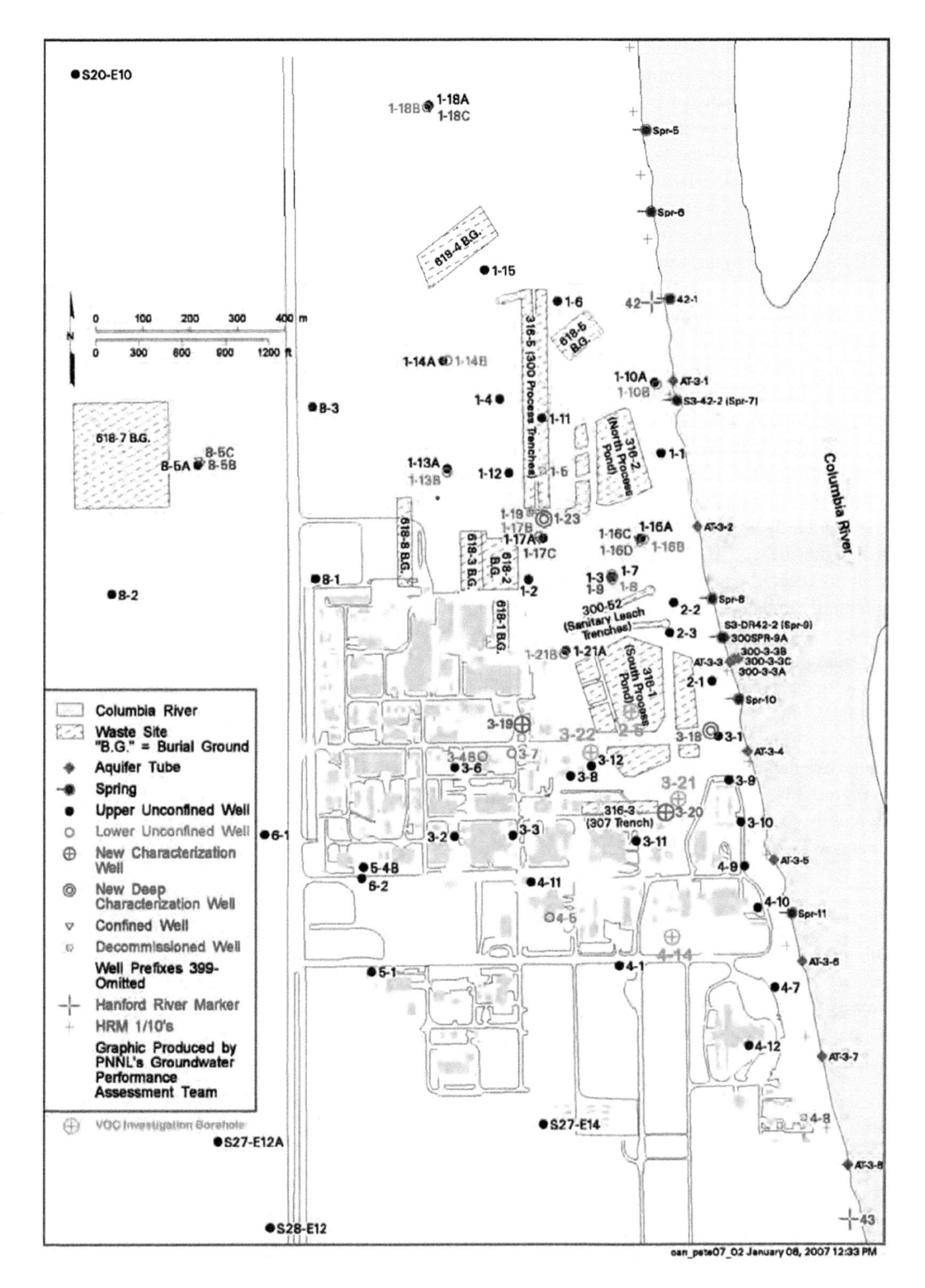


**FIGURE 3.9** Map showing locations of monitoring wells and shoreline sampling sites. Reprinted with permission from (PNNL-17034).

436.1, *Departmental Sustainability*; DOE O 458.1, *Radiation Protection of the Public and the Environment*) and the Hanford Site Air Operating Permit 00–05–006. Additional sampling under the Comprehensive Environmental Response, Compensation, and Liability Act (CERCLA) program monitors contaminants leaving the Hanford Site to offsite locations. The samples are collected annually along the 300 Area riverbank springs, nearshore river flow, and mainstream river flow (DOE-RL 2018). Several sample types including sediment, wildlife, and water are also provided to the Washington State Department of Health for comparative analysis as part of the quality assurance program (DOE-RL 2018).

Hanford Site groundwater monitoring data are added to the Hanford Environmental Information System database for convenient access, with records dating back to the 1950s. It is important to have data readily available for analysis and decision-making throughout the remediation process. To make the data easier to evaluate while integrating other data sources, the Department of Energy and Pacific Northwest National Laboratory (PNNL) partnered to create the PNNL-Hanford Online ENvironmental Information eXchange (PHOENIX; www.phoenix.pnnl.gov), where historical and annual groundwater monitoring data are available, along with several query, visualization, and analysis tools to help better interpret these data. This information, and the availability and visualization of this information, can be used to better inform remedial decisions by advising and improving the conceptual site model (CSM). The CSM represents a culmination of the site assessment and characterization by combining processes that control the concentration and transport of contaminants in the subsurface. If monitoring data show contaminant concentrations or movement outside of what is expected based on the CSM, it is a good indication that the CSM should be reevaluated and potentially updated. The CSM is meant to be updated and revisited over time, and progress from conceptual to quantitative as the project lifecycle advances (USEPA 2011). The 300 Area CSM can be found in Zachara et al. (2012).

Groundwater monitoring data are used to develop uranium plume maps; periodic updating of the maps allows changes to the plume shape to be visible over time (Figure 3.10). The size of the 300-Area uranium plume has fluctuated over time; available data from 2003 to 2020 show the plume extent between a low of 0.2 $km^2$ in 2019 and a peak of ~0.7 $km^2$ in 2012 (Cline 2020) as shown in Figure 3.11. Due to the continued source (uranium associated with sediments in a form that can be readily solubilized into groundwater) in the vadose zone, the concentration varies seasonally as shown in Figure 3.11, depicting the uranium plume during high and low river stages (DOE 2021).

## 3.5 AFTER CHARACTERIZATION

Due to the proximity to the Columbia River, the River Corridor of the Hanford Site has been a priority for cleanup efforts. The subsurface characterization information is used to inform decisions on remediation strategies for the site; however, choosing the optimal remediation strategy is not always straightforward. The path to remediation for the 300-Area uranium plume is discussed in Chapter 10.

## 3.6 CONCLUSIONS

Comprehensive environmental characterization and monitoring data have been crucial to decision-making about the evolving cleanup process of the Hanford Site's 300 Area. Characterization and monitoring included regular analysis of groundwater and river water through monitoring wells and aquifer tubes, sediment characterization pre- and post-remediation, and geophysical monitoring during and after active remediation. The transition from historical removal of soil and debris to monitored natural attenuation, and then to enhanced attenuation technologies (liquid polyphosphate solutions to form apatite), was conducted based on the changes in uranium in groundwater

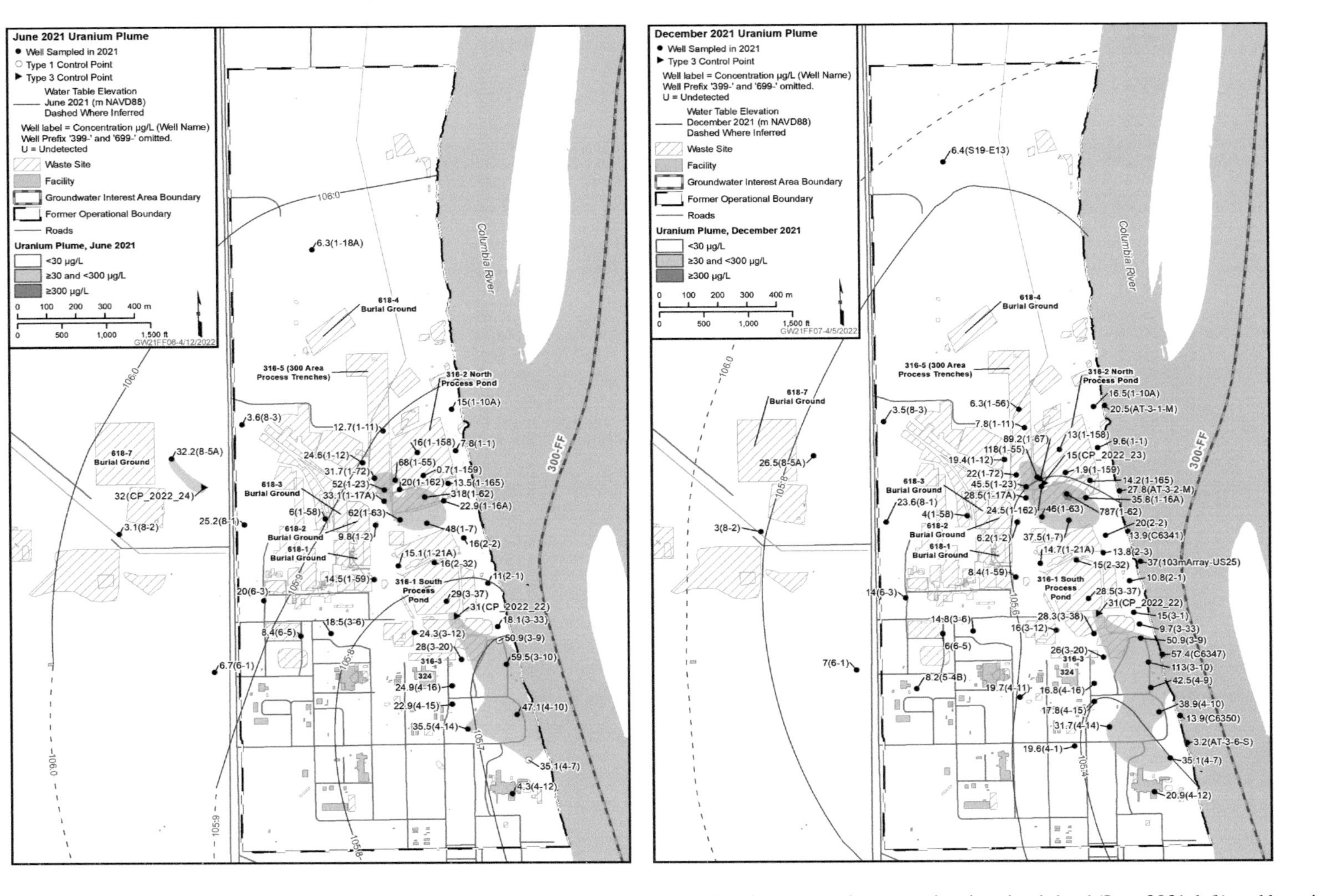


**FIGURE 3.10** Uranium plume maps demonstrating the change in shape and size between high river stage when water is migrating inland (June 2021, left) and low river stage when water is migrating toward the river (December 2021, right.) Reprinted with permission from DOE (2021).

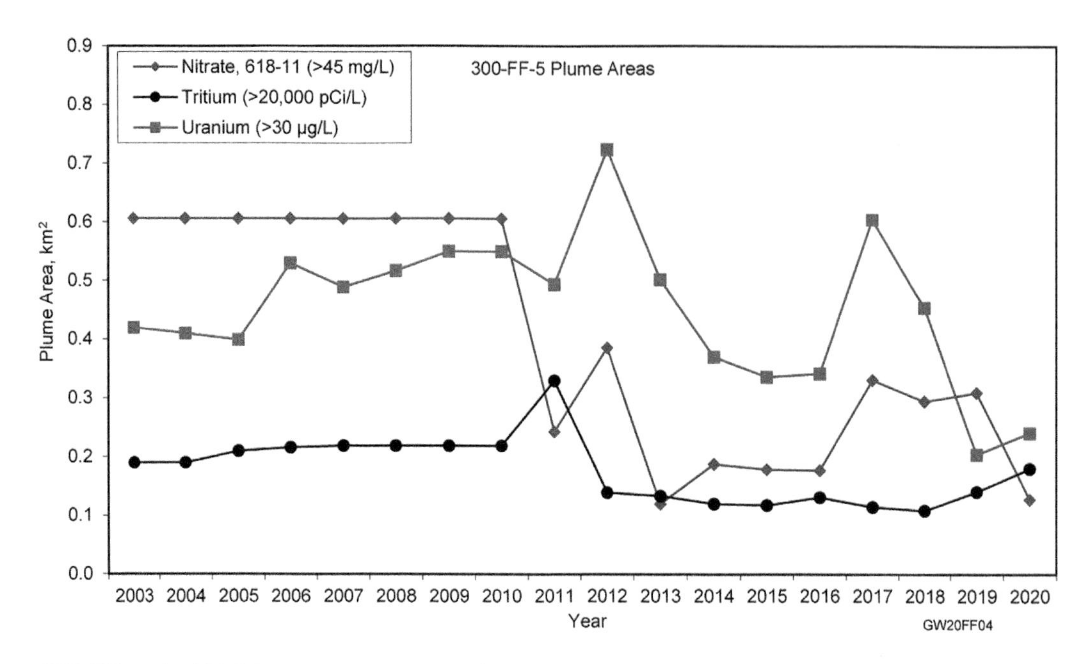


**FIGURE 3.11** 300-FF-5 plume areas (in $km^2$) between 2003 and 2020. Reprinted with permission from Cline (2020).

and Columbia River water over time. Moreover, real-time monitoring via geophysics allowed for process optimization between Stage A and B of the liquid polyphosphate injections for enhanced attenuation of uranium.

## REFERENCES

Briggs, J. D. 2001. *Historical Time Line and Information about the Hanford Site, Report.* PNNL-13524. Pacific Northwest National Laboratory, Richland, WA.

Brown, C. F. and R. J. Serne. 2008. Deep vadose zone characterization at the Hanford site: Accomplishments from the last ten years. *WM Symposia*, Tempe, AZ.

Cline, M. W. 2020. *300-FF-5 Operable Unit Enhanced Attenuation Uranium Sequestration Completion Report.* SGW-63113 Rev 0. No. 20-SGD-0071/SGW-63113. Department of Energy, Richland Operations Office, Richland, WA. https://pdw.hanford.gov/document/AR-04134.

DOE-RL. 1994a. *Phase I Remedial Investigation Report for the 300-FF-5 Operable Unit.* DOE/RL-93-21. U.S. Department of Energy, Richland Operations Office, Richland, WA.

DOE-RL. 1994b. *Phase II Report for the 300-FF-5 Operable Unit.* DOE/RL-93-22. U.S. Department of Energy, Richland Operations Office, Richland, WA.

DOE-RL. 2002. *300-FF-5 Operable Unit Sampling and Analysis Plan. DOE/RL-2002-11 Rev 2.0.* U.S. Department of Energy, Richland Operations Office, Richland, WA. https://pdw.hanford.gov/document/0095179

DOE-RL. 2013. *Hanford Site Groundwater Monitoring and Performance Report:* 2009. *Chapter 18.0: 300-FF-5 Operable Unit.* DOE/RL-2010-11, Rev. 1. U.S. Department of Energy, Washington, D.C. https://pdw.hanford.gov/document/0084237.

DOE-RL. 2018. *Hanford Site Environmental Monitoring Plan.* DOE/RL-91-50, Rev. 8. U.S. Department of Energy, Richland Operations Office, Richland, WA. https://www.hanford.gov/files.cfm/2018_EMP_estars.pdf.

DOEORL. 2019. *Hanford Site Groundwater Monitoring Report for 2018.* DOE/RL-2018-66, Rev. 0. U.S. Department of Energy, Richland Operations Office, Richland, WA. https://pdw.hanford.gov/document/AR-03138.

DOE-RL. 2021. *Hanford Site Groundwater Monitoring Report for 2021*. Rev. 0 pending. DOE/RL-2021-51. U.S. Department of Energy, Richland Operations Office, Richland, WA. https://www.hanford.gov/files.cfm/DOE-RL-2021-51_R0_Clean.pdf.

DOE-RL. 2022. *Hanford Site Environmental Monitoring Master Sampling Schedule for Calendar Year 2022*. DOE/RL-2021-54. U.S. Department of Energy, Richland Operations Office, Richland, WA. https://www.hanford.gov/files.cfm/MSA_Attachment.pdf.

DOE-RL. 2023. *Hanford Site Waste Management Units Report*. DOE/RL-88-30 Rev 32. Central Plateau Cleanup Company (CPCCo) for the Department of Energy Richland Operations Office (DOE-RL), Richland, WA. https://pdw.hanford.gov/document/AR-23391.

EPA. 2013. Hanford Site 300 Area Record of Decision for 300-FF-2 and 300-FF-5, and Record of Decision Amendment for 300-FF-1. Environmental Protection Agency, Washington, D.C.

Fritz, B. G. and E. V. Arntzen. 2007. Effect of rapidly changing river stage on uranium flux through the hyporheic zone. *Groundwater*, *45*(6), 753–760.

Gerber, M. S. 1992a. *Legend and Legacy*: *Fifty Years of Defense Production at the Hanford Site*. No. WHC-MR-0293-Rev. 2. Westinghouse Hanford Co., Richland, WA.

Gerber, M. S. 1992b. *Compilation of Historical Information of 300 Area Facilities and Activities*. No. WHC-MR-0388. Westinghouse Hanford Co., Richland, WA.

Gerber, M. S. 1993. *Multiple Missions*: *The 300 Area in Hanford Site History*. No. WHC-MR-0440. Westinghouse Hanford Co., Richland, WA.

Haney, W. A. 1957. *Dilution of 300 Area Uranium Wastes Entering the Columbia River*. No. HW-52401. General Electric Co. Hanford Atomic Products Operation, Richland, WA.

Harvey, D. W. 2000. *History of the Hanford Site*: *1943–1990*. PNNL-SA-33307. Pacific Northwest National Laboratory, Richland, WA. https://www.osti.gov/biblio/887452.

Jones, C. G. 2005. A review of the history of US radiation protection regulations, recommendations, and standards. *Health Physics*, *88*(6), 697–716.

Last, G. V. and D. G. Horton. 2000. *Review of Geophysical Characterization Methods Used at the Hanford Site*. PNNL-13149. Pacific Northwest National Laboratory, Richland, WA.

NCRP. 2009. *Ionizing Radiation Exposure of the Population of the United States*. Report No. 160. National Council on Radiation Protection and Measurements, Washington, DC.

Peterson, R.E., Rockhold, M.L., Serne, R.J., Thorne, P.D., and Williams, M.D. 2008. *Uranium contamination in the subsurface beneath the 300 Area, Hanford Site, Washington*. PNNL-17034. Pacific Northwest National Laboratory, Richland, WA.

Phillips, S. J., and J. R. Raymond. 1975. *Monitoring and Characterization of Radionuclide Transport in the Hydrogeologic System*. BNWL-SA-5494. Battelle Pacific Northwest Laboratory, Richland, WA. https://www.osti.gov/servlets/purl/4153593.

Stubbs, J. E., L. A. Veblen, D. C. Elbert, J. M. Zachara, J. A. Davis, and D. R. Veblen. 2009. Newly recognized hosts for uranium in the Hanford Site vadose zone. *Geochimica et Cosmochimica Acta*, 73(6), 1563–1576. https://doi.org/10.1016/j.gca.2008.12.004

Teel, S.S. and K.B. Olsen. 1990. *Final Report: Surface radiation survey for the Phase I remediation investigation of the 300-FF-1 Operable Unit on the Hanford Site*. EMO-1008. U.S. Department of Energy. Richland, WA.

Truex, M. J., J. E. Szecsody, N. Qafoku, C. E. Strickland, J. J. Moran, B. D. Lee, M. Snyder, A. R. Lawter, C. T. Resch, B. N. Gartman, and L. Zhong. 2017. *Contaminant Attenuation and Transport Characterization of 200-DV-1 Operable Unit Sediment Samples*. PNNL-26208; RPT-DVZ-AFRI-037. Pacific Northwest National Laboratory, Richland, WA.

U.S. DOE. 1996. Declaration of the Record of Decision, USDOE Hanford 300 Area, 300-FF-1 and 300-FF-5 Operable Units. U.S. Department of Energy, Washington, D.C.

USEPA. 2001. *US DOE Hanford Site First Five Year Review Report*. U.S. Environmental Protection Agency, Washington, D.C. https://www.hanford.gov/files.cfm/First_CERCLA_Review_-_Hanford_5-Year_Review_Final.pdf.

USEPA. 2011. *Cleanup Best Management Practices: Effective Use of the Project Life Cycle Conceptual Site Model*. U.S. Environmental Protection Agency, Washington, D.C.

USGS. 2000. Borehole Geophysics - New York Water Science Center, U.S. Geological Survey. https://www.usgs.gov/media/images/geophysical-logs-hydrogeologic-units-and-qw-and-well-construction

Walters, W. H., R. L. Dirkes, and B. A. Napier. 1992. *Literature and Data Review for the Surface-Water Pathway: Columbia River and Adjacent Coastal Areas*. PNL-8083-HEDR. Pacific Northwest National Laboratory, Richland, WA. https://pdw.hanford.gov/document/E0023909.

Yokel, J. and M. Priddy. 2010. *Uranium and Other Chemical Contaminants Entering the Columbia River from the South Columbia Basin Irrigation Outfalls: A Cooperative Study by the Washington State Departments of Ecology and Health.* Publication number 10-05-019. Washington State Department of Ecology, Nuclear Waste Program, Richland, WA.

Zachara, J. M., C. Brown, J. Christensen, J. A. Davis, E. Dresel, C. Liu, S. Kelly, J. McKinley, J. Serne, and W. Um. 2007. *A Site-Wide Perspective on Uranium Geochemistry at the Hanford Site*. PNNL-17031. Pacific Northwest National Laboratory, Richland, WA.

Zachara, J. M., M. D. Freshley, G. V. Last, R. E. Peterson, and B. N. Bjornstad. 2012. *Updated Conceptual Model for the 300 Area Uranium Groundwater Plume*. PNNL-22048. Pacific Northwest National Laboratory, Richland, WA.

# 4 Plutonium Finishing Plant Building and Subsurface Waste Release Characterization

*Calvin H. Delegard, Carolyn I. Pearce, Hilary P. Emerson, Andrea M. Hopkins, and Theodore J. Venetz*

The operations history of Hanford's Plutonium Finishing Plant (PFP) and subsequent characterization of the subsurface waste disposal facilities associated with the PFP are described in this chapter as an introduction and prelude to Chapter 9 on PFP demolition and subsurface disposal structure grouting. The assessment of the waste and contamination within the PFP building and released to the subsurface was important for developing a conceptual site model with which remediation decisions have been made for the PFP building demolition and are still being made for the subsurface.

## 4.1 HANFORD'S PLUTONIUM FINISHING HISTORY AND OPERATIONS

The Hanford Site (formerly known as the Hanford Engineer Works, Site W) in southeast Washington State was created by the Manhattan Project in 1943 to produce plutonium (Pu) for use in atomic weapons in World War II. This first kilogram-scale production of an artificial element was accomplished by irradiating natural uranium in graphite-moderated reactors and, using chemical precipitation processes, separating the Pu, present at about 220 g per metric ton of uranium, from the uranium and its fission products. A thick Pu nitrate paste product[1] was shipped from Hanford's 231 or 231-W (later, 231-Z) Building in 80- and 160-g (Pu) batches to the Los Alamos (Site Y, Los Alamos, NM[2]) Building D starting from February 5, 1945. Here, it was further purified by precipitation and solvent extraction processes and then converted to Pu metal, alloyed, and shaped for nuclear explosive cores. By the end of Building D's weapons part operations in September 1945, Pu metal cores for the Trinity device, the Nagasaki bomb, a second Pu-based bomb, and the first "composite" core using both enriched uranium and Pu had been produced. Starting from November 1945 and until 1949, Site Y's Delta Prime (DP) Site assumed weapons core production for national defense and continued Pu metal production, part fabrication, Pu processing, and research until mid-1978 (Section C of DuPont 1944; Section 2.1 of Gerber 1997; Chapters 1 and 2 of Christensen and Maraman 1969; Garde et al. 1982; Chapter 4 and Figure 2–6 of Widner et al. 2010).

Hanford's Plutonium Finishing Plant (PFP; the 234–5 or 234–5Z Building, Z Plant, or colloquially Dash-5), envisaged in 1946, designed in 1947, and constructed in 1947–1949, succeeded the DP Site on July 5, 1949, as the nation's center to convert Pu nitrate to metal and produce Pu metal parts for the nation's atomic weapons stockpile even as the DP Site continued Pu processing and fabrication of atomic weapons test items. The dilute waste handling and treatment facility (241-Z) for in-ground disposal was part of the original PFP construction. Later, the growth of the PFP mission prompted the addition of the Plutonium Reclamation Facility (PRF; 236-Z) to purify scrap Pu, a waste incinerator and Pu leaching facility (232-Z), and the PRF waste treatment and americium, Am, recovery system (242-Z) shown in Figure 4.1. Though weapons part production at PFP concluded in mid-December 1965, with these responsibilities transferred to the Rocky Flats Plant, routine Pu oxide and metal production continued at PFP until shutdown in June 1989.[3]

DOI: 10.1201/9781003329213-6

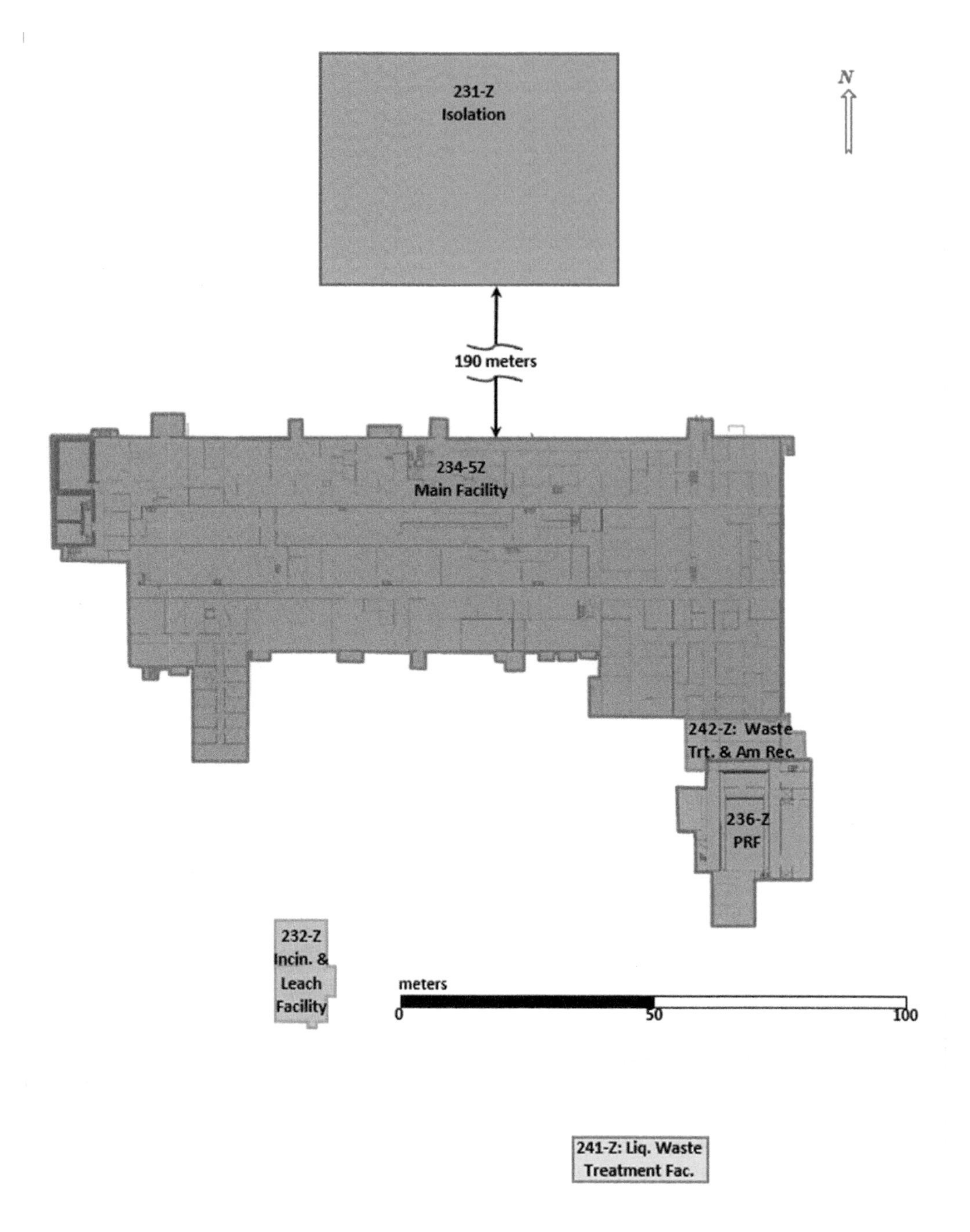


**FIGURE 4.1** Major process facilities at Hanford's Plutonium Finishing Plant.

### 4.1.1 PFP Operations

Initial Pu processing at the PFP adopted the chemical steps used at Los Alamos' DP Site to convert Hanford's Pu nitrate paste to Pu metal, and then alloy, precisely form, and coat the metal to the requisite weapons parts (Figure 4.2). At PFP, the operations were divided into "Tasks" (below) with different potential waste and scrap streams produced from each that would require different operations for treatment and disposal following characterization of the materials ultimately determined to be waste.

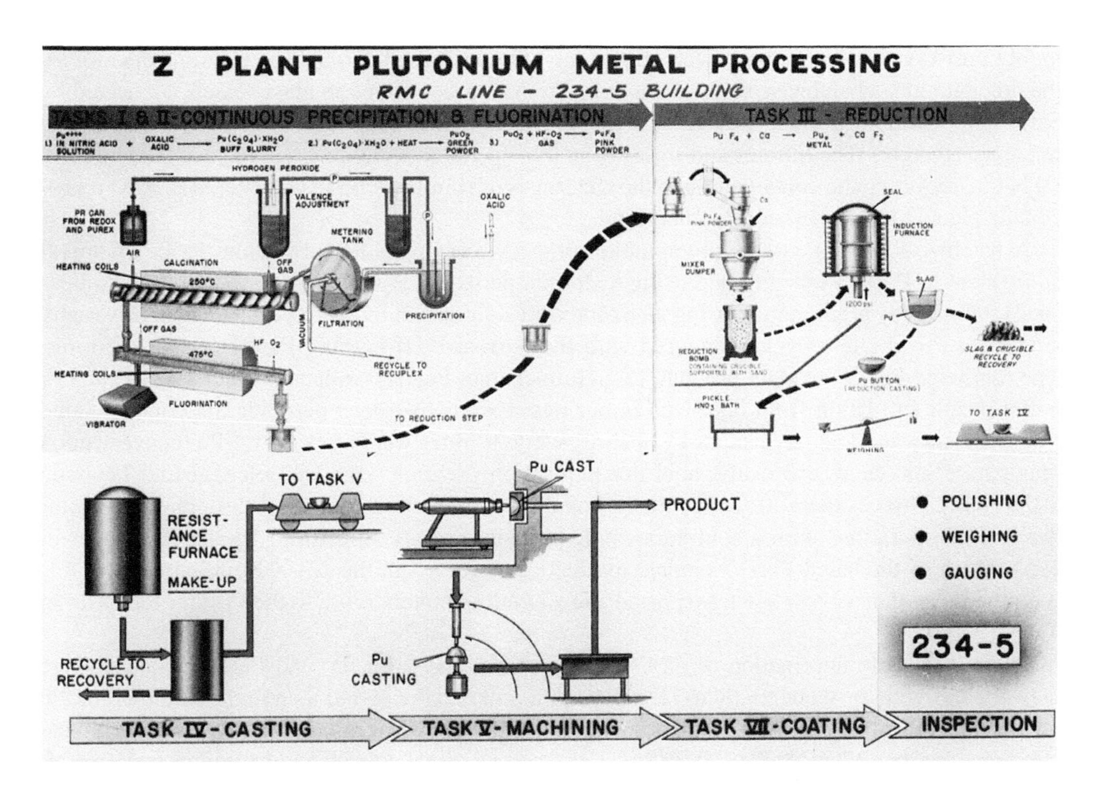


**FIGURE 4.2** Plutonium conversion, processing, finishing, and fabrication at Hanford's Remote Mechanical C Line. Reprinted with permission from Hanford (1961).

- **Task I:** Wet Chemistry (Pu oxalate precipitation for purification and to convert the solution to a solid intermediate on the way to usable metal)
- **Task II:** Hydrofluorination (also known as fluorination and dry chemistry to convert the oxalate to a Pu fluoride compound suitable for reduction to metal)
- **Task III:** Reduction (of Pu fluoride to Pu metal)
- **Task IV:** Casting (to approximate shape needed for weapons parts)
- **Task V:** Machining (shaping to exact dimensions)
- **Task VI:** Cleaning (chemical degreasing)
- **Task VII:** Coating (nickel coating deposited using nickel carbonyl)
- **Task VIII:** Final inspection (of Pu metal weapons parts)
- **Tasks IX–XIV:** Instrumentation, control, ventilation, conveyor system, equipment maintenance, and sampling

Due to its value, Pu in the PFP process waste and scrap streams was recovered to the extent technically achievable.

Chemical conversions were developed at Site Y to purify Hanford's initially relatively impure Pu nitrate paste through precipitation and solvent extraction (Carritt et al. 1946). By the time of PFP operations, the Pu nitrate paste purity had markedly improved such that these purification steps were no longer as extensive. As at Site Y's Building D and DP Site, the initial Task I Wet Chemistry step at PFP was conversion of the Pu nitrate paste to trivalent Pu, Pu(III), oxalate. This was accomplished by dissolving the paste (later, the solution concentrate) containing tetravalent Pu, Pu(IV), and hexavalent Pu, Pu(VI), in nitric acid, $HNO_3$; chemically reducing the Pu to Pu(III) using hydriodic acid, HI; and then mixing that solution with oxalic acid, $H_2C_2O_4$, solution. Large, teal-colored, and readily rinsed crystals of Pu(III) oxalate chemically precipitated. Because HI is highly corrosive to the stainless-steel apparatus otherwise used throughout Hanford operations, the precipitation, a batchwise settle-decant operation, was conducted in glass vessels connected by Saran™ (polyvinyl chloride) and tantalum tubing (Section 3.5.4 of Delegard and Peterson 2020 and references therein). The filtrates and rinses from this operation were fumed to remove the corrosive HI before recycling the contained Pu to the end stages of Hanford's B, T, and, later, REDOX[4] reprocessing plants for recovery and purification.

A batchwise Pu(IV) oxalate precipitation process was implemented in June 1952. Chemical adjustment to Pu(IV) was attained using hydrogen peroxide, $H_2O_2$, solution (benign to stainless steel) followed by precipitation using an oxalic acid solution. In terms of performance, this vastly simplified Pu filtrate recycle compared with the corrosive HI-bearing Pu(III) oxalate filtrates. The tan/beige Pu(IV) oxalate product, though finer than Pu(III) oxalate, provided equivalent or better decontamination from dissolved impurities. Excess hydrogen peroxide disproportionated to innocuous water and oxygen gas by heating, while filtrates were recycled for Pu recovery upon manganese ion–catalyzed oxalic acid decomposition (Section 3.5.5 of Delegard and Peterson 2020 and references therein). Task I Pu(IV) oxalate precipitation continued through the remainder of PFP operations with a continuous precipitation process supplanting the batch process in 1957. Much of the batch Pu(IV) oxalate production occurred in the 231-Z Building (Figure 4.1) with the Pu oxalate cakes then transported a few hundred meters south to the PFP for the ensuing Tasks.

In Task II Hydrofluorination or Dry Chemistry, the Task I Pu(III) and Pu(IV) oxalates were converted directly or sequentially to olive green Pu dioxide, $PuO_2$, and then to the salmon pink Pu tetrafluoride, $PuF_4$, intermediate by heating and exposure to hydrogen fluoride gas, HF. Specialized corrosion-resistant filter boats used in Task I also served as calcination and hydrofluorination containers in the early Task II batch processes. The boats' filters and drains used to separate the precipitated Pu(IV) oxalate from the mother solution in Task I also directed hot HF gas in Task II through the drying and calcining Pu oxalate/oxide cake to form $PuF_4$.

A mechanized apparatus integrated the Task I and II operations into a continuous process. It was installed in 1957, partially fulfilling an original PFP goal to reduce the intensive "hands-on" batchwise processing formerly intrinsic to these steps.

Throughout PFP operations, reduction to Pu metal, Task III, took place by the batchwise reaction of chemically excess calcium metal, Ca, with $PuF_4$ in magnesium oxide, MgO, crucibles within sealed, metallic pressure-tight "bombs." The reaction is exothermic such that once it initiates upon induction heating, the charge rapidly self-heats to ~1,500°C–2,000°C and goes to completion with the dense, molten Pu metal product sinking through the molten calcium fluoride, $CaF_2$, slag to the crucible bottom. A "booster" of elemental iodine, $I_2$, was used to increase reaction heat with the product calcium iodide, $CaI_2$, concurrently lowering the $CaF_2$ slag melting point. Initial operations based on the Site Y design produced 500-g Pu metal "buttons" with 800- and 2,000-g buttons later produced using larger crucibles, charges, and pressure vessels. After reaction and cooling, the Pu metal buttons were broken free from the frangible MgO crucibles and adhering solidified $CaF_2$ slag. The buttons were "pickled" in $HNO_3$ solution to dissolve slag traces, and then sampled for purity. Buttons not meeting purity specifications were recycled by burning to oxidize Pu metal to $PuO_2$, followed by dissolution in $HNO_3$ with hydrofluoric acid, HF, as a catalyst and complexant, and finally purified. Electrolytic button dissolvers also were used. Though 99% conversions of $PuF_4$ to Pu metal generally were obtained, with appreciable decontamination from impurities, occasional failures to fire or coalesce occurred. Slags and crucibles containing scrap Pu were collected in any case for Pu recycle. Task III equipment and technique improvements decreased worker handling, especially for the high neutron dose $PuF_4$.

Task IV involved metal blending and the vacuum casting of approximate metal shapes in molds. Alloying the Pu metal with gallium also could be done in this step to obtain the machinable delta (δ) phase. The cast shapes were broken from the molds, and the Pu metal adhering to the molds, known as "skulls," was burnt to $PuO_2$ and dissolved with $HNO_3$/HF for Pu recovery.

The cast Pu metal parts were machined to close tolerances in Task V using "lard oil," a solution of hog lard in carbon tetrachloride, $CCl_4$, as a lubricant and coolant. The relatively pure machining chips and turnings were collected, degreased with fresh $CCl_4$, compressed, and returned to the Task III or IV (reduction and casting) operations or burnt to $PuO_2$ and the lard oil filtered to collect Pu fines.

Task VI cleaned/degreased the machined Pu metal alloy parts, and Task VII coated the parts with nickel using nickel carbonyl gas. This coating inhibited corrosion of the part and sloughing of Pu contamination. Task VIII was Final Inspection for integrity and dimension. The as-implemented Task I through VIII sequence, which took place from 1957 until 1965, and diagrams of the mechanized Tasks I, II, and III, which continued until the PFP shutdown in 1989, are shown in Figure 4.2.

### 4.1.2 The Rubber Glove Line

To lower operator radiation exposure and improve process consistency, the original PFP design was to have a remotely operated and mechanized apparatus achieve Tasks I, II, and III. However, the Atomic Energy Commission accelerated the planned end-of-1949 PFP startup one year to January 1, 1949, and, under Cold War pressures, chose to forgo mechanized remote processing and opt for a completely manual process, similar to that used at the DP Site. The Task I through VIII manual operations were conducted sequentially within a 180-ft (55-m)-long line of 28 hermetically connected and windowed stainless-steel gloveboxes, fitted with heavy rubber gloves. Workers manually performing the process steps thus were separated from direct contact with the radioactive Pu in this so-called Rubber Glove (RG) Line. Hot startup was July 5, 1949, but operational difficulties and radioactive contamination leaks delayed full operations until December 1949. The RG Line continued full operation until March 1953, when it was limited to Task II and III operations, and completely ceased operations in September 1953.

### 4.1.3 The Remote Mechanical A Line

The Remote Mechanical A Line, that is, the RMA Line, commenced Task II–VIII operations in March 1952 at PFP, with the Pu(IV) oxalate Task I feed for the RMA Line being prepared batchwise in the nearby 231-Z Building. The similar Remote Mechanical B (RMB) Line, installed with cold shakedown testing occurring in May 1952, proved obsolete even before hot operations commenced due to RMA Line process improvements. Further improvements included continuous auger-driven Task II hydrofluorination furnaces installed in RMA in May 1955 and remotely operated lathes emplaced in 1955–1956 for Task V. Integrated continuous Pu(IV) oxalate precipitation, calcination of the Pu(IV) oxalate to form $PuO_2$, and hydrofluorination of the $PuO_2$ to form $PuF_4$ were implemented in July 1957 with equipment upgraded based on process experience. Details on continuous Tasks I–III process parameters are given by Crocker and Hopkins (1963). RMA's weapons part manufacture (Tasks IV through VIII) concluded in December 1965 with RMA then briefly placed on hold. RMA produced some Pu metal in 1966. In 1968, the hydrofluorination channel in the Task II apparatus was converted to a secondary calciner to produce $PuO_2$. From 1968 until a final stabilization run in 1979, the RMA Line only produced $PuO_2$ product, including some for Fast Flux Test Facility fuel.

### 4.1.4 The Remote Mechanical C Line

The Remote Mechanical C, RMC, Line began operations in October 1960 to supplement production by the RMA Line. By November 1960, all Task I and II operations and 80% of PFP's Task III operations occurred in the RMC. By 1961, the RMC Line accomplished all Task I–III operations, while the RMA focused on Pu metal part fabrication. By 1962, the RMC Line increasingly undertook the fabrication operations as well. With RMA Line production ramped down and closing at the end of 1964, RMC Line production increased. Although part fabrication equipment in both RMA and RMC Lines was laid up by March 1966 following the cessation of weapons part production in mid-December 1965, the RMC Line continued producing Pu metal through August 1973. After a 12-year pause, the refurbished RMC Line undertook two metal production campaigns in 1985 and 1986 with a final metal campaign occurring from July 1988 to June 1989.

### 4.1.5 Plutonium Scrap Recovery Operations

The original mission at Hanford's 231-Z Building was the purification and concentration of bismuth phosphate process Pu nitrate solution product through two sequential precipitations as Pu peroxide in the "Isolation Process" (Section 3.5.2 of Delegard and Peterson 2020). In early operations, off-specification Pu nitrate and Pu peroxide filtrates from 231-Z were recycled to the lanthanum fluoride, $LaF_3$, precipitation stages in Hanford's 224-T and 224-B Buildings. Off-specification Pu nitrate also could be kept at 231-Z for further purification via Pu peroxide precipitation.

With PFP operations beginning in 1949, Pu flows for recovery (from the Task I through V scrap streams – oxalate, oxide, fluoride, metal, filtrates, metal foundry, and machining residues) increased greatly in magnitude but still relied on the return of the Pu scrap, first dissolved in $HNO_3$, to 231-Z, to the 224-T and 224-B Buildings, and, from 1952, to the REDOX reprocessing plant for purification and recycle.

RECUPLEX, a dedicated in-house Pu scrap recovery facility, eliminated cross-site Pu solution shipments and operated within the 234–5 Building from 1955 until its abrupt shutdown because of a nuclear criticality on April 7, 1962. The PRF, in the 236-Z Building adjoined to 234–5, succeeded RECUPLEX in 1964 for Pu scrap recovery and purification and ran until 1994. Both the RECUPLEX and PRF processes purified Pu(IV) by solvent extraction from $HNO_3$ solution into tributyl phosphate (TBP) extractant diluted in $CCl_4$. Polishing to remove trace Pu from PRF raffinates was done by dibutyl butyl phosphonate (DBBP) solvent extraction in $CCl_4$ diluent in the 242-Z Building, located between the 236-Z and 234–5 Buildings. A cation exchange process in 242-Z to recover americium-241 ($^{241}$Am) in-grown from $^{241}$Pu decay operated from 1964, just after the start of PRF, until August 30, 1976, when the ion exchange column exploded and permanently shut down this system.

**THE EXPLOSION AT 242-Z**

Removal and disposal of the highly contaminated debris and wreckage remaining from the August 30, 1976 explosion at 242-Z were key challenges in decommissioning the facility. The explosion was caused by radiolytic and chemical degradation of the ion exchange resin, made worse by leaving 130 grams of Am loaded onto about 13 L of resin for five months during a labor strike. The resin suffered $2.7\times10^9$ rad ($2.7\times10^7$ Gray) exposure over this interval and thus extensive radiolytic damage; a nearly identical resin suffered 20% capacity loss at 10% of the dose in a much less oxidizing hydrochloric acid medium (Kajanjian and Horrell 1975). In the leadup to the accident, the compromised resin was washed with 0.3 M $HNO_3$ and then eluted according to routine procedure using 7 M $HNO_3$, producing a dark-colored solution indicating resin damage. Elution was stopped while eluate analyses were performed. In this period, the high concentration $HNO_3$ continued oxidizing the radiation-damaged resin in an autocatalytic and self-accelerating manner. About 3 hours after introduction of the 7 M acid, Harold McCluskey, the chemical operator, reached into the glovebox, observed brown fumes so dense he could not read a pressure gauge within the glovebox, felt unusual heat, heard "hissing," and observed liquid leakages. Before he could fully disengage from the glovebox, the 15 cm diameter by ~90-cm-long stainless-steel ion exchange column exploded, failing by hoop stress, blowing out glovebox gloves and windows, including the overhead window, jetting solution and staining the ceiling, shattering the leaded glass shielding, and piercing McCluskey with contaminated glass shards and splashing him with Am-bearing acid and resin (Figure 4.3). The dose rates immediately in front of the glovebox were 3 rad/hour and 0.5 rad/hour 2.5 m away, indicating significant Am dispersal. Limited cleanup and stabilization of the process spaces occurred, and the area was isolated from further access and operations until PFP closure and remediation occurred as described in Chapter 9 (BNWI 1976). McCluskey, 64 years old at the time of the accident, underwent 79 days of isolation in a radiological hospital facility with medical treatment, mostly for lacerations and acid burns, and decontamination, including internal and external chelation therapy and scrubbing with mild detergents. He then resided in an adjacent custom camping trailer with continued treatments. He began short daily visits home after ~103 days and returned to living at home in January 1977, about five months after the accident. McCluskey died eleven years after the accident of congestive heart failure arising from pre-existing coronary heart disease. Post-mortem examination showed 540 kBq 241Am (4.6 μg), ~94% of which was in bone surface and bone marrow and ~5% in the liver (Carbaugh 2016).

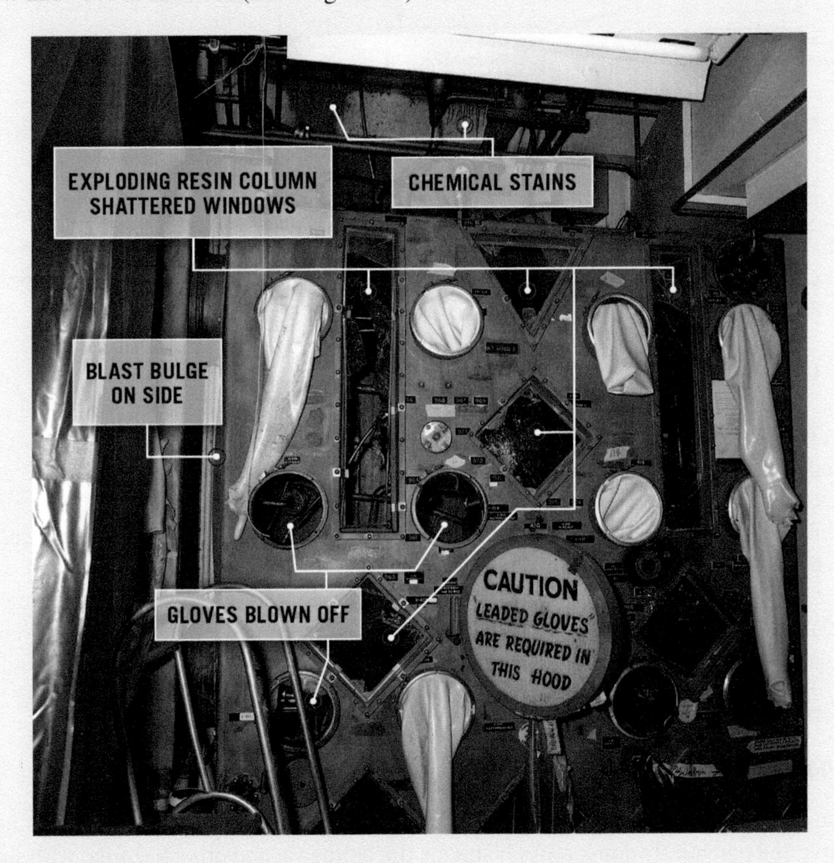


**FIGURE 4.3** Image of the McCluskey glovebox following the explosion.

Scrap fed to RECUPLEX and PRF included Pu nitrate solutions derived by dissolving Task I through V residues, as well as glovebox sweeps and ash from burning Pu-bearing materials – filters, plastics, gloves – in the 232-Z incinerator and leach facility (1962–1973). Solid Pu-bearing scrap from other national weapons material sites, most prominently Rocky Flats, also was processed at PRF. Details on RECUPLEX and PRF/242-Z operations and the scrap feeds are described elsewhere (Delegard et al. 2019 and references therein). The PFP process cycle of Tasks I–VIII; RECUPLEX, PRF, and 242-Z recovery; and 232-Z are diagrammed in Figure 4.4 and associated Pu materials are shown in Figure 4.5.

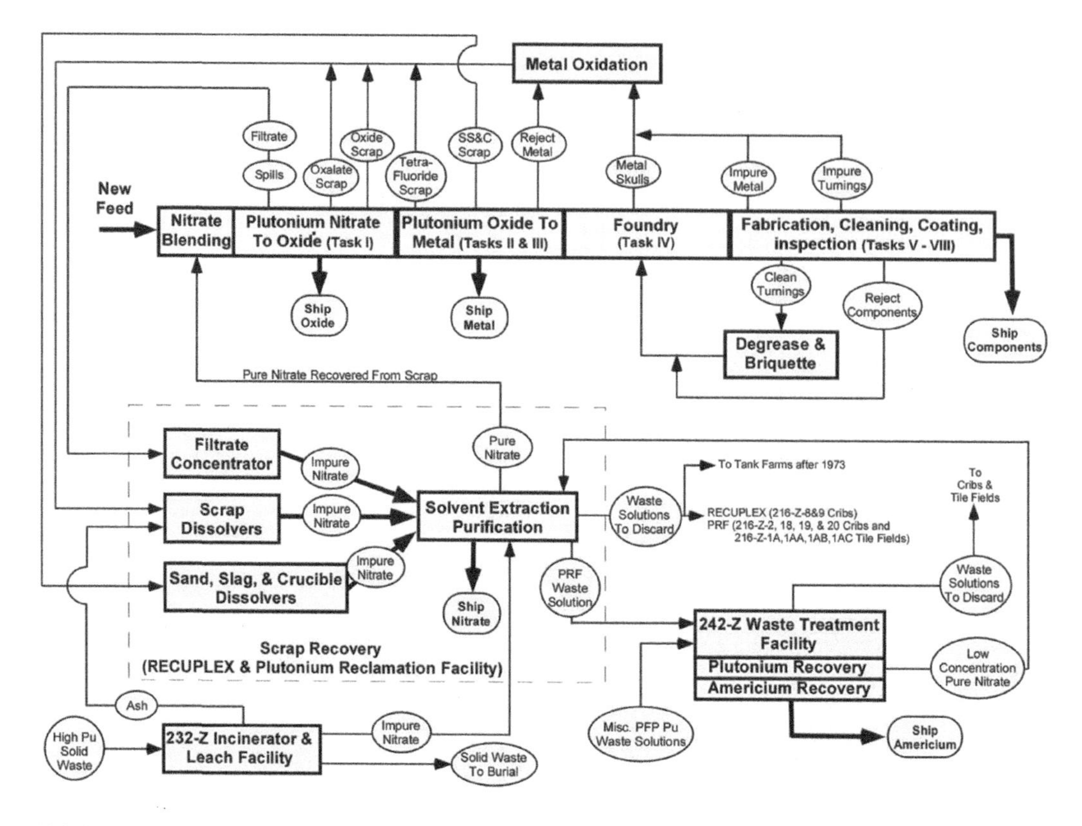


**FIGURE 4.4** Plutonium processing cycle at the PFP. Reprinted with permission from Hoyt and Teal (2004).

At the effective end of its processing and production mission in 1989, the PFP was left in operational standby, still holding significant Pu-bearing scrap and product-quality Pu oxide and metal inventories, as well as process chemicals. From that time, the process lines fell into disrepair, while radiological control, Pu inventory maintenance, environmental monitoring, and physical security measures necessarily remained in place with their attendant costs. The DOE issued a letter in October 1996 calling for PFP's shutdown, thus taking the first step in its decommissioning (Hebdon et al. 2003). Early characterization of the PFP building's chemical and radiological inventory and the associated subsurface structures' inventories, dispositions, and behaviors of Pu and Am are discussed in the following sections.

### 4.1.6 PFP Building Complex Closure Inventory Assessment

Initial assessments were completed in 1997–1998 to identify risks and vulnerabilities associated with reactive and potentially dangerous chemicals (including Pu and other hazards) with subsequent assessments of wastes. Assessments of chemical hazards were conducted on active areas of the PFP in 1999 and primarily inactive systems in the early 2000s (Hopkins et al. 2003). The characterization process included a team for physical inspection, as well as a team that compiled the vessel inventory, which included the following components:

- Documenting the evaluation criteria and processes used
- Listing process equipment (including tanks, piping, and waste lines) that might contain residual chemicals
- Evaluating the hazards of process equipment containing residual chemicals
- Assessing the hazards of remaining chemicals, their hazard characteristics, and the quality of characterization data

**FIGURE 4.5** Solid plutonium materials produced and scrap processed at the PFP. (Courtesy of T Venetz.)

- Identifying risks associated with changes to chemicals due to aging, evaporation, or leakage and inadvertent combination with chemicals in associated systems
- Categorizing items relative to risk to human health and the environment
- Determining which items require a response action prior to final deactivation
- Providing a schedule for response actions required prior to final deactivation

Multiple types of data were involved – physical inspection and characterization, engineering evaluations, research, interviews, and historical documentation in several different types of databases – all of which were compiled into a single database for simplified control and management. Ultimately, relative risks were assigned based on ranking factors by severity of the hazard, the likelihood of occurrence of hazard, and hazard group. Hazard groups included (i) explosive, unstable reactive, unstable over time (because of aging in storage or contamination during use), and organic peroxides (29 CFR 1910.1200); (ii) pyrophoric, water-reactive, flammable gas, fissile materials (29 CFR 1910.1200); (iii) corrosive and highly toxic chemicals (29 CFR 1910.1003, 1017, 1044–50); and (iv) all other materials (generally not very reactive) including flammable and toxic substances (Hopkins et al. 2003). Example descriptions of ranking factors for severity and likelihood of risks are shown in Table 4.1.

**TABLE 4.1**
**Example Descriptions of Ranking Factors for Severity and Likelihood of Risks**

| Evaluation Factor | Value of Severity and Likelihood Factors | | |
|---|---|---|---|
| | **Severity** | | |
| A1. Physical injury potential to humans | Capable of damaging a limb, vision, or hearing. Capable of initiating a heart attack to a surprised person who is prone to heart attacks. Capable of second- or third-degree skin burns. Victim would need help getting to an eyewash or safety shower | Victim would require nothing beyond first aid treatment. Victim could reach nearest eyewash or safety shower unassisted | Essentially none |
| A2. Exposure potential to humans | Could release gases, mist, or powder that threatens lives of those in the same air space if inhaled. Also includes release of irritants | Could release gases, mist, or powder that should not be inhaled but are not irritants or life-threatening | Essentially none |
| A3. Significance of secondary impact | Capable of damaging safety systems or other equipment, not necessarily causing catastrophic failure of equipment. Cleanup is complicated | Capable of releasing chemicals that corrode other systems slowly. Cleanup requires personnel protective equipment but is not really difficult | Any released chemical(s) do not corrode other vessels. Cleanup is routine |
| | **Likelihood** | | |
| B1. Design | Bad material of construction. Unvented tank that could generate high pressures. Vented tanks where venting obviously is not adequate | Vented tank that could generate gases and heat rapidly, but vent is expected to be adequate | Heat and pressure generation very improbable in planned use |
| B2. Operation | Space occupied frequently. Chemical(s) more reactive and/or react(s) more vigorously | Space occupied infrequently. Chemical(s) not very reactive and/or not as vigorously reactive | Vessel or space not used at all or very seldom. Chemical(s) cannot start any dangerous reaction |
| B3. Containment vessel condition | Container integrity is questionable. High pressure could be generated | Container integrity is adequate. High pressure cannot be generated | Container integrity is good and expected to stay that way |
| B4. Emergency planning and safety basis | Difficult or slow to take countermeasures when abnormal conditions are noticed | Countermeasures available and easy to implement | Not expected to generate any kind of emergency response |
| B5. Maintenance and inspection | Inspections are needed at least weekly. Repairs needed more often (months apart) because of the other factors. Violation of Washington (State) Industrial Safety and Health Administration/Occupational Safety and Health Administration | Inspections are needed monthly-quarterly. Repairs expected to be needed seldom (years apart) | Routine inspections will be done but repairs are not expected to be needed |

The PFP complex contained an inventory of approximately 3,600 kg (7,900 pounds) of reactive, plutonium-bearing materials, which were grouped into the following categories for the initial environmental impact statement: (i) plutonium-bearing solutions; (ii) oxides, fluorides, and process residues; (iii) metal alloys; and (iv) polycubes and combustibles (DOE/RL 2003). In addition, approximately 50–75 kg (110–165 pounds) of plutonium-bearing materials were estimated, largely by non-destructive analyses, to have accumulated within plant hardware (e.g., ventilation, process equipment, and piping).

Several areas were difficult to inspect due to limited physical access and/or visibility. Therefore, evaluations were conducted based on historical documents, including cleanout records for the different areas of the facility, as well as interviews. Following the final assessment, no items were found to pose a significant risk (relative risk was a maximum of 15 out of 100); therefore, no action items required immediate attention and were deferred. It should be noted that the relative risk scores identified in the 1999 assessment were significantly decreased due to (i) new knowledge on the potential for chemical interactions, (ii) conditions for reaction confirmed to no longer be present, or (iii) vessels in difficult-to-inspect areas were confirmed to be empty. These rankings were then used to prioritize activities moving forward into three categories (Hopkins et al. 2003):

1. **High Priority**. Items determined to warrant actions (i.e., cleanup or movement into a safe configuration and/or passive controls) in the near term based on relative risk values and/or engineering judgment; due to the relatively low risk determined overall, these did not represent significant risks and were deferred (six items)
2. **Other/Work Scheduled**. Items that have been removed, are awaiting shipment, or have been mitigated to a safe configuration until deactivation and decommissioning; includes relative risk scores from 3 to 6 (84 items)
3. **Deferred**. Items that pose minimal relative risk and can be safely deferred; includes relative risk scores <6 (219 items)

Plans were developed for the PFP based on these characterization and assessment activities. As detailed in Chapter 9, the de-inventory, deactivation, decommissioning, and demolition process for the PFP building complex was accomplished in pieces with some iteration on the plan due to unforeseen challenges. Although the characterization process sought to identify all potential risks or pitfalls, cleanup of complex sites like the PFP often presents additional challenges along the way, resulting in an iterative approach as opposed to a linear, step-by-step process.

## 4.2 CONTAMINATION OF THE SUBSURFACE SURROUNDING THE PLUTONIUM FINISHING PLANT

The PFP discharged Pu-bearing wastes, in waste solution and as solution-entrained solids, to various in-ground structures and facilities from 1949 until 1973 (Gerber 1997; Owens 1981). Among the most prominent, the 1955–1962 RECUPLEX operations discharged about 4.0 million liters of Pu particle-bearing aqueous raffinates and 0.1 million liters of organic solution to the concrete-ceilinged 216-Z-9 (Z-9) trench. The acidic, high-salt, aqueous wastes were partially neutralized to about pH 2.5 and contained 0.48 M $Al(NO_3)_3$, 0.36 M $AlF(NO_3)_2$, 0.33 M $Mg(NO_3)_2$, 0.17 M $Ca(NO_3)_2$, 2.25 M $NaNO_3$, and an estimated 106 kg Pu (earlier estimates were lower). With so much Pu present, renewed concern for inadvertent nuclear criticality in Z-9 arose in the early 1970s, leading to assays of Pu distribution and concentrations by non-destructive (e.g., gamma and neutron) and destructive (e.g., microscopy; sampling and analysis) techniques. To allay the criticality concern, the top 30 cm (~50 $m^3$) of Z-9 soil, containing about 58 kg Pu, was "mined" from 1976 to 1978 and packaged in 5,222 10-L canisters, still leaving about 48 kg Pu in the trench. Efforts to recover Pu from the retrieved soil were entertained but, because of relatively low concentration compared with other Pu-bearing scrap and difficulties in processing back into operations, never were implemented (Sections 3.0, 4.1, and 6.0 of Delegard et al. 2019). The mining was not considered to be a remediation action but an important safety action.

In a parallel manner, ensuing PRF operations released about 5.2 million liters of high-salt aqueous waste underground to the 216-Z-1A tile field (1964–1969), partially neutralized to pH ~2.5, with about 57 kg Pu in 0.2 M $Al(NO_3)_3$, 0.3 M $AlF(NO_3)_2$, 0.3 M $Mg(NO_3)_2$, 0.2 M $Ca(NO_3)_2$, and

1.1 M $NaNO_3$ (Delegard et al. 2019). Similar high-salt PRF wastes containing about 23 kg Pu were discharged (1969–1973) from PRF to the 216-Z-18 crib (Owens 1981).

Low-salt and pH-neutral aqueous PFP wastes also went to ground disposal. A major low-salt waste composite consisted of an incinerator offgas scrubber and hydrofluorinator aspirator solutions, laboratory wastes, and cooling waters. These were passed through the 241-Z Liquid Waste Treatment Facility, collected in the 241-Z-361 settling tank, and sent downstream to the 216-Z-12 crib. Process records show that this system received 273 million liters of solution containing about 25 kg Pu during its operational lifetime from March 1959 until May 1973. However, sampling and analyses later found about 32 kg in the Z-361 settling tank with an estimated 9 kg Pu sent to the 216-Z-12 crib (Section 3.3 of Delegard et al. 2019 and references therein). The 216-Z-1 and 216-Z-2 cribs received process cooling waters, steam condensates, and occasional PRF wastes for a ~8 kg Pu total, similarly determined by process records. The 216-Z-1 and 216-Z-2 tile fields received ~7 kg Pu as pH 8–10 process, analytical, and development laboratory wastes with 216-Z-3 receiving ~6 kg Pu. From 1947 to 1966, the 216-Z-7 crib received 2 kg Pu as process and laboratory wastes from 231-Z. Other ground disposal sites received less than 1 kg Pu total (Owens 1981).

As shown in Table 4.2, about 184 kg of Pu still remains in PFP operations ground disposal sites (Owens 1981). The inventories were estimated by multiplying the discharge volumes and Pu concentrations recorded in historical records over the disposal site lifetimes. Because of entrained solids settling enroute, the entire reported inventory may not actually have been delivered to the respective disposal site. On the other hand, entrained solids may have been under-accounted. Beginning in 1973, the PFP no longer disposed Pu-rich wastes to the ground but instead disposed to Hanford waste tanks 241-TX-105, 241-TX-109, and 241-TX-118 and SY-102.

Americium, from beta decay of $^{241}Pu$,[5] is also present in Pu-bearing waste solutions and solids discharged to the in-ground structures. The total Am in the PFP ground disposal can be estimated as follows:

1. Based on 216-Z-9 trench disposal estimates (Table 4 in Delegard et al. 2019), Am quantities from the weapons-grade RECUPLEX operations were about 2.6 mass % of the 106 kg Pu at discharge, or about 2.8 kg, thus leaving ~1.2 kg after the mining operation.[6]
2. The PRF, in turn, processed both weapons-grade and higher exposure Pu, thus containing higher $^{241}Am$ concentrations than RECUPLEX feeds, though much of the $^{241}Am$ was recovered from the PRF raffinates at 242-Z. Wastes from PRF went to the 216-Z-1A tile

**TABLE 4.2**
**Estimated Pu Inventories Disposed to PFP Waste Solution Ground Disposal Sites (Owens 1981)**

| PFP Disposal Site | Pu Inventory, kg[a] | Structure or Facility | Pu Inventory, kg[a] |
|---|---|---|---|
| 216-Z-1 & 216-Z-11 ditch | 8.22 | 216-Z-9 trench | 48.0 |
| 216-Z-1 and 216-Z-2 cribs | 7.03 | 216-Z-10 reverse well | 0.050 |
| 216-Z-1AA tile field | 30.0 | 216-Z-12 crib | 25.1 |
| 216-Z-1AB tile field | 16.6 | 216-Z-13 French drain | Unknown, but low |
| 216-Z-1AC tile field | 10.8 | 216-Z-14 French drain | Unknown, but low |
| 216-Z-3 crib | 5.7 | 216-Z-15 French drain | Unknown, but low |
| 216-Z-4 trench | 0.002 | 216-Z-16 crib | 7.16 |
| 216-Z-5 crib | 0.034 | 216-Z-17 trench | 0.0502 |
| 216-Z-6 crib | 0.005 | 216-Z-18 crib | 22.9 |
| 216-Z-7 crib | 2.0 | 216-Z-19 ditch | 0.03–0.06 |
| 216-Z-8 French drain | 0.0484 | **Total** | ~184 |

[a] Pu inventory estimates, such as for 216-Z-9, may differ from information presented in other studies.

field (June 1964 until May 1969) and then to the 216-Z-18 crib (until 1973). Sediment analyses showed the $^{239,240}Pu$:$^{241}Am$ Curie concentration ratio to be 1.0 in Z-1A (Price et al. 1979). For the $^{239,240}Pu$-specific activity in the processed 88.2% $^{239}Pu$ and 10.8% $^{240}Pu$ (Table 5–9 of Hoyt and Teal 2004), the Am mass thus would be about 2.35% that of the Pu. Assuming equivalent behavior in Z-18, the $^{241}Am$ originally disposed in PRF's Z-1A and Z-18 sites would be 1.9 kg.[7]

3. The total Am in the PFP ground disposal sites would have increased since then through subsequent $^{241}Pu$ decay from the ground-disposed Pu. Because the latest ground discharges were nearly 50 years ago, over three $^{241}Pu$ half-lives, nearly all $^{241}Pu$ would have decayed to $^{241}Am$. The $^{241}Pu$ constituted about 0.13 mass % of the RECUPLEX weapons-grade Pu at processing and about 0.5% of the varied weapons and higher exposure Pu processed through PRF solvent extraction (Table 5–9 of Hoyt and Teal 2004). Therefore, the $^{241}Am$ present in the RECUPLEX Z-9 trench and the PRF Z-1A tile field and Z-18 crib from subsequent $^{241}Pu$ decay would be an additional 0.5 kg.[8]
4. The total $^{241}Am$ present in the Z-9, Z-1A, and Z-18 ground disposal sites thus would be about 3.6 kg[9] with ~86% introduced to the sediment as "free" Am not arising from $^{241}Pu$ decay in the ground.
5. Because these three sites represent about 70% of the Pu found in all of the PFP ground disposal sites,[10] the total Am in the PFP ground disposal sites could be about 5.1 kg.[11]

Aside from the 216-Z-9 trench "mining" operation (Section 4.2.2), the Pu and Am discharged from PFP and 231-Z to the ground disposal sites remain in the ground awaiting final remediation actions. Extensive well-logging and sampling measured Pu and Am concentration distributions within the 216-Z-9 trench (Price and Ames 1976) in anticipation of interim remedial action (i.e., Pu mining) and then at the 216-Z-1A tile field (Price and Ames 1976; Kasper et al. 1979; Price et al. 1979) and the 216-Z-12 crib (Kasper 1982). Particulate $PuO_2$ filtration was observed for all three sites by soils at the discharge level. Multiple wells afforded excellent maps of Pu and Am plumes for Z-1A and Z-12. Attenuation of Pu and Am concentration with depth was considerable for acidic wastes, decreasing by a factor of 1,000 with 2 m depth for Z-1A and Z-9 (Price and Ames 1976) with Pu and Am concentration increases observed within silt layers underlying Z-1A (Price et al. 1979). Pu and Am distributed similarly on soils from the low-salt, pH-neutral Z-12 wastes, with activity decreasing rapidly with distance below discharge, falling about a factor of 1,000 within 3 m depth and a further factor of 1,000 at 12 m depth (Kasper 1982). Trends in contaminant distribution in the Z-9 trench are illustrated in Figure 4.6.

Further background specific to PFP operations and waste disposal practices and sites is combined in a previous review (Delegard et al. 2019). Perspectives from historical characterization work pertaining to the PFP ground disposal sites, to Pu and Am ground disposal behaviors, germane to the PFP sites, and more recent studies are summarized in the following sections.

Impacts of PFP operations on the subsurface are not limited to disposed Pu and Am, acidic solutions, and the associated inorganic salts. Substantial $CCl_4$ and degraded TBP, DBBP, and lard oil also were disposed in various combinations from PFP to the 216-Z-9 trench, 216-Z-1A tile field, and 216-Z-18 crib from RECUPLEX and PRF operations. The estimated $CCl_4$ discharged to those waste sites ranges from 577 to 922 metric tons (Rohay et al. 1994).

### 4.2.1 Assessment of Subsurface Mobility: Early Plutonium and Americium Soil Uptake and Leaching Studies

Early (1950s–1960s) laboratory-scale studies of Pu and Am uptake onto, and leaching from, Hanford sediments help illuminate the underlying complex mechanisms at play and define parameters that limit Pu and Am movement to the underlying groundwater in waste solution disposal or process leak situations. Questions with regard to Pu and Am behavior in the existing subsurface

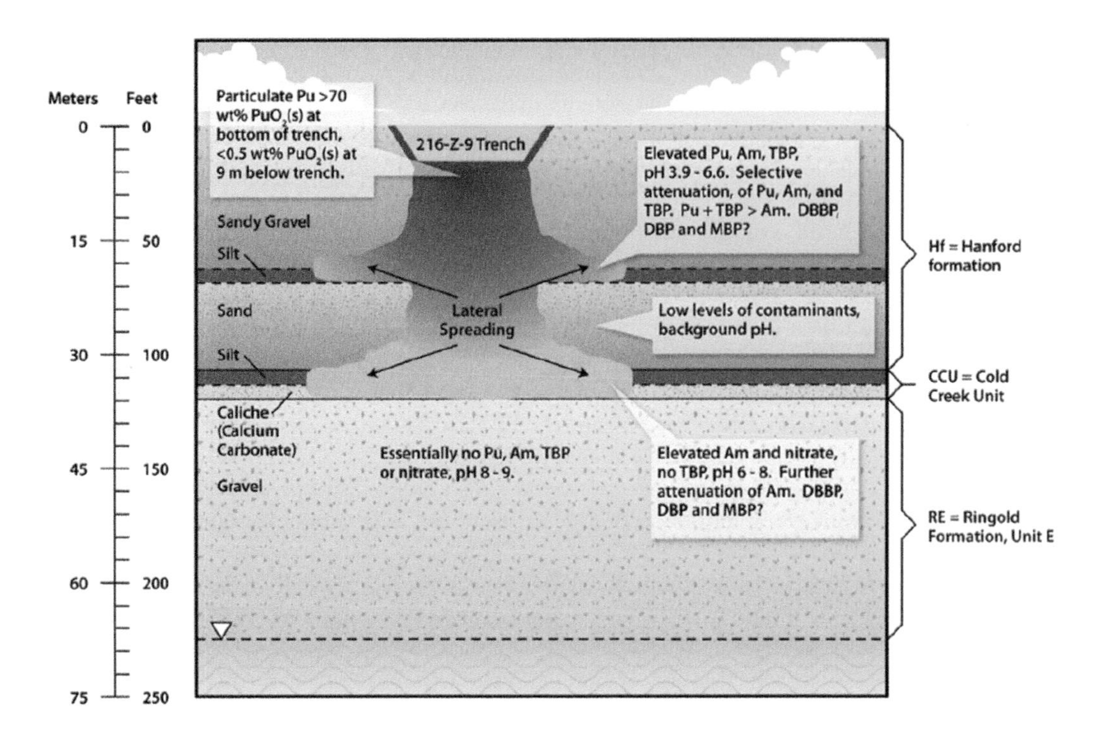


**FIGURE 4.6** Conceptual Pu and Am distribution beneath the 216-Z-9 trench. Reprinted with permission from Cantrell and Riley (2008).

structures for solvent extraction wastes arose during disposal operations and motivated work at this time (Hajek 1966), due to concerns with Pu and Am mobilization from their deposits by solution intrusion (e.g., surface spills or uncommonly high rain or snowfall) and/or diffusion, which could concentrate Pu or Am leading to criticality concerns or mobilize Pu or Am downward toward groundwater. Examination of these findings can help inform remedial action decisions and improve a general understanding of Pu and Am mobilities in the Hanford subsurface.

Generic studies of Pu uptake onto Hanford soils and sediments were undertaken to investigate the roles of dissolved salt content, time, and pH. These studies showed high uptake that increases with time and favors distribution to particles of high specific surface area and high exchange capacities (Rhodes 1952, 1957a, b, and references therein). Ammonium, sodium, nitrate, and phosphate ions had little effect on Pu uptake, but acetate inhibited Pu uptake to a limited degree, suggesting the formation of a Pu acetate complex.

A series of studies of the effects of sediments, waste compositions, and sorption/desorption of Pu and Am was undertaken from 1965 to 1969 for the 216-Z-9 trench (RECUPLEX) and the chemically and geologically similar 216-Z-1A crib (PRF) systems at Hanford. The initial studies were to support the start of the PRF and delivery of its wastes to Z-1A (Knoll 1965; Hajek and Knoll 1966), while later studies examined Pu and Am leaching by simulated groundwater and 1 M sodium nitrate from contaminated Z-9 sediments (Hajek 1966) and generic leaching of Pu and Am from contaminated sediments by several organic solvents (Knoll 1969). Further details on each of these studies are provided in the following paragraphs.

Thus, for Z-1A, Knoll (1965) prepared a high-salt simulated PRF waste solution also containing lard oil, DBBP, and TBP with $CCl_4$ as partly miscible organic compounds. The simulant waste solution was spiked with actual PRF waste; Pu and Am uptakes were measured by passing the solution through packed columns containing sediments collected from the Z-1A spoil pile (sieved to <2 mm to remove larger particles). Retention on the sediment was low for Pu (50% breakthrough at 1.5–4 column volumes) and even less for Am (50% breakthrough at <0.01 column volumes). Disposal of

PRF wastes to Z-1A sediments was further studied by Hajek and Knoll (1966) using the same waste simulant spiked with actual PRF waste. Soil titrated with acid showed the neutralization capacity greatest at the 140- to 155-ft depth, the horizon where $CaCO_3$ is found (Price et al. 1979). Batch experiments, both for the acid waste itself with soil and for wastes adjusted to pH 2 and pH 3, also showed low sorption: 2.4–2.9 mL/g distribution coefficients ($K_d$s) for Pu and <1 mL/g for Am. When the wastes were made alkaline, closer to the natural pH of the Hanford subsurface, the Pu $K_d$ became unmeasurably large because of Pu precipitation, and the Am $K_d$ rose to 212 mL/g. Adding a "slug of organic" (20% DBBP in $CCl_4$) to these pH-neutralized waste/sediment mixtures decreased the Pu $K_d$ to 1.4 and Am to 42 mL/g. Note that DBBP was routinely dumped into the cribs when its performance degraded rather than purified for reuse.

Column leaching of contaminated sediments from Z-9 with groundwater showed Pu removal plateauing at ~0.1% and 0.05% for two different Z-9 sediments (containing 0.6 and 0.4 g Pu/L, respectively), while the respective removals for Am were 7.5% and 6%. Leaching was markedly greater with high-salt solutions (1 M $NaNO_3$), removing 3.5% of the Pu and 33% of the Am. Diffusion rates for Pu and Am were found to be too low to be of concern.

More tests of Pu and Am uptake and leaching were conducted by Knoll (1969). In these tests, fresh sieved Hanford sediment was "contaminated" with Pu and Am dissolved/dispersed in tap water at sediment-controlled pH (~8, natural Hanford subsurface pH). The contaminated sediments then were leached in column tests by various organic solvents. In one test, 130 column volumes of 20% TBP in normal paraffin hydrocarbon (NPH, similar to kerosene) were passed through a column of the Pu- and Am-contaminated sediment and removed 5% of the sorbed Pu and 10% of the Am. A similar experiment using 80 column volumes of 30% DBBP in $CCl_4$ removed 40% of the Pu and 15% of the Am, while another experiment with 30 column volumes of fab oil (Pu metal machining oil, i.e., 25% lard in $CCl_4$) removed <4% of Pu and <1% of Am. Thus, the three principal organic constituents of RECUPLEX and PRF wastes increasingly leached Pu and Am from pH-neutral sediments in the order lard oil<TBP<DBBP. Tests with more acidic organic leachants showed Am was rapidly removed with 0.4 M D2EHPA (di-2-ethylhexyl phosphoric acid) and 0.2 M TBP in NPH but <30% of the Pu was removed. Hydroxyacetic acid, at pH 3.5–3.8, removed all of the Am quickly and all the Pu, but about half as fast. In uptake experiments, clean soil absorbed Pu from 20% TBP in $CCl_4$ with 50% breakthrough in 30 column volumes and 50% Am breakthrough in 9 column volumes.

Together, these early studies show limited Pu and Am uptake (or high subsurface mobility) from the high-salt acid wastes produced in the RECUPLEX and PRF processes, although uptake would increase as the acidic waste solutions were neutralized with depth in the subsurface. Though initial sorption was low, leaching from already contaminated sediments likewise was low unless high salt concentrations, low pH, and/or complexing organic leachants were used. Results of later studies are summarized in Sections 4.2.2 and 4.2.3.

### 4.2.2 Mining and Removal of Pu- and Am-Contaminated Sediments from 216-Z-9 and Related Studies

Even early in RECUPLEX operations, concern was expressed that disposal of Pu to the 216-Z-9 trench could pose a nuclear criticality hazard (Section 15.2 of Gerber 1997). Sampling of trench sediments at four locations showed as much as 1.5 g Pu/L (Reisenauer 1959). Criticality concerns continued in the 1960s and early 1970s as Pu measurements and estimates widely varied (27.5–196 kg Pu) with the lower figure based on Pu accountability and the higher figure based on conservative projections from disparate sample analyses. A 1972 mission to recover Pu from PFP wastes and from soil made the Z-9 trench an attractive potential Pu resource (Section 15.2 of Gerber 1997). Therefore, retrieval of the top 12 inches (~30 cm) of the Z-9 sediments was undertaken, but not until further measurements and the addition of cadmium to the trench as a neutron absorber improved criticality safety assurance. The mining operation from the ~9 × 18-m trench bottom (Figure 4.7) took place from August 1976 to July 1978 (Delegard et al. 2019).

**FIGURE 4.7** Sign at Hanford's plutonium mine. Reprinted with permission from Hanford Site photo archive image (72698-29CN).

The sediments, tested for Pu recovery by chemical leaching, provided ancillary characterization data (Panesko 1972; Swanson 1973). Ultimately, because of process difficulties anticipated for this dilute (<0.1 wt%) Pu scrap feed, no Pu recovery was performed. The 58 kg of Pu in the ~51 $m^3$ of retrieved sediment, packaged into 5,222 10-L canisters, still remains at Hanford Site's T Plant (Delegard et al. 2019).

More detailed Z-9 sediment characterizations by autoradiography, electron microscopy, and electron microprobe[12] showed Pu disposition and how the acidic wastes altered the native minerals (Ames 1974; Price and Ames 1976). The Pu concentrations were as high as 28 g/L, and particulate $PuO_2$ was found in the uppermost layers where the sediment itself acted as a filter. The Pu concentrations decreased steeply with depth, falling by a factor of 100 at 0.5 m depth below the disposal horizon and by a factor of 100,000 at 9 m (Price and Ames 1976). Mineralogical studies of the shallow Z-9 sediments showed marked chemical attack, particularly for glassy basalt, with leaching of alkali and alkaline earth elements (e.g., Na, K, Mg, and Ca) and aluminum to leave silica-rich gels (Ames 1974).

Further examinations of Z-9 samples and a 216-Z-12 crib sediment sample from pH-neutral PFP wastes measured their Pu leachabilities as functions of pH (Rai et al. 1980). Similar tests for Am leachability from Z-9 also were done (Rai et al. 1981). The soil samples first were water-washed to remove soluble salts before being contacted with 0.0015 M $CaCl_2$ and allowed to equilibrate. For the Pu tests, parallel tests were run with 5 mg of crystalline $PuO_2$ added as a spike. Significantly, Pu concentrations in tests with and without the $PuO_2$ spike were in the range observed for tests run with $PuO_2$ only, thus giving evidence that Pu concentrations were limited in Z-9 and Z-12 by $PuO_2$ solubility. A correlation of linearly decreasing log-Am concentration with increasing pH was found (Rai et al. 1981) with a similar correspondence found for sediments taken 9 m below the discharge line in the Z-1A tile field in batch tests with a variety of leachants (Delegard et al. 1981). Curiously, column leach tests in the latter study failed to follow as faithfully the log[Am]-vs-pH correlation though Am concentrations decreased with increasing pH.

### 4.2.3 Recent Studies of Pu and Am Behaviors in Support of Understanding PFP Disposal to the Ground

Despite the large body of research presented here, questions remain around the mechanisms responsible for Pu and Am migration into the deep subsurface at Hanford, and the possibility for this Pu and Am to be remobilized and reach groundwater. The subsurface behaviors of Pu and Am at Hanford disposal sites were examined and reviewed (Cantrell and Riley 2008). In a further comprehensive review of Pu and Am geochemistry at Hanford, Cantrell and Felmy (2012) identified key research challenges whose resolution would support remediation decisions to achieve scientifically defensible end states. These research challenges, which have been the focus of recent studies, are as follows: (i) determining the transformation of Hanford Site sediments in response to changes in waste/vadose zone porewater composition; (ii) assessing the impact of this transformation on solubility and acidic adsorption of Pu and Am, and the potential for pseudocolloid formation that could facilitate migration; (iii) evaluating the role of organic complexants and/or non-aqueous solvents in the transport of Pu and Am in the deep subsurface; and (iv) understanding the potential for lateral transport above low transmissivity layers. These challenges are being addressed using advanced imaging and spectroscopy techniques including high-resolution transmission electron microscopy (TEM) with electron diffraction, nanometer-scale secondary-ion mass spectrometry (nanoSIMS), and Pu $L_2$ X-ray absorption fine structure spectroscopy (EXAFS).

Felmy et al. (2010) conducted Pu $L_2$ EXAFS to show that the chemical form of Pu in Hanford sediments is far more variable than just the expected $PuO_2$, and that it differs between surface sediments and deep subsurface sediments. They concluded that identifying the distribution of different chemical forms of Pu in the sediments beneath the PFP would provide information on subsurface transport mechanisms.

In a review of Pu transport in the environment, Kersting (2013) examined sediment samples from beneath the Z-9 trench with a high spatial resolution (50–100 nm) using nanoSIMS. Pu was found to be associated with primary silicate and oxide minerals (e.g., quartz, feldspar, biotite, and clay) with iron oxide coatings, ranging in size from colloids to tens of micrometers. As many of these mineral grains are too large to have migrated through the vadose zone, it is more likely that the organic complexants and/or non-aqueous solvents in the waste played a significant role in the transport of Pu, and that Pu sorbed to inorganic colloids did not facilitate vertical transport.

Subsequently, Buck et al. (2014) used TEM and Pu $L_2$ EXAFS to demonstrate the formation of nano-sized mixed plutonium-iron phosphate hydroxide, structurally related to the rhabdophane group minerals, in sediments from under the Z-9 trench. They hypothesized that the formation of these phases may depend on the local microenvironment and phosphate availability and that the distribution of these minerals may control long-term migration of plutonium in the soil.

Recent studies have focused on the behaviors of both Pu and Am in systems akin to, or taken from, PFP waste solution disposals to the ground (Emerson et al. 2021, 2022). These studies used the carefully curated 1973 Z-9 sediment samples and some 2006 samples retrieved from a slant borehole drilled under the Z-9 trench at greater depth (0–0.15, and ~12 and ~25 m below waste discharge, respectively). The 2021 work also used pH-neutral Z-12 crib samples. Findings from the former study (Emerson et al. 2021) and internally cited data showed Pu and Am in contaminated sediments could be mobilized, but at low concentrations, by an artificial groundwater even as the pH for the shallower Z-9 sediments decreased leachant pH by 1–2 units. The Pu and Am solubility-controlling phases remained to be identified. Significantly, both Pu and Am had low mobilities.

Further insights were gained in the latter study (Emerson et al. 2022), which investigated solubilities of $PuO_2$ prepared by different methods relevant to synthesis procedures utilized historically at the Hanford Site (from calcined oxalate and burnt Pu metal) having varied particle sizes. Solubilities, with time, tended to those expected for amorphous hydrated $PuO_2$ (i.e., $PuO_{2\ am,hyd}$). For Pu-contaminated sediments that are low in phosphate, $PuO_{2\ am,hyd}$ likely controls the Pu concentration. However, both Pu phosphate and Am phosphate phases, predicted by solubility index calculations and discovered by high-resolution TEM, control solubilities in high-phosphate and low-pH conditions. The phosphate likely is being slowly released to the Z-9 and similar Z-1A and Z-18 sediments by radiolysis and hydrolysis of TBP. Sediments from the slant borehole deeper beneath the Z-9 trench are less acidic and contain correspondingly lower Pu concentrations. Here, the Pu concentration likely is controlled by $PuO_{2\ am,hyd}$. The Am concentrations in deeper sediments potentially are controlled by adsorption or release from $PuO_2$ after Am-241 ingrowth. Shallow sediments have both Pu and Am in the colloidal fraction, though colloidal Pu is not found in tests with the synthetic $PuO_2$, thus suggesting the presence of Pu/Am pseudocolloids, that is, Pu and Am associated with mineral colloids. Neither Pu nor Am was associated with colloids for the deep slant borehole samples.

Additional interpretation of Am behavior in Z-1A sediments has been performed (Delegard et al. 2023). Based on Delegard et al. (1981) findings, examination of unpublished 1980s data from the same study, and an assessment of related research, a fraction of the Am in sediments interacts by surface complexation, likely onto the amorphous silica abundantly present in the waste-altered sediments. This mechanism explains the observed linearly decreasing log-Am solution concentration with increasing pH. Most of the Am, however, remains incorporated, and thus immobile, within dynamically precipitating amorphous silica. Similar Am behavior is probable in the analogous Z-9 sediments.

Work to understand and predict the behavior of Pu and Am in complex subsurface environments continues.

## 4.3 CONCLUSIONS

The Hanford Site's PFP was a workhorse, providing 40 years of service to the nation in purifying, recovering, and finishing Pu for U.S. defense needs at the height of the Cold War. Processes adopted from Los Alamos were improved, upscaled, and made more efficient with lower radiological dose rates to the operators over the plant lifetime. Descriptions of the plant processes and residues remaining within the PFP as provided in this chapter help inform the operations leading to the demolition of the PFP as described in Chapter 9.

Process waste solution disposal from the PFP into a number of engineered ground disposal facilities (trenches, cribs, ditches, French drains) occurred until 1973, leaving a legacy of nearly 200 kg of Pu and about 5 kg of Am in the surrounding PFP sediments from high-salt and acidic wastes to low-salt and pH-neutral wastes. Hundreds of tons of $CCl_4$ with accompanying organic extractants also were discarded into the sediments. Early testing showed relatively low sorption of Pu onto the sediments from the acid waste, though further study shows more nuanced behaviors with Pu interacting with sediment minerals and with phosphate generated by TBP breakdown all occurring as acid is neutralized.

While more studies are required to better understand long-term Pu, and associated Am, interactions with sediment minerals, the extant data are pivotal to our current understanding of Pu and Am transport in the Hanford Site's vadose zone and are being used to inform the conceptual site model for Pu and Am transport and to make subsurface remediation decisions.

## NOTES

1 The Hanford product initially contained up to 3 wt% process contaminants, primarily lanthanum but also uranium and stainless-steel corrosion products (iron, chromium, nickel).
2 During the Manhattan Project, the existence of war-related activities at Los Alamos was held as secret with the location known only as Site Y. The name Los Alamos Scientific Laboratory was adopted on January 1, 1947, and became Los Alamos National Laboratory in 1981.
3 Information here and in the remaining narrative is taken from Gerber (1997, 2002) unless otherwise noted.
4 The B and T Plants employed successive chemical carrier precipitations in bismuth phosphate and lanthanum fluoride to recover plutonium from irradiated uranium fuel. This technology was utilized from 1944 until 1956. The REDOX Plant was the fourth of five Hanford reprocessing plants. Running 1952–1967, the REDOX (REDuction-OXidation) Plant used solvent extraction in methyl isobutyl ketone to recover uranium and plutonium from dissolved irradiated uranium metal fuel.
5 $t_{1/2} = 14.3$ years.
6 $(0.026 \times 48 = \sim 1.2)$.
7 $(0.0235 \times 80.3 = 1.9)$.
8 $(0.0013 \times 48 + 0.005 \times 80.3 = .5)$.
9 $(1.2 + 1.9 + 0.5 = 3.6)$.
10 $[100\% \times ((48 + 80.3)/184) = \sim 70\%]$.
11 $(3.6/0.7 = 5.1)$.

12 Autoradiography is a technique where radioactive emissions are allowed to interact with a photographic film and produce a visible pattern indicating the distribution of radioactivity in a sample. Electron microscopy techniques are used for collecting high-resolution images by interaction of a beam of high-energy electrons with a sample, while microprobe techniques measure X-rays emitted after interaction of the beam with a sample to identify the elements present in that sample.

## REFERENCES

Ames LL. 1974. *Characterization of Actinide Bearing Soils: Top Sixty Centimeters of 216-Z-9 Enclosed Trench.* BNWL-1812, Battelle Pacific Northwest Laboratories, Richland, WA. https://www.osti.gov/biblio/4313915-characterization-actinide-bearing-soils-top-sixty-centimeters-enclosed-trench.

BNWI. 1976. *Explosion of Cation Exchange Column in Americium Recovery Service, Hanford Plant, August 30, 1976. BNWI-1006*, Battelle-Northwest, Richland, WA. https://doi.org/10.2172/6017334.

Buck EC, DA Moore, KR Czerwinski, SD Conradson, ON Batuk, and AR Felmy. 2014. "Nature of Nano-Sized Plutonium Particles in Soils at the Hanford Site." *Radiochimica Acta* 102(12):1059–1068. https://doi.org/10.1515/ract-2013-2103.

Cantrell KJ and AR Felmy. 2012. *Plutonium and Americium Geochemistry at Hanford: A Site-Wide Review.* PNNL-21651, Pacific Northwest National Laboratory, Richland, WA. https://www.pnnl.gov/main/publications/external/technical_reports/pnnl-21651.pdf.

Cantrell KJ and RG Riley. 2008. *A Review of Subsurface Behavior of Plutonium and Americium at the 200-PW-1/3/6 Operable Units.* PNNL-SA-58953, Pacific Northwest National Laboratory, Richland, WA. https://www.pnnl.gov/main/publications/external/technical_reports/PNNL-SA-58953.pdf.

Carbaugh EH. 2016. "The Atomic Man: Case Study of the Largest Recorded $^{241}$Am Deposition in a Human". USTUR Special Session 61st Annual Meeting of the Health Physics Society Spokane, WA, July 19, 2016. https://s3.wp.wsu.edu/uploads/sites/1058/2017/07/TAM-A.2-Carbaugh-EH.pdf

Carritt DE, CE Hagen, and AC Wahl. 1946. *Purification Processes: D-Building Plutonium Purification.* LA-405, Los Alamos Scientific Laboratory, Los Alamos, NM. https://sgp.fas.org/othergov/doe/lanl/lib-www/la-pubs/00350288.pdf.

Christensen EL and WJ Maraman. 1969. *Plutonium Processing at the Los Alamos Scientific Laboratory.* LA-3542, Los Alamos Scientific Laboratory, Los Alamos, NM. https://www.osti.gov/biblio/4794689-plutonium-processing-los-alamos-scientific-laboratory.

Crocker HW and HH Hopkins Jr. 1963. *Conversion of Plutonium Nitrate to Plutonium Tetrafluoride via the Continuous Oxalate Fluoride Method.* HW-SA-2880 and HW-76621, Hanford Atomic Products Operation, General Electric Company, Richland, WA. https://www.osti.gov/scitech/biblio/4089211.

Delegard CH, HP Emerson, KJ Cantrell, and CI Pearce. 2019. *Generation and Characteristics of Plutonium and Americium Contaminated Soils Underlying Waste Sites at Hanford.* PNNL-29203, Pacific Northwest National Laboratory, Richland, WA. https://www.osti.gov/biblio/1682309-generation-characteristics-plutonium-americium-contaminated-soils-underlying-waste-sites-hanford.

Delegard CH, SA Gallagher, and RB Kasper. 1981. "Saturated Column Leach Studies: Hanford 216-Z-1A Sediment." Presented at *International Symposium on the Migration in the Terrestrial Environment of Long-Lived Radionuclides*, 27–31 July 1981 and in the proceedings of that meeting, *Environmental Migration of Long-Lived Radionuclides*, International Atomic Energy Agency, Proceedings Series STI/PUB/597, 1982; also as RHO-SA-210, Rockwell Hanford Operations, Richland, WA. https://www.osti.gov/servlets/purl/6279493.

Delegard CH, CI Pearce, and HP Emerson. 2023. "On Silica's Roles in Controlling Americium Migration in Contaminated Sediments". *Applied Geochemistry* 154:105690. https://doi.org/10.1016/j.apgeochem.2023.105690.

Delegard CH and RA Peterson. 2020. "Precipitation and Crystallization Processes in Reprocessing, Plutonium Separation, Purification, and Finishing, Chemical Recovery, and Waste Treatment." Chapter 3 of *Engineering Separations Unit Operations for Nuclear Processing*, RA Peterson, ed., 51–144. CRC Press, Taylor & Francis Group, Boca Raton, FL.

DOE-RL. 2003. *Environmental Assessment, Deactivation of the Plutonium Finishing Plant, Hanford Site, Richland, WA, Predecisional Draft.* DOE/EA-1469. U.S. Department of Energy, Richland Operations Office, Richland, WA.

DuPont. 1944. *Hanford Engineer Works Technical Manual.* HW-10475 ABC, General Electric, Hanford Atomic Products Operation, Richland, WA. https://pdw.hanford.gov/arpir/pdf.cfm?accession=0076415H.

Emerson HP, CI Pearce, CH Delegard, KJ Cantrell, MMV Snyder, M-L Thomas, BN Gartman, MD Miller, CT Resch, SM Heald, AE Plymale, DD Reilly, SA Saslow, W Nelson, S Murphy, M Zavarin, AB Kersting, and VL Freedman. 2021. "Influences on Subsurface Plutonium and Americium Migration." *ACS Earth and Space Chemistry* 5(2):279–294. https://doi.org/10.1021/acsearthspacechem.0c00277.

Emerson HP, SI Sinkov, CI Pearce, KJ Cantrell, CH Delegard, MMV Snyder, M-L Thomas, DD Reilly, EC Buck, L Sweet, AJ Casella, JC Carter, JF Corbey, IJ Schwerdt, R Clark, FD Heller, D Meier, M Zavarin, AB Kersting, and VL Freedman. 2022. "Solubility Controls on Plutonium and Americium Release in Subsurface Environments Exposed to Acidic Processing Wastes." *Applied Geochemistry* 153:105241. https://doi.org/10.1016/j.apgeochem.2022.105241.

Felmy AR, KJ Cantrell, and SD Conradson. 2010. "Plutonium Contamination Issues in Hanford Soils and Sediments: Discharges from the Z-Plant (PFP) Complex." *Physics and Chemistry of the Earth* 35:292–297. https://doi.org/10.1016/j.pce.2010.03.034.

Garde R, EJ Cox, and AM Valentine. 1982. *Los Alamos DP West Plutonium Facility Decontamination Project, 1978-1981*. LA-9513-MS, Los Alamos National Laboratory, Los Alamos, NM. https://www.osti.gov/servlets/purl/6633819.

Gerber MS. 1997. *History and Stabilization of the Plutonium Finishing Plant (PFP) Complex, Hanford Site*. HNF-EP-0924, Fluor Daniel Hanford, Inc., Richland, WA. https://www.osti.gov/biblio/325360-history-stabilization-plutonium-finishing-plant-pfp-complex-hanford-site.

Gerber, MS. 2002. *"Plutonium Finishing."* Section 5 *in The Hanford Site Historic District: History of the Plutonium Production Facilities, 1943–1990,* TE Marceau, DW Harvey, DC Stapp, SD Cannon, CA Conway, DH Deford, BJ Freer, MS Gerber, JK Keating, CF Noonan, and G Weisskopf, eds., 2-5.1–2-5.29. Battelle Press, Columbus, OH. https://www.osti.gov/scitech/biblio/807939.

Hajek BF. 1966. *Plutonium and Americium Mobility in Soils*. BNWL-CC-925, Pacific Northwest Laboratory, Richland, WA. https://reading-room.labworks.org/Files/GetDocument.aspx?id=D8616997.

Hajek BF and KC Knoll. 1966. *Disposal Characteristics of Plutonium and Americium in a High Salt Acid Waste*. BNWL-CC-649, Pacific Northwest Laboratory, Richland, WA. https://www.osti.gov/biblio/4485088-disposal-characteristics-plutonium-americium-high-salt-acid-waste.

Hanford. 1961. *Z Plant Plutonium Processing & 2345 Operations*. Diagrams 26934-1 [N1389757] and 26934-3 [N1389882]. Hanford Atomic Products Operation, Richland, WA.

Hebdon J, J Yerxa, L Romine, AM Hopkins, R Piippo, L Cusack, R Bond, O Wang, and D Willis. 2003. "Collaborative Negotiations a Successful Approach for Negotiating Compliance Milestones for the Transition of the Plutonium Finishing Plant (PFP), Hanford Nuclear Reservation, and Hanford, Washington." *WM'03 Conference*, 23–27 February 2003, Tucson, AZ. https://archivedproceedings.econference.io/wmsym/2003/pdfs/520.pdf.

Hopkins, AM, AR Sherwood, LR Fitch, and DG Ranade. 2003. *The Chemical Hazards Assessment Prior to D&D of the Plutonium Finishing Plant, Hanford Nuclear Reservation*. HNF-14264, Rev 0. Fluor Hanford, Richland, WA.

Hoyt RC and JA Teal. 2004. *Plutonium Finishing Plant Operations Overview (1949-2004): Contamination Events and Plutonium Isotope Distributions of Legacy Holdup Material in Process Systems*. HNF-22064, Rev. 0, DEL, Fluor Hanford, Richland, WA.

Kajanjian AR and DR Horrell. 1975. *Radiation Effects on Ion-Exchange Resins - Part III: Alpha Irradiation of Dowex 50W*. RFP-2298, Rocky Flats Plant, Golden, CO. https://www.osti.gov/biblio/4215563.

Kasper RB. 1982. *216-Z-12 – Transuranic Crib Characterization: Operational History and Distribution of Plutonium and Americium*. RHO-ST-44, Rockwell Hanford Operations, Richland, WA.

Kasper RB, SM Price, MK Additon, RM Smith, GV Last, and GL Wagenaar. 1979. *Transuranic Distribution beneath a Retired Underground Disposal Facility, Hanford Site*. RHO-SA-131, Rockwell Hanford Operations, Richland, WA.

Kersting A. 2013. "Plutonium Transport in the Environment." *Inorganic Chemistry* 52:3533–3546. https://pubs.acs.org/doi/10.1021/ic3018908.

Knoll KC. 1965. *Reaction of High Salt Aqueous plus Organic Waste with Soil – Interim Report*. BNWL-CC-313, Pacific Northwest Laboratory, Richland, WA. https://www.osti.gov/biblio/4397154-reaction-high-salt-aqueous-plus-organic-waste-soil-interim-report.

Knoll KC. 1969. *Reactions of Organic Wastes and Soils*. BNWL-860, Pacific Northwest Laboratory, Richland, WA. https://www.osti.gov/biblio/4826993-reactions-organic-wastes-soils.

Owens KW. 1981. *Existing Data on the 216-Z Liquid Waste Sites*. RHO-LD-114, Rockwell Hanford Operations, Richland, WA. https://www.osti.gov/servlets/purl/5703558.

Panesko JV. 1972. *Plutonium Recovery from Soil*. ARH-2570, Atlantic Richfield Hanford Company, Richland, WA.

Price SM and LL Ames. 1976. "Characterization of Actinide-Bearing Sediments Underlying Liquid Waste Disposal Facilities at Hanford". *Paper IAEA-SM-199/87, Transuranium Nuclides in the Environment, STI/PUB/410*, pp. 191–211, International Atomic Energy Agency, Vienna, Austria. Also published as ARH-SA-232. https://inis.iaea.org/collection/NCLCollectionStore/_Public/07/238/7238004.pdf?r=1&r=1.

Price SM, RB Kasper, MK Additon, RM Smith, and GV Last. 1979. *Distribution of Plutonium and Americium beneath the 216-Z-1A Crib: A Status Report*. RHO-ST-17, Rockwell Hanford Operations, Richland, WA.

Rai D, RJ Serne, and DA Moore. 1980. "Solubility of Plutonium Compounds and Their Behavior in Soils." *Soil Science Society of America Journal* 44(3):490–495. https://doi.org/10.2136/sssaj1980.03615995004400030010x.

Rai D, RG Strickert, DA Moore, and RJ Serne. 1981. "Influence of an Americium Solid Phase on Americium Concentrations in Solutions." *Geochimica et Cosmochimica Acta* 45:2257–2265. https://doi.org/10.1016/0016-7037(81)90075-2.

Reisenauer AE. 1959. *216-Z-9 Core Sampling Data*. HW-61787, Hanford Laboratory Operations, Richland, WA.

Rhodes DW. 1952. *Preliminary Studies of Plutonium Adsorption in Hanford Soil*. HW-24548, Hanford Works, Richland, WA.

Rhodes DW. 1957a. "The Effect of pH on the Uptake of Radioactive Isotopes from Solution by a Soil." *Soil Science Society of America Journal* 21:389–392. https://doi.org/10.2136/sssaj1957.03615995002100040009x.

Rhodes DW. 1957b. "Adsorption of Plutonium by Soil." *Soil Science* 84:465–472. https://journals.lww.com/soilsci/Citation/1957/12000/ADSORPTION_OF_PLUTONIUM_BY_SOIL.5.aspx.

Rohay VJ, KJ Swett, and GV Last. 1994. *1994 Conceptual Model of the Carbon Tetrachloride Contamination in the 200 West Area at the Hanford Site*. WHC-SD-EN-TI-248, Westinghouse Hanford Company, Richland, WA.

Swanson JL. 1973. *Nature of Actinide Species Retained by Sediments at Hanford: Interim Progress Report*. BNWL-B-296, Battelle Pacific Northwest Laboratories, Richland, WA. https://www.osti.gov/biblio/4374249-nature-actinide-species-retained-sediments-hanford-interim-progress-report.

Widner T, J Shonka, R Burns, S Flack, J Buddenbaum, J O'Brien, K Robinson, and J Knutsen. 2010. *Final Report of the Los Alamos Historical Document Retrieval and Assessment (LAHDRA) Project*. Centers for Disease Control and Prevention, Washington, DC. https://wwwn.cdc.gov/LAHDRA/Content/pubs/Final%20LAHDRA%20Report%202010.pdf.

# 5 Tank Waste Characterization
## *History, Challenges, and Successes*

*Emily Campbell, Carolyne Burns, and Richard Daniel*

The gap in tank waste characterization information became abundantly clear after the U.S. Department of Energy (DOE), Washington State Department of Ecology, and U.S. Environmental Protection Agency came together in 1989 to sign the Hanford Federal Facility Agreement and Consent Order; more commonly known as the Tri-Party Agreement (see Chapters 1 and 2). This precipitated shortly after the formation of the Defense Nuclear Facilities Safety Board (DNFSB) in 1988 by Congress to comply with the mandate bestowed under the Atomic Energy Act to provide safe oversight into nuclear weapons complex operated by the DOE (Rasmussen, 2017). The legally binding Tri-Party Agreement spells out how Washington State and the federal government will cooperate to ensure cleanup of the chemical and radioactive nuclear waste at Hanford in compliance with the Resource Conservation and Recovery Act (RCRA) and the Comprehensive Environmental Response, Compensation, and Liability Act (CERCLA). The Tri-Party Agreement provides the tasks and schedules for the final closure of high-level waste tanks, including the transfer, treatment, and final disposition of the high-level and low-level wastes for the eventual remediation and restoration of the Hanford Site.

The chemical and radioactive Hanford tank waste was generated by three distinct chemical separation processes (bismuth phosphate, REDOX, and PUREX), and further complicated by waste volume reduction operations and two tank waste processing campaigns. Initially, extensive records were kept describing the placement of waste in specific tanks. However, evolution of the processes during plant operation was not well documented, the records for tank contents were incomplete, and numerous tank-to-tank transfers occurred (Simpson et al., 1996). While the intent of this book is not to present a technical history of Hanford, understanding the historical events and evolution of the technologies employed at Hanford contributing to the tank wastes' present complexity is important. Presented below is a summary of the historical events and Hanford's technological history, but for further details, the reader is encouraged to consult Gephart (2003).

The first plutonium production reactors (labeled B, D, and F) began construction in 1943 on the isolated desert plains of Hanford, Washington. B Reactor achieved its first nuclear chain reaction on September 26, 1944, just shy of a year after the start of construction. The bismuth phosphate process at the Hanford Site's B and T Plants was used to separate the plutonium (Pu) from the irradiated fuel (1944–1956). As in all Pu separation processes at Hanford, the fuel cladding first was dissolved, leaving the bare irradiated uranium metal. The uranium metal then was dissolved in strong nitric acid. The bismuth phosphate process relied on successive chemical precipitations, centrifuge separations, and dissolutions using bismuth phosphate and lanthanum trifluoride as the precipitation reagents to carry or leave in solution, in alternative successive steps, the desired plutonium. For both bismuth phosphate and lanthanum fluoride, the plutonium was carried when in its lower (tetravalent) oxidation state, leaving some undesired constituents in solution for discard. The plutonium was not carried when in its higher (hexavalent) oxidation state, leaving other undesired constituents with the precipitating carrier for discard. The succeeding third step separated most of the lanthanum from the plutonium to produce carrier-free plutonium peroxide product through two successive precipitations followed by decomposition and dissolution in nitric acid. The plutonium


DOI: 10.1201/9781003329213-7

nitrate product dissolved from the second plutonium peroxide precipitation was concentrated by heating to a thick paste and transported to Los Alamos where it was dissolved, further purified, and processed into 160- to 500-g metallic Pu "buttons." The Pu metal then was alloyed, formed, and provided the chain-reacting cores in the first nuclear explosion in the July 16, 1945, Trinity Test, and in the "Fat-Man," the nuclear weapon detonated over Nagasaki, Japan, August 9, 1945. Weapons-grade plutonium production at Hanford accelerated from 1950 and peaked ~1965 during the Cold War before ceasing in the early 1970s although "fuel-grade" reactor production continued. Weapons-grade production briefly resumed during the 1980s although not at the levels achieved earlier.

## PERSPECTIVES: CARY SEIDEL

Chemist Cary Seidel supported the Hanford Site's cleanup mission for 33 years, first at the contractor Washington River Protection Solutions and then as the DOE Analytical Laboratory Manager.

### WHAT WAS THE MAIN TECHNICAL CHALLENGE YOU FACED IN YOUR CAREER AT HANFORD?

"Back at the start, the most challenging aspect was how to handle waste outside the tank. I was involved in the first shipment of high-level waste (HLW) sludge to Savannah River and Idaho National Lab. In the early nineties, a core truck was used to retrieve sludge from the tank using a push and rotary action, like soil core samples. A total of 22 core segments, approximately 19″ long and 1.5 ″ in diameter were transferred to 222-S to extrude the solids out of the core. To ship the HLW and supernate samples to Savannah River, 8–10 L went by semi-truck and an environmental assessment of the highways was performed prior to transport. It was found that the HLW could not be in close proximity to another batch and the waste had to come back after analysis. The knowledge and limitation of this shipping event culminated in the development and certification of the original Hedgehog, a multi-configuration shielded Type A packaging for radioactive solids and liquids."

### WHAT TECHNICAL CONTRIBUTION ARE YOU MOST PROUD OF DURING YOUR TIME AT HANFORD?

"Improving the 222-S characterization lab by integrating waste characterization was something I took pride in. I interfaced with Westinghouse and Lockheed Martin and directed a team of 20 chemists to execute the work. Originally, 222-S was a support lab for REDOX and a site support lab for production facilities. When I stepped in as DOE Analytical Laboratory Manager, I helped establish analytical equipment and partnered with Pacific Northwest National Laboratory (PNNL) to develop a laser ablation mass spectrometer that could be used in the hot cell on tank waste samples."

### WHAT WAS THE MOST CHALLENGING ASPECT OF CHARACTERIZATION?

"The most challenging aspect was budget and schedule. In the early days, one tank sample took 6 months of analytical work from the time the sample was taken until the analysis was complete. Additionally, the radiation from the tanks created its own problems with shuffling the sampling crew to be cognizant of personnel dose."

The acidic waste from the Hanford bismuth phosphate process separations, which contained all the uranium, minor actinides, fission products, and separation process reagents, was made alkaline with sodium hydroxide and sodium carbonate and then discharged to mild-steel–lined underground storage tanks. Certain tanks were selected to contain the valuable uranium-rich fraction for its later recovery.

Storage of waste was an ongoing concern at Hanford. Therefore, industrial-scale methods for chemically separating plutonium from the irradiated fuel (itself containing residual uranium, other actinides, and fission products) were pioneered at Hanford to minimize waste generation and increase production rates. Thus, the second-generation plutonium separation process used methyl isobutyl ketone to solvent-extract both uranium and plutonium from acid-dissolved irradiated fuel in the Reduction-Oxidation (REDOX) Plant (1952–1967). Aluminum nitrate was used as a salting-out reagent in the process, leading to the discharge of relatively large quantities of this element to the waste tanks. The third and final generation of plutonium separation was PUREX (1956–1972; 1983–1990). In PUREX, the irradiated fuel was dissolved in acid and the uranium and plutonium extracted using tributyl phosphate diluted in kerosene. The salting-out reagent in this case was nitric acid itself, which could be distilled and recycled. The waste was an acidic high-level solution, which contained trace amounts of plutonium and uranium, fission products, and actinide metals, intermixed with non-radioactive organic solvent breakdown products, process reagents, and fuel cladding metals (aluminum for eight of the Hanford reactors and zirconium for the ninth). For both REDOX and PUREX, the acidic wastes again were made alkaline before disposal to the mild-steel–lined underground tanks. All told, 96,900 MT of irradiated uranium fuel was processed at Hanford: ~70% by PUREX, ~25% by REDOX, and ~5% by bismuth phosphate (Gephart 2010).

In 1952, Hanford's U Plant, built but not used as a bismuth phosphate plutonium processing canyon for World War II, was retrofitted as the Metal Recovery Plant (1952–1958) to recover uranium from the reserved bismuth phosphate waste as the scarcity of high-grade uranium supplies became a concern (Harvey, 2000). The plant used the newly developed tributyl phosphate solvent extraction technique (simplified PUREX) to separate uranium from the acidified wastes. During this recovery mission, ferrocyanide salts were added to the waste stream to precipitate cesium-137 and the resulting decontaminated waste was then either evaporated or discharged to the ground, freeing tank storage space (Gerber, 2001; Boomer et al., 2012). Additionally, ferrocyanide was added directly to certain waste storage tanks, again to scavenge cesium-137, allowing the cesium-denuded solutions to be discharged to the ground, likewise freeing tank storage space. The ferrocyanide added further complex challenges to Hanford's tank waste cleanup efforts. Non-radioactive strontium and calcium nitrate were added to scavenge radioactive strontium-90 by in-tank precipitation. Later, B Plant was refurbished to employ technologies to separate cesium-137 by ion exchange and strontium-90 by precipitation. Known as waste fractionization, these operations utilized organic complexants (principally ethylenediaminetetraacetate – EDTA; hydroxyethyl ethylenediaminetriacetate – HEDTA; citrate; and glycolate), which, when added to the waste tanks, later caused concerns for generation of retained and expelled gases, some flammable (Strachan et al., 1993).

Adding to the tank waste complexity were discharges from the Plutonium Finishing Plant (see Chapter 3). This plant received the plutonium product from the B, T, REDOX, and PUREX separation plants and processed it into forms suitable for weapons and fuel with the valuable scrap recycled by purification via tributyl phosphate solvent extraction. Until 1973, wastes from the Pu purification processes were discharged to the surrounding sediments (see Chapter 4); beginning in 1973, these acidic wastes were made alkaline and disposed of to the underground tanks. Importantly, the wastes contained small quantities of solid plutonium oxide and even metal, as well as trace coprecipitated Pu, that together demand special attention for criticality safety management in the tank wastes.

**PERSPECTIVES: DAN HANSEN**

Dan Hansen dedicated 35 years to work at the Hanford Site, first at the production plant's analytical laboratories and finally at the 222-S Laboratory, where he performed organic analysis of tank waste.

**WHAT WAS THE MAIN TECHNICAL CHALLENGE YOU FACED IN YOUR CAREER AT HANFORD?**

"The most challenging technical problem faced in my career was development of organic analysis techniques on HLW, which had never been done before. Liquids were the biggest challenge where large volumes were required for EPA methods, however the style of analysis, RCRA vs. CERCLA was not defined and was a moving landscape that the regulators weren't even sure the correct path."

**WHAT TECHNICAL ADVANCEMENT WOULD BE A BREAKTHROUGH FOR CHARACTERIZATION EFFORTS?**

"The Tri-Party Agreement shoehorned characterization into a certain box with analyzing for a whole slew of things that weren't expected to be there. Why analyze for a nitrobenzene? Why waste the time? There are a lot of organics in the tanks and we were not figuring out what those organics were, but were analyzing for what they were not. The biggest gap in knowledge with respect to organics is what organics are present in the supernate?"

The wastes from plutonium separation, either by precipitation or by solvent extraction, from plutonium finishing (beginning 1973), and the treated wastes generated from uranium recovery and waste fractionization were stored in carbon steel–lined single-shell tanks (SSTs) meant as a temporary solution with a projected design expectancy of less than 25 years. That lifetime has long been exceeded for all SSTs. The 149 SSTs are grouped into farms of between 2 and 18 tanks that were originally solely associated with specific plants. SSTs have operating capacities of up to 1 Mgal. Chapter 6 provides further details on the tanks' construction.

As noted, all waste was neutralized to a basic (alkaline) pH prior to storage tank disposal; mild steel corrodes severely in acid. The resulting alkalinity precipitated many waste constituents as a "sludge" of solid polyvalent metal hydroxides (e.g., of iron, bismuth, aluminum) or low-solubility sulfates and phosphates (Wells et al., 2011). The pH adjustment was accomplished by adding sodium hydroxide and sodium carbonate, which as industrial-grade chemicals also contain other sodium salts, principally chloride. Further, to improve utilization of the available tank space, waste solutions were concentrated by evaporation to the extent that the dissolved sodium salts were crystallized. The crystallized salt waste, known as saltcake, contains sodium nitrate, aluminate, carbonate, sulfate, fluoride, and phosphate, and some sodium double salts of these anions (Herting et al., 2015). Thus, the tank wastes are comprised of saltcake, sludge, and supernatant and interstitial solutions highly concentrated in sodium salts.

By 1956, the first leak was suspected from a single-shell tank, which, with the need for additional waste storage capacity, prompted the design and construction of double-shell tanks (DSTs) starting in 1968 (Gephart, 2003). Construction of 28 DSTs, also mild-steel–lined, was completed in 1986, each with a nominal million-gallon waste capacity. As part of the SST interim stabilization program completed in 2005, the free liquids for SSTs have been transferred to the DSTs (Boomer et al., 2012). Eventually, the remaining solids (i.e., sludge and saltcake) and interstitial liquid in the SSTs will also be retrieved and transferred to DSTs for subsequent processing and disposal.

As a result, there are about 54 million gallons of sludge, supernate, and saltcake waste (Rodgers, 2023) with over 130 million curies of radioactivity stored in the Hanford underground waste tanks. Hanford tank waste exhibits significant chemical and physical complexity in terms of composition and behavior and leads to challenges for its sampling and analysis as described later. Details on the chemical and physical complexities are published extensively elsewhere and are available through online tools described later. One significant report that describes the waste and, importantly, the historical origins of its constituents is by Kupfer et al. (1999). Table 5.1 lists the primary chemical constituents in Hanford tank waste, aside from water and the hydr(oxide)s associated with low-solubility metals, for example, Fe, Mn, and Si. The key radiochemical constituents of Hanford tank waste are listed in Table 5.2. Both inventory sets, derived from the online tool described later, illustrate the complexity of the waste. It is seen that sodium salts make up most of the waste (e.g., sodium nitrate, nitrite, and carbonate alone comprise ~80% of the listed mass), while cesium-137, strontium-90, and their daughter products barium-137 and yttrium-90, respectively, constitute most of the radioactivity.

Again, the flowsheets for the waste-generating processes are well known at the start of a process, but institutional knowledge and paperwork were lost over decades of operation of the waste tank farms. In addition, process vessel corrosion and impurities in industrial-grade process chemicals can dramatically affect the nature of the waste stream. The active concentration of tank waste supernates in six different evaporator campaigns spanning the 50 years of processing has not been

**TABLE 5.1**
**Primary Chemical Constituents of Hanford Tank Waste**

| Chemical Constituent[a] | Total Mass (Metric Tons) | Proportion of Total Mass (%) |
|---|---|---|
| Aluminum | 8,006 | 5.42 |
| Bismuth | 525 | 0.36 |
| Calcium | 245 | 0.17 |
| Chloride | 804 | 0.54 |
| Carbon (within organic compounds) | 1,199 | 0.81 |
| Carbonate | 9,656 | 6.53 |
| Chromium | 549 | 0.37 |
| Fluoride | 1,227 | 0.83 |
| Free hydroxide | 4,257 | 2.88 |
| Iron | 1,144 | 0.77 |
| Mercury | 2 | <0.01 |
| Potassium | 952 | 0.64 |
| Lanthanum | 33 | 0.02 |
| Manganese | 155 | 0.10 |
| Sodium | 46,036 | 31.16 |
| Nickel | 88 | 0.06 |
| Nitrite | 11,969 | 8.10 |
| Nitrate | 50,745 | 34.34 |
| Lead | 73 | 0.05 |
| Phosphate | 4,940 | 3.34 |
| Silicon | 723 | 0.49 |
| Sulfate | 3,382 | 2.29 |
| Strontium | 44 | 0.03 |
| Uranium | 615 | 0.42 |
| Zirconium | 393 | 0.27 |

[a] Other constituents, not listed, comprise, in total, 1% or less of the total mass.

**TABLE 5.2**
**Radiochemical Constituents (>0.01% of Total) of Hanford Tank Waste**

| Radionuclide | Radioactivity (Ci) | Proportion of Total Radioactivity |
|---|---|---|
| Americium-241 | 1.21E5 | 0.09% |
| Barium-137 (metastable) | 2.87E7 | 21.57% |
| Cesium-137 | 3.04E7 | 22.85% |
| Europium-154 | 2.24E4 | 0.02 |
| Nickel-63 | 1.19E5 | 0.09% |
| Plutonium-238[a] | 2.22E3 | <0.01% |
| Plutonium-239[a] | 4.27E4 | 0.03% |
| Plutonium-240[a] | 9.34E3 | 0.01% |
| Plutonium-241[a] | 4.59E4 | 0.03% |
| Plutonium-242[a] | 1.38E0 | <0.01% |
| Samarium-151 | 2.97E6 | 2.24% |
| Strontium-90 | 3.53E7 | 26.52% |
| Technetium-99 | 2.52E4 | 0.02% |
| Yttrium-90 | 3.53E7 | 26.52% |

[a] Total plutonium inventory in tank waste is 731 kg.

thoroughly documented. There are many uncertainties in the tank transaction histories for these campaigns. The waste heating during evaporator campaigns may also have accelerated chemical reactions, changing waste properties from those described in the flowsheets or in early sample analysis. The waste transactions associated with the removal of tank waste for the two major tank waste processing campaigns (uranium recovery in the 1950s and waste fractionization campaign in the 1960–1970s) are incomplete (Simpson et al., 1996). These processes were based on assumed waste characteristics, not actual waste characteristics.

As a result, in 1993, it was noted by the DNFSB that there was insufficient tank waste technical information to ensure that Hanford tank waste could be safely stored, handled, and characterized for future disposal. Due to these concerns, DNFSB transmitted Recommendation 93–5 (DNFSB, 1993) to the U.S. DOE. The DOE responded with an implementation plan that was accepted in March 1994 (Recommendation 93-5 Implementation Plan, 1996). One of the tasks within the implementation plan was to improve data accessibility, data control, and data readability, as well as establish "standard inventory estimates for all tanks" (Recommendation 93-5 Implementation Plan, 1996).

**PERSPECTIVES: GARY COOKE**

Gary Cooke pioneered phase characterization at the Hanford Site.

**WHAT WAS THE MAIN TECHNICAL CHALLENGE YOU FACED IN YOUR CAREER AT HANFORD?**

"Putting aside the highly radioactive materials, dealing with the sludges was the most difficult technical challenge. Every tank had a different sludge that had never been seen before. Very challenging and very rewarding. It was frustrating to characterize 3% of the material, but chemistry and sequential leaches were a powerful tool for chemical characterization."

**WHAT WAS THE MOST CHALLENGING ASPECT OF CHARACTERIZATION?**

"A nearly insurmountable task is to extrapolate the contents of a million-gallon tank from a milligram sample. The statistical analysis of the sample results to the bulk of the tank is a difficult sell."

**WHAT TECHNICAL CONTRIBUTION ARE YOU MOST PROUD OF DURING YOUR TIME AT HANFORD?**

"On the way to retirement, I was granted 3 million dollars' worth of equipment including a Raman, Fourier Transform Infrared Spectrometer, Thermogravimetric analyzer, and a state-of-the-art X-ray diffractometer (XRD)." Cooke considered this a crowning achievement in his career at Hanford.

## 5.1 TANK SAMPLING PROCESSES AND CHALLENGES

The Tank Waste Characterization Project mission was to provide information and waste sample material necessary for Tank Waste Remediation System (TWRS) to define and maintain safe interim storage and to process waste into stable forms for final disposal (Brown et al., 1998). As part of this effort, the Tank Waste Characterization Basis was developed to determine the priority of tank sampling and characterization efforts where those tanks with flammable gases, organic fuel, ferrocyanide, and vapor generation were placed as highest priority for characterization and treatment.

The preparation and chemical and radiochemical analyses of Hanford tank waste samples can be performed with standard laboratory equipment and instruments as relatively routine processes that are not particularly challenging. Rather, the main challenges of tank waste characterization are associated with radiological dose and sampling limitations. Accurate, representative, and effective sampling techniques are difficult with the waste tanks because they were not designed for routine sampling. There are a finite number of sampling locations for each tank based on riser positioning, depth, and the operational functionality of the sampling riser. For example, in one recently emptied SST, there was one riser that was found to have had concrete dumped down it, thereby eliminating that sampling port. Additionally, the waste within the tank, especially true for the saltcake and sludge, is not homogeneous. The ability to adequately mix a million-gallon DST is a concern for data reproducibility. Another real challenge that must be managed for sampling single-shell tanks is that saltcake waste in a compromised (leaking) SST cannot be dissolved so a representative sample can be extracted. These physical constraints mean that uncertainty in the representativeness of samples must be considered when applying analytical results to the bulk contents of the tank.

The tank waste is highly radioactive and thus can only be handled initially by facilities that can receive samples into concrete-shielded hot cells having mineral oil and leaded-glass (i.e., radiologically shielded) windows with remote operation (see Figure 5.1). The shielding protects the worker from the radiological dose, while remotely operated manipulators enable the samples to be handled. At Hanford, analytical laboratories with these hot cell capabilities are limited to the PNNL and the main Hanford operations support laboratory, 222-S Laboratory. Because of their high radioactivity (and high chemical concentrations), samples generally must be sufficiently diluted to facilitate their analysis outside of a shielded cell. In some cases, this means some accuracy must be compromised to complete the analysis beyond that normally encountered for non-radioactive material.

The 222-S Laboratory initiated operations in 1951 and is expected to provide the analytical characterization support to tank farm operations in the future. The laboratory contains 11 hot cells and receives around 10,000 waste samples annually with ~120 analyses performed per working day.

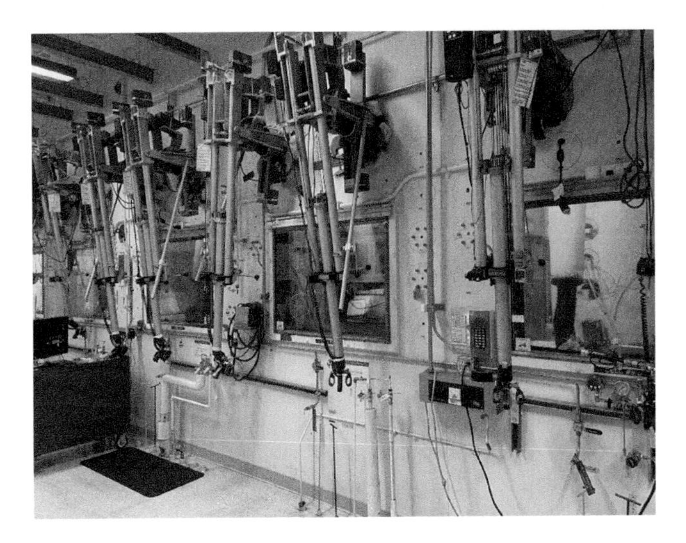

**FIGURE 5.1** Six hot cells as part of the Shielded Analytical Laboratory at Pacific Northwest National Laboratory in Richland, Washington.

DOE Office of River Protection (ORP) relies on 222-S to test waste compatibility and physical characteristics to support tank-to-tank waste transfers, support tank corrosion inhibition through corrosion rate studies, and provide input to the engineering specification for operating the tank farm.

In the very early phases of characterization, the effort to sample and analyze SSTs was done in collaboration with the Radiochemical Processing Laboratory (RPL) at PNNL. Due to its close proximity, ability to receive and handle highly radioactive samples, and technical expertise in analytical chemistry, the RPL worked closely with 222-S to develop procedures and protocol to analyze tank waste samples in the mid-nineties during the characterization campaign.

The 222-S Laboratory is equipped with standard analytical instruments including inductively coupled plasma atomic emission spectrometry (ICP-AES) for metal analysis, ion chromatography (IC) to measure anions, titrations for hydroxide concentration, standard gamma counting, etc. Table 5.3 highlights the analytical instrumentation available at the 222-S Laboratory in support of the Hanford Site, including analytes measured and the number of instruments owned by the laboratory. Note there is more than one of every instrument type (except for the titrator). It is necessary to have a back-up instrument in the event instruments are down due to maintenance or recalibration. Table 5.4 highlights the instruments 222-S employed for physical property testing on the tank waste samples along with the quantity of those instruments that support the characterization mission.

A Tank Sample and Analysis Plan (TSAP) was created for each tank and waste type to detail the characterization plan for that sample. Post analysis, the 222-S Laboratory was responsible for generating Tank Characterization Reports (TCRs) to report and evaluate data collected from the tanks. The TCRs were electronically filed, and the characterization data was collated in one location.

When a sample is received at the laboratory and the packaging is removed, the sample is visually inspected, and photographs are taken to document the observations. Figure 5.2 is a flowchart of the characterization methods outlined to support waste retrieval operations at the 222-S Laboratory as described by Berry (2018). Treatment and characterization of liquid samples are relatively straightforward; the liquid is subaliquoted for measurement of specific gravity, anions by ion chromatography, and TIC/TOC using the silver-catalyzed persulfate method. An aliquot of the liquid undergoes acid digestion and is analyzed by ICP-AES for metals. The analytical characterization and physical property characterization of solid samples are much more involved. First, the bulk density of the solid is measured, and then, a whole slew of characterization techniques may be employed. Most of the analytical methods required the sample to be present as a liquid; thus, acid digestion, fusions,

**TABLE 5.3**
**Analytical Instrumentation with Associated Sample Preparation Utilized at the 222-S Laboratory to Support Characterization of Tank Waste Samples**

| Instrumentation | Analytes Measured | Preparation Steps | Quantity |
|---|---|---|---|
| Liquid scintillation counting (LSC) | $^{3}H$, $^{14}C$, $^{63}Ni$, $^{79}Se$, $^{93}Zr$ | Analyte-specific separation required prior to analysis | 3 |
| | $^{59}Ni$, $^{60}Co$, $^{106}Ru$, $^{125}Sb$, $^{134}Cs$, $^{144}Ce$, $^{152}Eu$, $^{154}Eu$, $^{155}Eu$, $^{226}Ra$ | Radio-isotopic acid/fusion digest | |
| | $^{241}Pu$ | Pu and Am separation | |
| Total inorganic compound/total organic compound (TIC/TOC) | Organic carbon and inorganic carbon ($CO_3^{2-}$), semi-volatile organic compounds (SVOCs), volatile organic compounds (VOCs) | Persulfate – liquids<br>Combustion – solids<br>SVOCs – separation followed by organic concentration | 5 |
| Inductively coupled plasma atomic emission spectrometer (ICP-AES) | Al, Bi, Ca, Cr, Fe, K, Mn, Na, Ni, Pb, Si, Sr, Zr | Radio-isotopic acid/fusion digest | 2 |
| Inductively coupled plasma mass spectrometer (ICP-MS) | $^{237}Np$, $^{233}U$, $^{234}U$, $^{236}U$, $^{238}U$, $^{63}Ni$, $^{99}Tc$, $^{126}Sn$, $^{229}Th$, $^{232}Th$, $^{239}Pu$, $^{135}Cs$ | | 3 |
| Gamma emission analyzer (GEA) | $^{129}I$, Gamma fission products | | 9 |
| Alpha emission analyzer (AEA) | $^{238}Pu$, $^{239/240}Pu$, $^{241}Pu$, $^{241}Am$, $^{242}Cm$, $^{243/244}Cm$ | Radio-isotopic acid/fusion digest followed by Pu and Am separation | 30 |
| | $^{232}Th$ | $^{232}Th$ separation | |
| Ion chromatography (IC) | Ammonia ($NH_3$) | Distillation | 5 |
| | $Cl^-$, $F^-$, $NO_2^-$, $NO_3^-$, $PO_4^{3-}$, $SO_4^{2-}$ | Water digest | |
| Cold vapor atomic absorption (CVAA) | Mercury (Hg) | | 2 |
| Gas proportional counter (GPC) | Radiochemical analysis | | 13 |
| Gas chromatograph (GC) – mass spectrometer (MS) | PCBs, SVOCs, VOCs | Extraction followed by organic concentration for PCBs and SVOCs | 11 |
| Titrator | Hydroxide ($OH^-$) | | 1 |

**TABLE 5.4**
**Instrumentation Used at 222-S Laboratory to Measure the Physical Properties of Tank Waste Samples**

| Instrument | Use | Quantity |
|---|---|---|
| Differential scanning calorimeter (DSC) | Measures how physical properties of a sample change with temperature over time | 2 |
| Thermal gravimetric analysis (TGA) | Measures % water | 1 |
| X-ray diffractometer (XRD) | Solid phase characterization | 4 |
| Scanning electron microscopy (SEM) | | 6 |
| Polarized light microscopy (PLM) | | 1 |
| Rheometer | Measures physical properties of liquids | 3 |

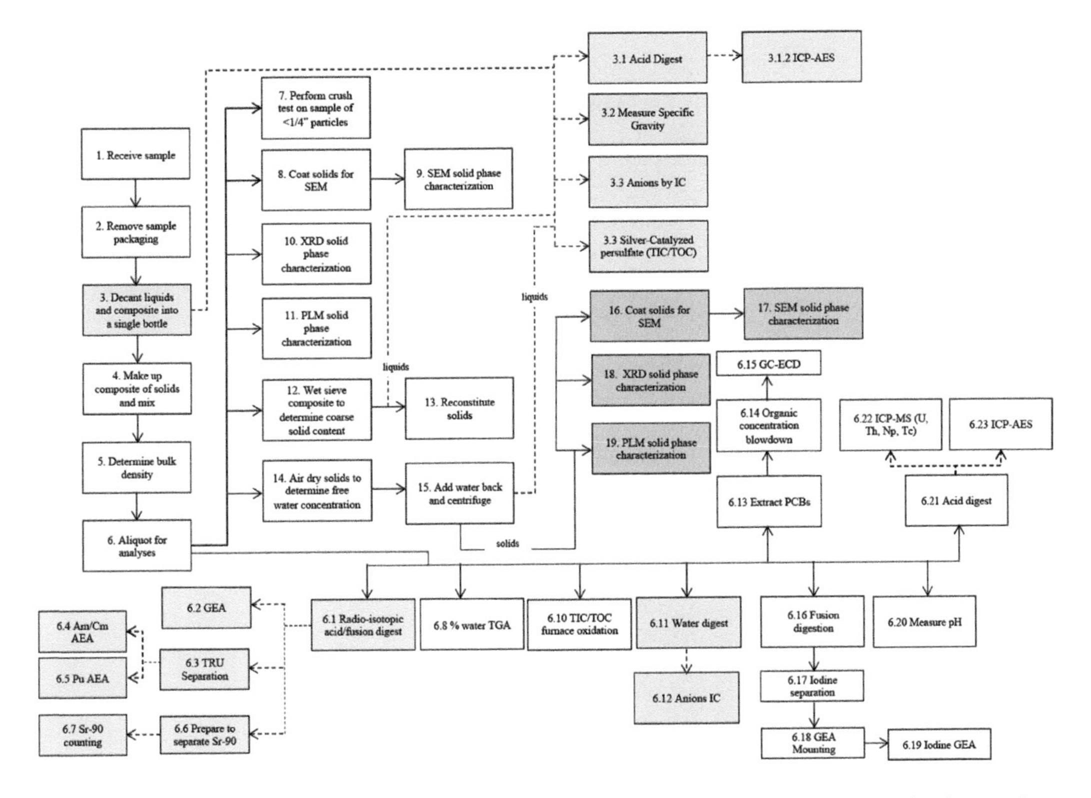


**FIGURE 5.2** Example of the characterization pathway to support Hanford tank waste retrieval operations at the 222-S Laboratory.

and water leaching of the solids are done to quantify the analytes leached/dissolved in the liquid phase. As for the physical characterization of the solids, XRD, SEM, and PLM techniques can be utilized to identify phase characterization.

## 5.2 HANFORD WASTE PHYSICAL PROPERTY CHARACTERIZATION

Historically, characterization of Hanford waste liquids, solids, and slurries has targeted the following physical properties to enable engineering assessment of tank inventories and the feasibility of waste pumping and treatment operations:

- physical appearance
- liquid, solid, and slurry density
- solid content (including total, undissolved, and dissolved solids)
- solid settling rate
- liquid viscosity and slurry and sludge rheology
- sludge and slurry yield stress
- solid particle size distribution

Standard measuring techniques (Table 5.5) for all properties itemized above typically require 10–500 mL of test material, with the exception of particle size determination, which can be reliably done with as little as 0.1–1 g of solid waste particulate. As such, waste characterization campaigns requesting the full suite of physical properties typically receive a total of 1–10 L of the target Hanford tank waste. Prior to characterization, individual tank waste samples must be composited, homogenized, and then refractionated to provide representative (well-mixed)

**TABLE 5.5**
**Typical Physical Property Targets for Hanford Waste Characterization (and the Equipment Needed to Effect Characterization)**

| Physical Property | Equipment and Materials |
|---|---|
| Physical appearance | • Reference color chart<br>• Digital camera |
| Settling rate and percent settled solids of slurries and suspensions | • Volume-graduated labware of appropriate size to accommodate sample<br>• Calibrated analytical balance, sensitivity based on sample size and data needs<br>• Rod, spatula, vortex mixer, or gas sparger (for mixing sample as needed)<br>• Centrifuge |
| Density or specific gravity of sludges, solutions, and slurries | • Graduated centrifuge cone, volumetric flask, or graduated cylinder<br>• Centrifuge with swing rotor (for centrifuged method)<br>• Calibrated analytical balance, sensitivity based on sample size and data needs |
| Determination of volume percent, weight percent, and densities of centrifuged solids and supernatants from sludges and slurries | • Graduated centrifuge cone with cap<br>• Centrifuge with swing rotor<br>• Calibrated analytical balance, sensitivity based on sample size and data needs<br>• Graduated cylinder<br>• Transfer pipette |
| Determination of weight percent total solids, dissolved solids, and total oxides | • Ceramic crucibles<br>• Glass vials<br>• Calibrated analytical balance, sensitivity based on sample size and data needs<br>• Desiccator containing dry indicating desiccant<br>• Muffle furnace (100–1,000°C)<br>• Drying oven with calibrated thermocouple or thermometer |
| Determination of viscosity and rheology | • Viscometer or rheometer<br>• Certified Newtonian viscosity standards<br>• Constant temperature bath/circulator |
| Determination of yield stress/shear strength (vane method) | • Viscometer<br>• Shear vane |

subplots for separate physical characterizations (Fiskum et al., 2008, 2009; Edwards et al., 2009; Lumetta et al., 2009; Snow et al., 2009).

Documentation of the sample physical appearance is relatively straightforward and involves recording liquid and sludge morphology and color. For samples handled in shield facilities, a direct visual inspection may be complicated by distortions and sample obscuration caused by shielded facility glass, which is thick and can be leaded or have an oil fill layer to moderate dose. Use of in-cell and out-of-cell reference color charts or in-cell cameras can be used to aid inspection and overcome color distortion caused by the shielded facility glass. Care must be taken when inspecting samples with in-cell digital cameras, as visual inspection can be hampered by baseline camera and monitor color accuracy. Radiation damage may further alter camera color accuracy (over the life of the camera) and may eventually lead to digital camera failure.

Determination of waste solid content, density, settling rates, viscosity, and rheology is governed by ASTM C1752-21 (2021). Density measurements of liquids and slurries require 10–50 mL of material and are done using standard laboratory balances and volumetric glassware (typically Grade A) or graduated centrifuge cones. For settling slurries, the ASTM method documents steps to determine gravimetrically settled and centrifuged solid density. For determination of solid content, the ASTM method relies on gravimetric methods where samples of liquids, slurries, and centrifuged solids are partitioned, weighed, and then heated at low temperature (90–105°C) to drive off water. The mass of solids recovered is then ratioed to the initial sample mass and subsequently interpreted to determine the dissolved, undissolved, and total solid content of the originating samples.

Generic methods for determining liquid viscosity and slurry viscosity/rheometry and yield stress are outlined in ASTM C1752-21 (2021). These methods rely on conventional bench-top rheometers and viscometers, which can represent a substantive (but not significant) capital investment. For measurements in shielded facilities (such as hot cells), the sensitive electronic equipment housed in modern viscometers/rheometers must be shielded or moved outside the radiological enclosure to prevent premature equipment failure (due to radiation damage) and associated replacement costs for even modest term (e.g., 1-year) characterization efforts. The volume of the test material needed for determination of viscosity and rheology typically ranges from 10 to 100 mL of material (and depends on the geometry used to characterize the stress response of the material). Determination of yield stress, particularly as a function of the sample hold period (where, e.g., the increase in settled sludge strength is characterized as a function of time), requires samples of sufficient size to meet the geometric requirements outlined in ASTM C1752-21 (2021). Here, up to 500 mL of waste material may be needed to accurately determine yield stress/shear strength.

Measurement of Hanford waste particle size distribution largely relies on optical microscopy (either using standard OM methods or using automatic flow image analyzers) and laser diffraction. As most Hanford waste solids derive from high-salt (5–10 M Na), high-pH (>14) liquid media, accurate assessment of size distribution (as it exists in the tanks or Hanford process lines) requires wet/chemical-cell analysis methods that can accept solids suspended in their originating supernatant. Likewise, because Hanford solids contain dense materials (iron-, zircon-, and transuranic metal-bearing solids), size analysis must typically be coupled with some form of active flow or mixing to prevent size bias because of particle settling. For this reason, Hanford waste size analyses have typically used commercial off-the-shelf laser diffraction analyzers with built-in chemical-resistant flow cells. Particle dispersion is accomplished through the use of active flow during size measurement, along with sonication of the sample before or during measurement (with the latter accomplished by an in-cell ultrasonic probe). Laser diffraction analyzers operate in the single-scattering regime and have active measuring paths on the millimeter length scale; the volume of dispersed material needed to effect representative measurement is low (typically 0.01–0.2 vol%). Modern analyzers have small volume dispersion units, requiring as little as 6–20 mL of test dispersion. As such, single measurements of size analysis can be accomplished with ~0.1 g of solids, allowing size distribution measurements to generally be completed outside of shielded facilities and in radiologically controlled fume hoods. Laser diffraction analysis can be limited by the relatively large volume of liquid waste generated because of flow-cell cleaning. A typical single measurement using a 20 mL dispersion generates 60–100 mL of liquid waste.

## 5.3 KEEPING THE RECORDS STRAIGHT

The Best-Basis Inventory Maintenance tool, or BBI (https://twins.hanford.gov/twinsdata/best-basis-inventory), was initially developed in FY 1999 at the culmination of the Tank Waste Characterization Project to fulfill DFSNB Recommendation 93–5. The justification for developing it came from the realization that it was an impossible task to keep 177 paper documents up to date in the face of tank farm operations including incorporation of new sample data, updating sampling events, and addressing mistakes/errors in existing calculations (Rasmussen, 2017). During the first year of deployment (FY 2000), BBI underwent several upgrades, including the option to (i) separate the tank waste in (up to) six waste phases: supernate, saltcake solids, saltcake liquids, sludge solids, sludge liquids, and retained gas; (ii) identify sampling events used to update BBI; and (iii) update waste volumes consistent with sampling or transfer events. The current (2022) BBI estimates are presented on a tank-by-tank basis and a global (total) basis. The tank-by-tank inventory includes 25 chemical components and 46 radionuclides. The standard analytes account for 99% of the chemical inventory (not including water and hydroxide), and the 46 radionuclides account for 99% of the total radioactivity. The global inventory originated from key historical records, whereas the tank-by-tank inventory is calculated from tank sample reports; thus, these estimates are independent of one another.

On a broader scope, the Tank Waste Information Network System (TWINS; https://twins.hanford.gov/twinsdata/home) is a universal location for documents pertaining to tank farm operations and characterization. The database includes the most recent reports for sample analysis, tank transfers, vapor documentation, tank levels and temperatures, and the BBI.

The DOE partnered with PNNL to develop the PNNL-Hanford Online Environmental Information Exchange (PHOENIX), a web-based tool for accessing current and historical data associated with Hanford tank waste. PHOENIX was launched to the public in 2015 and provides access to:

- leak status and other basic information about SSTs and DSTs
- historical and current in-tank sensor data: temperature, surface level, interstitial liquid level
- breakdown of tank/farm waste volumes by phase (sludge, supernate, saltcake) and source
- tank-specific inventory of radionuclides
- tank volume and waste transfer history

in a visually appealing and user-friendly manner. The website to explore the tank waste information is http://phoenix.pnnl.gov/.

PHOENIX displays the tanks in "Map View" or "Tank Plot" view. Figure 5.3 shows the tank plot view where the tanks are color-coordinated to separate SSTs, DSTs, and the capacity of said tanks.

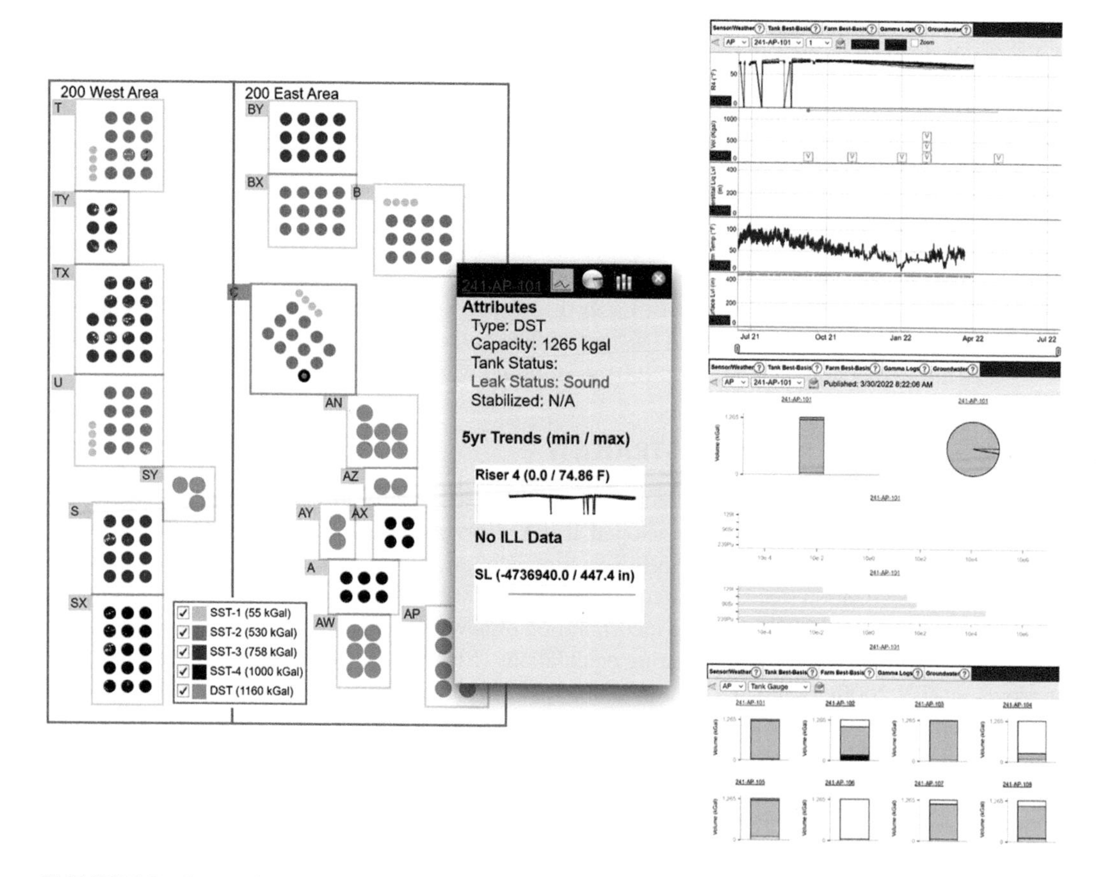


**FIGURE 5.3** Interactive display of the tank farms in the PHOENIX tool to depict tank waste characterization including tank temperature over time, sampling times and dates, radionuclide breakdown, and phase characterization.

The tanks are also broken up and labeled into the tank farms. Tank 241-AP-101 was selected as an example in Figure 5.3, and the three figures on the right show the sensor data, including up-to-date temperature readings, as well as indications of tank transfers/sampling events, followed by the Best-Basis Inventory data icon, where a waste volume is displayed as a bar chart, pie chart, and bar graph with the most prevalent radionuclides and their respective concentrations. Finally, there is an option to display the Farm Best-Basis Inventory data where the tanks are displayed as bar graphs with color-coordinated volumes of the different waste phases present in those tanks. PHOENIX is an excellent interactive tool for tank waste/Hanford education without the need to mine the data from different sources and references.

## REFERENCES


ASTM C1752-21. 2021. *Standard Guide for Measuring Physical and Rheological Properties of Radioactive Solutions, Slurries, and Sludges*. Standard, ASTM International, West Conshohocken, PA.

Berry J. 2018. *Direct Feed Low-Activity Waste 222-S Laboratory Operations Research Model Basis and Assumptions Document*. RPP-RPT-59477, Rev. 1. Washington River Protection Solutions, Richland, WA.

Boomer KD, JB Johnson, TJ Venetz, and DJ Washenfelder. 2012. *Overview of Enhanced Hanford Single-Shell Tank (SST) Integrity Project - 12128*. WRPS-51713-FP, Rev. 0. Washington River Protection Solutions, Richland, WA.

Brown TM, JW Hunt, and LJ Fergestrom. 1998. *Tank Characterization Technical Sampling Basis*. HNF-SD-WM-TA-164, Rev. 4. Lockheed Martin Hanford. Corp., Richland, WA.

DNFSB. 1993. *Recommendation 93-5 to the Secretary of Energy*. Defense Nuclear Facilities Safety Board, Washington, DC.

Edwards, MK, JM Billing, DL Blanchard, EC Buck, AJ Casella, AM Casella, JV Crum, RC Daniel, KE Draper, SK Fiskum, LK Jagoda, ED Jenson, AE Kozelisky, PJ MacFarlan, RA Peterson, RW Shimskey, LA Snow, and RG Swoboda. 2009. *Characterization, Leaching, and Filtration Testing for Tributyl Phosphate (TBP, Group 7) Actual Waste Sample Composites*. PNNL-18119, Rev. 0. Pacific Northwest National Laboratory, Richland, WA.

Fiskum SK, JM Billing, JV Crum, RC Daniel, MK Edwards, RW Shimskey, RA Peterson, PJ MacFarlan, EC Buck, KE Draper, and AE Kozelisky. 2009. *Characterization, Leaching, and Filtrations Testing of Ferrocyanide Tank Sludge (Group8) Actual Waste Composite*. PNNL-18120, Rev.0. Pacific Northwest National Laboratory, Richland, WA.

Fiskum SK, EC Buck, RC Daniel, KE Draper, MK Edwards, TL Hubler, LK Jagoda, ED Jenson, AE Kozelisky, GJ Lumetta, PJ MacFarlan, BK McNamara, RA Peterson, SI Sinkov, LA Snow, and RG Swoboda. 2008. *Characterization and Leach Testing for REDOX Sludge and S-Saltcake Actual Waste Sample Composites*. PNNL-17368, Rev. 0. Pacific Northwest National Laboratory, Richland, WA.

Gephart RE. 2003. *A Short History of Hanford Waste Generation, Storage, and Release*. PNNL-13605, Rev. 4. Pacific Northwest National Laboratory, Richland, WA.

Gephart RE. 2010. "A Short History of Waste Management at the Hanford Site". *Physics and Chemistry of the Earth, Parts A/B/C* 35(6–8):298–306. https://doi.org/10.1016/j.pce.2010.03.032.

Gerber MS. 2001. *History of Hanford Site Defense Production (Brief)*. HNF-5041-FP, Rev. 0. Fluor Hanford, Richland, WA.

Harvey DW. 2000. *History of the Hanford Site: 1943-1990*. United States. Pacific Northwest National Laboratory, Richland, WA. https://doi.org/10.2172/887452.

Herting DL, GA Cooke, JS Page, and JL Valerio. 2015. *Hanford Tank Waste Particle Atlas*. LAB-RPT-15-00005. Washington River Protection Solutions LLC, Richland, WA. https://www.osti.gov/biblio/1423895.

Kupfer MJ, AL Boldt, KM Hodgson, LW Shelton, BC Simpson, RA Watrous, MD LeClair, GL Borsheim, RT Winward, BA Higley, RM Orme, NG Colton, SL Lambert, and DE Place. 1999. *Standard Inventories of Chemicals and Radionuclides in Hanford Site Tank Wastes*. Report No. HNF-SD-WM-TI-740, Rev. 0 and 0C. Lockheed Martin Hanford Company, Richland, WA.

Lumetta GJ, EC Buck, RC Daniel, K Draper, MK Edwards, SK Fiskum, RT Hallen, LK Jagoda, ED Jenson, AE Kozelisky, PJ MacFarlan, RA Peterson, RW Shimskey, SI Sinkov, and LA Snow. 2009. *Characterization, Leaching, and Filtration Testing for Bismuth Phosphate Sludge (Group 1) and Bismuth Phosphate Saltcake (Group 2) Actual Waste Sample Composites*. PNNL-17992, Rev. 0. Pacific Northwest National Laboratory, Richland, WA.

Rasmussen JH. 2017. *Guidelines for Updating Best-Basis Inventory*. RPP-7625, Rev. 13. Washington River Protection Solutions, Richland, WA.

Rodgers MJ. 2023. *Waste Tank Summary Report for Month Ending June 30, 2023*. HNF-EP-0182, Revision 426. Washington River Protection Solutions, Richland, WA.

Simpson BC, SJ Eberlein, TM Brown, CH Brevick, and SF Agnew. 1996. *Characterization of Hanford Waste and the Role of Historic Modeling*. WHC-SA-3029-FP. Westinghouse Hanford Company, Richland, WA.

Snow LA, EC Buck, AJ Casella, JV Crum, RC Daniel, KE Draper, MK Edwards, SK Fiskum, LK Jagoda, ED Jenson, AE Kozelisky, PJ MacFarlan, RA Peterson, and RG Swoboda. 2009. *Characterization and Leach Testing for PUREX Cladding Waste Sludge (Group 3) and REDOX Cladding Waste Sludge (Group 4) Actual Waste Sample Composites*. PNNL-18054, Rev. 0. Pacific Northwest National Laboratory, Richland, WA.

Strachan DM, WW Schulz, and DA Reynolds. 1993. *Hanford Site Organic Waste Tanks: History, Waste Properties, and Scientific Issues*. PNL–8473. Pacific Northwest Laboratory, Richland, WA. https://www.osti.gov/biblio/6699333.

US Department of Energy. Hanford Tank Waste Retrieval, *Treatment, and Disposition Framework*. September 24, 2013. US Department of Energy, Washington, DC.

US DOE. 1996. *Recommendation 93-5 Implementation Plan*. DOE/RL-94-0001, Rev. 1. United States Department of Energy, Richland, WA.

US DOE 2009. *Categorical Exclusion for the Facility Upgrades at 222-S Laboratory and Complex Conducted under the American Recovery and Reinvestment Act*. DOE/CX-00004. U.S. DOE, Office of River Protection, Richland, WA.

Wells BE, DE Kurath, LA Mahoney, Y Onishi, JL Huckaby, SK Cooley, CA Burns, EC Buck, JM Tingey, RC Daniel, and KK Anderson. 2011. *Hanford Waste Physical and Rheological Properties: Data and Gaps*. PNNL-20646, EMSP-RPT-006. Pacific Northwest National Laboratory, Richland, WA. https://www.pnnl.gov/main/publications/external/technical_reports/PNNL-20646.pdf.

# 6 Waste Tank Structural Assessment

*Christopher Grant and Naveen Karri*

Legacy industrial sites may contain process waste containment systems that were designed and constructed to less rigorous standards than would be employed today. Additionally, such systems may be required to fulfill roles for much longer than they were originally designed while the site is remediated. The liquid and sludge waste from plutonium production at the Hanford Site is a significant environmental liability not least because it is stored in underground tanks that are well beyond their design life; some have leaked in the past. Tanks were constructed to serve production facilities as part of the Manhattan Project during World War II, and others were added during the Cold War. They are grouped into "farms" with several connected as cascades so liquid waste could flow from one to another as they filled. The farms typically contain the primary storage tanks and smaller receiver or catch tanks that collected liquids from the containment structures of pumps and valves. This chapter focuses on the primary storage tanks. The farms are labeled by the production plant with which they were associated, with a "241" designator added to indicate the functional position in the production process. For example, the T-farm has tanks associated with one of the first production plants, T-plant, so the first tank is referred to as 241-T-101. Subsequent farms associated with T-plant then received another letter on the label, such as the TX for the second farm. The Hanford tanks can primarily be categorized into two types, single-shell and double-shell, according to their containment structure. While the primary function of tanks is to contain the nuclear waste during their active operation and sustain the thermal and operating loads, they must also maintain their integrity even after they are decommissioned to support retrieval and closure activities. Given the longevity of the mission to remediate the Hanford Site, understanding the structural integrity of the legacy storage tanks is of utmost importance.

## 6.1 SINGLE-SHELL TANKS

### 6.1.1 Single-Shell Tank Description

Most tanks at the Hanford Site are single-shell tanks (SSTs), which are currently inactive and do not contain any liquid waste. A total of 149 SSTs grouped into 12 farms, consisting of 4–18 tanks each, were constructed between 1943 and 1964. The original SST design included a reinforced concrete shell, dome, and bottom slab, with mild carbon steel lining the bottom and the cylindrical shell (typically referred to as tank wall). The liner, which extends close to the top of the wall, essentially defines the maximum volume of material that can be stored in an SST. Although the waste originating in the production plants was acidic, mild steel was selected for cost considerations during the war. Therefore, sodium hydroxide was added to the waste prior to discharge to make it alkaline and so compatible with the tank material. Mild steel continued to be used after the war to avoid introducing complexity into the waste management system. Over time, while the basic design remained essentially the same, the thickness of the steel liner within the concrete shell was increased with corresponding increases in operating depth and storage capacity. Minor design improvements were made in the tank bottoms and footing (the location where the cylindrical shell meets the bottom).

DOI: 10.1201/9781003329213-8

**FIGURE 6.1** Photograph of a single-shell tank farm under construction.

The SSTs were constructed in four batches, as the need for storage grew. A photograph of B tank farm under construction is shown in Figure 6.1 with 12 storage and 4 smaller receiver tanks. The first 64 tanks were built between 1943 and 1944 as part of the Manhattan Project. Another 42 were constructed between 1946 and 1949, and 39 more between 1950 and 1955 as plutonium production increased commensurate with the increasing probability of military conflict during the Korean War. The fourth batch of tanks, built between 1963 and 1964, contained only four tanks. The SSTs ranged in volume from 55,000 to 1 million gallons and are buried roughly 10 ft below grade. The first tanks were designed with 18-ft high carbon steel liners and operating volumes of approximately 530,000 gal. The next tanks were built first with 24-ft liners and operating volumes of approximately 758,000 gal, and finally with liner heights of 32 ft and operating capacities of approximately 1 million gallons.

The schematic diagram in Figure 6.2 illustrates a typical SST with its associated instrumentation for leak detection, level and temperature measurement, and equipment such as a central pump, camera, and filter. The lateral leak detection drywells indicated in the figure are physically located approximately 10 ft beneath the tank. The drywells consist of pipes through which probes can be inserted to monitor for gamma radiation that could indicate waste leakage.

Physical phenomena such as radiolytic heat and corrosion have contributed to failure of the containment in several SSTs as described by Field et al. (2014) and consequent leakage of waste into the ground beneath them. The effect of radiolytic heat was most dramatically illustrated in 1965 when likely vaporization of liquid between the floor liner and concrete separated a portion of the liner from the floor and wall of SST A-105 to create a ~100,000-gallon bulge covering half the tank floor. The first SST leaks were confirmed in 1959 (Tanks TY-106 and U-101), although anecdotal evidence had suggested a leak from U-104 in 1956. Since 1959, additional SST leaks have been declared, resulting in the current list provided by Hay (2023) of 58 tanks categorized as confirmed or assumed leakers. As a result, all SSTs were administratively removed from service in November 1980. The tanks were deactivated by pumping most of the supernatant liquid from the tanks using the installed high-volume turbine pumps. The remaining supernatant and interstitial liquid in the saltwell were removed with a jet pumping system, referred to as saltwell pumping, as part of an isolation and stabilization program. Nonetheless, waste can still leak from some tanks at low rates, especially if water infiltrates tanks and increases interstitial liquids. Several tank farms have had asphalt surface barriers installed to limit intrusion of rainwater into stabilized SSTs.

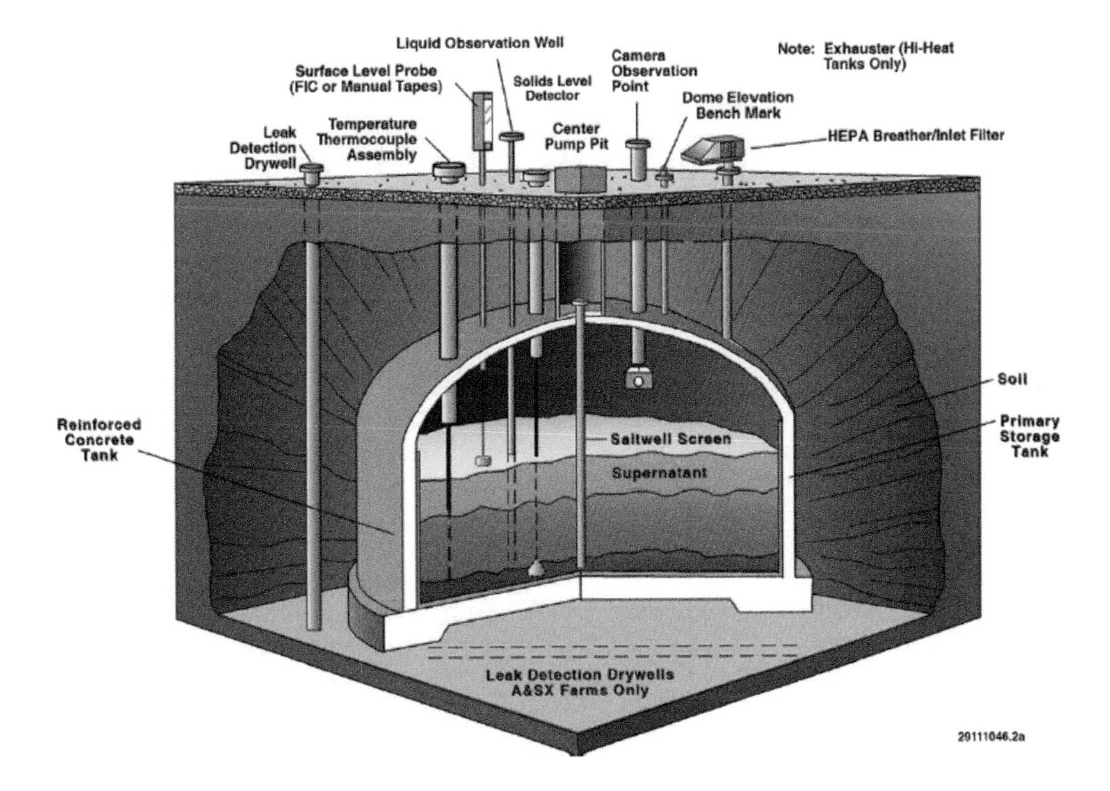


**FIGURE 6.2** Schematic diagram of a single-shell tank. Reprinted with permission from Parker and Barton (2007).

### 6.1.2 SST Structural Analysis of Record (AOR) Project

Retrieval, treatment, and final disposal of the waste contained in the SSTs is a multi-decadal process that began in the early 2000s and is ongoing. Therefore, the SSTs will be required to store residual waste (saltcake and sludge) for many years to come. As a result, the Single-Shell Tank Integrity Program (SSTIP), described by Rast et al. (2012) was initiated in 2009 to better assess and maintain the integrity of the SSTs. A panel of experts from industry and academia was created to recommend project activities in four main areas: structural integrity, liner degradation, leak integrity and prevention, and contamination migration in the surrounding soil. One of the panel's primary recommendations was to perform a detailed structural integrity analysis of the SSTs. This analysis conducted for four different tank designs bounding all the SSTs is described by Johnson et al. (2015) and summarized below. The Type-I through Type-IV SST tank designations correspond to 55 Kgal, 530 Kgal, 758 Kgal, and 1 Mgal tanks, respectively.

Finite element analysis was utilized to estimate how the SSTs would react structurally to the historical thermal and operational loads, in addition to seismic loads based on design parameters. To incorporate the thermal deterioration of concrete modulus and strength, as well as cracking caused by differential thermal expansion under in situ static loading conditions, bounding thermal histories were established from temperature records of waste and integrated into the model. The tanks' dynamic response to a design basis earthquake was assessed, incorporating the decreased stiffness of the reinforced concrete in the seismic analysis. The response to both static and seismic loads was then assessed against the design criteria of ACI-349-06, the American Concrete Institute standard for nuclear safety–related concrete structures.

For each of the four major SST designs, thermal and operating load analyses (TOLA) and seismic analyses were conducted to assess the structural integrity of tanks subject to historical thermal, static, and dynamic loads. The analyses were conducted with rigor to meet the American Society of Mechanical Engineers' Nuclear Quality Assurance (ASME-NQA-1) standards. The analyses of the bounding thermal, operating, and seismic loads for each SST design indicate that there are no issues that could compromise the Hanford SSTs' structural integrity. While some of the available tank temperature history data indicate that the bottom slabs of the Type-IV, million-gallon SSTs (SX, A, and AX tanks) likely sustained damage due to exposure to high temperatures, the analyses showed that the predicted damage due to high temperature degradation of concrete in the slabs is not severe enough to impact overall stability or structural integrity of those tanks. Analysis with detached slabs inward of the footing showed that the contribution of tank bottom (slab) to the overall structural integrity of an SST is minimal. The Type-II (530k gallon) and Type-III (758k gallon) SST did not experience very high temperatures. The historical peak temperatures were found to be in the 300°F–310°F range where concrete degradation is not significant.

Thermocouple readings available from some of the SSTs (AX tanks), which experienced very high tank bottom temperatures, showed that the tanks' walls, outer footing, and dome sections did not surpass the temperature of boiling waste (max 250°F or 121°C). This significant finding indicates that the concrete strength and stiffness degradation are restrained in the outer footing, wall, and dome sections, as the temperatures were moderate in those areas. This finding aligns with the outcome of the coring of the sidewall of tank A-106, the tank that exhibited the highest temperature among the SSTs. Based on laboratory analyses of these sampled cores, compressive strengths well above the 20 MPa design strength were measured throughout the height of the wall core (Misiak et al. 2015).

The SST analysis of record (Johnson et al. 2015) evaluated tank demands, which are the net section forces and moments under applied loads, against the design capacities of reinforced concrete per ACI-349 standards. The tank demands could be static or dynamic. Static forces and moments in the tank sections are due to long-term and permanent loads such as waste hydrostatic, waste thermal, soil overburden, and lateral soil loads. Dynamic loads are temporary and from equipment or seismically induced vertical and lateral soil loads. While the thermal and operating load analysis (TOLA) estimates static loads, the seismic analysis captures inertial effects from the dynamic loads. The structural integrity evaluations are conducted by combining the demands from TOLA and seismic analysis.

The results from structural analyses are presented in the form of demand-to-capacity (D/C) ratios, which are the ratio of combined static plus seismic forces to moment demand and ACI-349 prescribed capacity. The D/C ratios are evaluated at several sections in the dome, wall, footing, and slab regions of a tank as shown in Figure 6.3. The evaluations are performed in both meridional (axial) and hoop (circumferential) directions as the tanks experience different forces and moments in each of these directions at any given section. The D/C ratio for a section is estimated based on load-moment interaction diagram typically used in civil engineering to represent the variable strength of a reinforced concrete column under various combinations of forces and moments. Figure 6.4 shows a typical force-moment diagram along with the definition of the demand/capacity ratio. The D/C ratio is defined as the ratio of the vector length from the origin to the force-moment demand coordinate to the vector length from the origin to the capacity curve assuming the same ratio of force to moment. A D/C ratio exceeding 1.0 in any section indicates that the section doesn't meet the force-based ACI design criterion and may necessitate additional detailed analysis.

The primary factor contributing to the static tank demands is the soil overburden, which exerts the most substantial load on the tank. The hydrostatic waste loads in the tank and the concentrated live loads present on the soil surface (such as equipment loads) have only secondary effects. Thermal loads were particularly significant in the bottom slab and lower wall areas, where the temperatures were highest. The AOR considered several combinations of soil and concrete material properties to account for the variation and uncertainty in the material properties such as elastic modules and compressive/tensile strengths of soil and concrete. The combination of an upper-bound concrete modulus and a lower-bound soil modulus typically resulted in bounding section demands under both TOLA and seismic analysis.

Soil with a softer composition (low stiffness) tends to generate greater seismic demands overall (Johnson et al. 2015). However, the seismic demands are minor in comparison with the TOLA

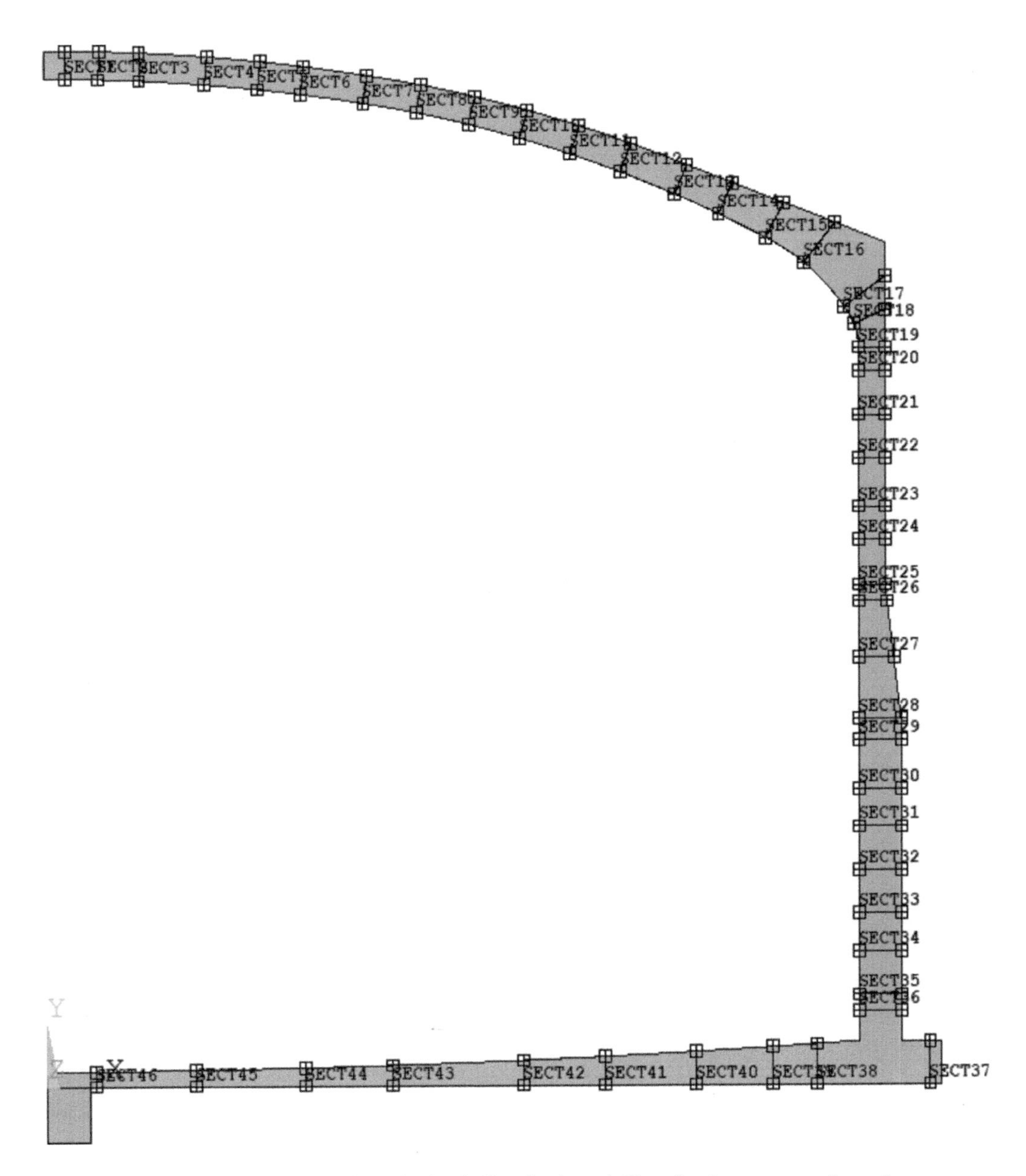


**FIGURE 6.3** Location of sections in a single-shell tank where ACI evaluations were performed.

demands, even when accounting for the estimated 20%–40% conservatism in the seismic input accelerations. The haunch (dome-to-wall transition region) usually experiences the highest seismic loads, while the seismic demands on the floor are insignificant compared to other locations. The dome's sensitivity is primarily determined by the soil configuration, while it remains largely insensitive to the waste level and properties. Waste configuration and soil stiffness have the greatest impact on the wall demands. The highest seismic demands in the wall occur when the tank is nearly full, and the waste is stiffer. The lowest seismic demands in the wall occur when the tank is empty. In the wall, the in-plane shear forces exhibit more sensitivity to the soil stiffness than other seismic demands. Through-wall shear forces typically peak near the haunch and the footing. The hoop (circumferential) forces in the wall exhibit more sensitivity to the waste height and stiffness than other seismic demands (viscous with high shear modulus).

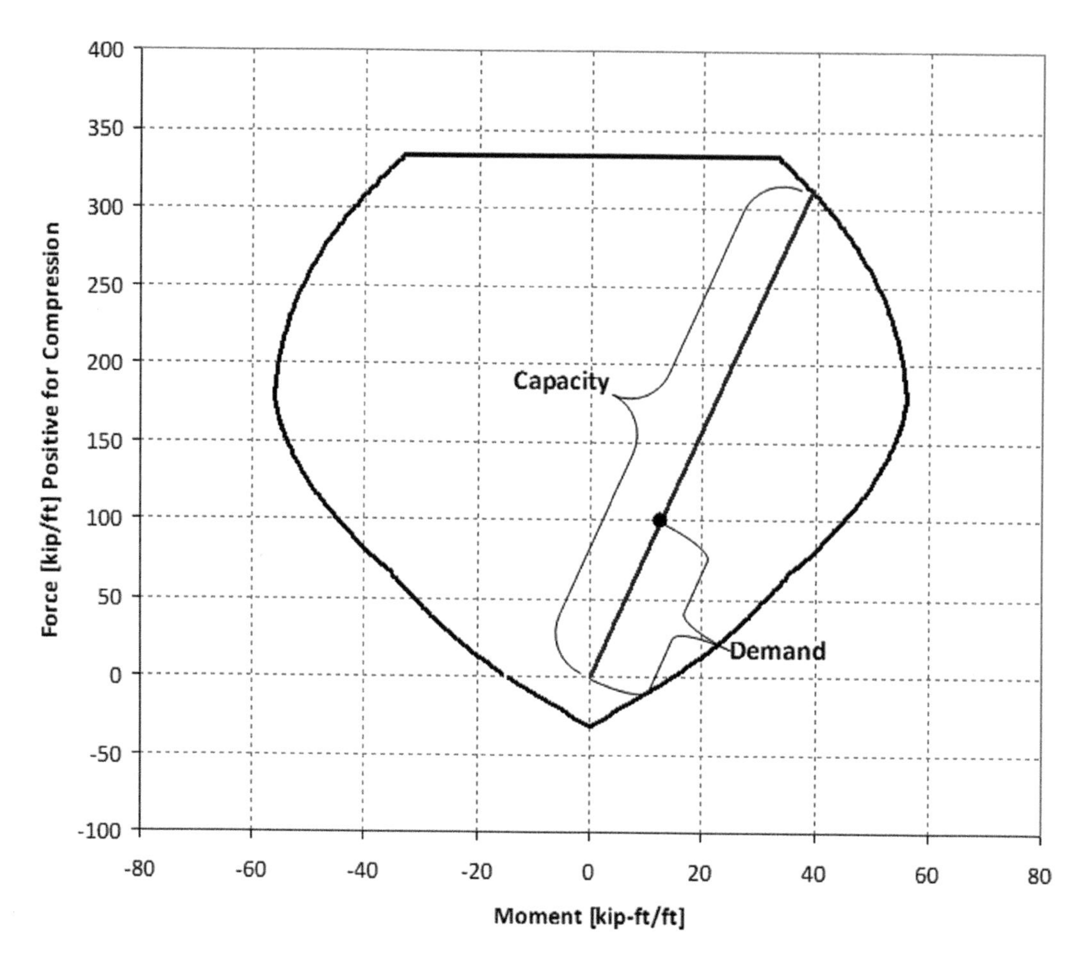


**FIGURE 6.4** ACI force-moment interaction diagram for evaluating reinforced concrete sections. The vector lengths defining the demand/capacity ratio are also shown.

When tanks are closely spaced in a tank farm, additional considerations are required from a seismic perspective. The influence of tank-to-tank interaction (TTI) on seismic demands is dependent on several factors. The tank spacing, differences in waste height, and dissimilar concrete properties between two interacting tanks all have an observable effect on concrete seismic demands. In some cases, the variation in demands between a full and empty tank can be as crucial a variable as tank spacing. While the differences in TTI seismic demands are significant on a percentage basis, they remain small in comparison with the thermal and operating loads (which also incorporate TTI effects). When calculating the demand/capacity (D/C) ratios, the impact of the TTI seismic differences is minimized by the higher TOLA loads. The TTI investigation demonstrated that the D/C ratios for two interacting tanks are comparable to those for a single tank, with the most significant differences arising near the bottom of adjacent tank walls.

## 6.2 DOUBLE-SHELL TANKS

### 6.2.1 Double-Shell Tank Description

The need for additional tank space and an increased radiolytic heat load led to a decision by the U.S. Atomic Energy Commission (predecessor to the U.S. Energy Research and Development Administration, and subsequently DOE) in the 1960s to initiate construction of double-shell tanks

(DSTs). The DSTs had improved design, materials, and construction over the SSTs. The construction of the DSTs began in 1968, with the sixth farm being completed in 1986. All of the DSTs have a nominal million-gallon waste capacity, with design lives of 20–50 years Campbell et al. 2021.

The 28 DSTs were built in six tank farms from 1968 to 1986. These provided improved liquid containment, greater waste storage capacity, and better access for monitoring and sampling over the SSTs. As their name implies, the DSTs have two steel liners separated by about 3 ft of space called the annulus. This provides a margin of safety if the inner liner leaks. Four tanks have a capacity of 1 million gallons, and 24 have a capacity of 1.16 million gallons.

As shown in Figure 6.5, the DSTs consist of a primary steel tank inside of a secondary steel liner. Both the primary tank and secondary liner are built of the same specification carbon steel (ASTM A537 Class-1). In each DST, the primary tank was post-weld heat-treated to reduce residual stresses from fabrication and the propensity for stress corrosion cracking failures. The secondary steel liner is encased by a reinforced concrete shell. The primary tank rests on a refractory concrete slab used to thermally insulate it from the secondary liner and concrete foundation. This refractory slab also provides air circulation/leak detection channels under the primary tank bottom plate. An annular space of 2.5–3 ft exists between the secondary liners and primary tanks. This annular space also allows for visual surface and ultrasonic volumetric inspections of the primary tank walls and secondary liners.

Given the improved integrity of the DSTs over the SSTs and ongoing leak issues associated with the latter, waste from the SSTs has been progressively removed and transferred into the DSTs. The DSTs provide interim storage of the waste before it can be treated and permanently dispositioned. The limited DST storage space and delays to initiate treatment have motivated work to increase their allowable service capacity by increasing the level to which they can be filled. A key part of that work has been the structural analysis described in Section 6.2.2.

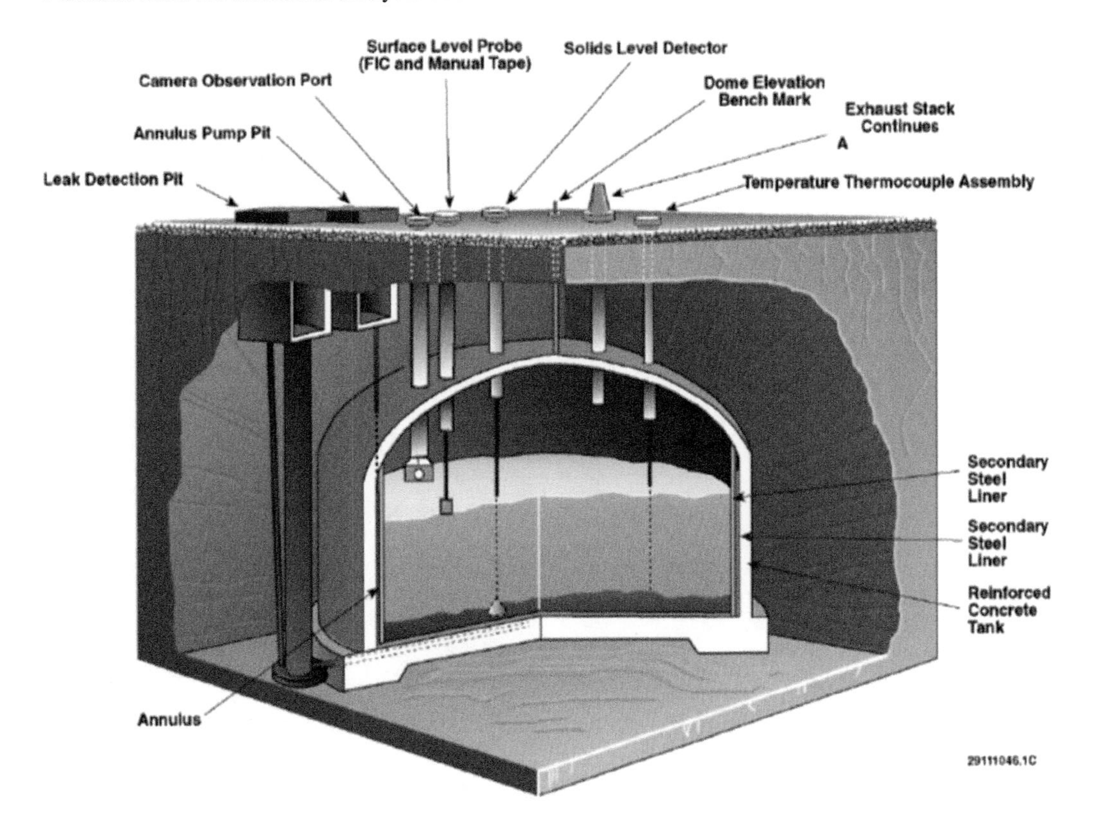


**FIGURE 6.5** Schematic diagram of a double-shell tank. Reprinted with permission from Campbell et al. (2021).

### 6.2.2 Double-Shell Tank Structural Analysis

The DSTs in the AP tank farm are the newest (initial service in 1986) followed by those in the AN (1981) and AW (1980) farms. These tanks were designed with a 50-year service life, higher strength concrete, primary tank steel, and thicker primary tank walls than the earlier tanks in the AY, AZ, and SY farms. A detailed structural analysis completed by Deibler et al. (2008) justified increasing the allowable waste level in the newest AP tanks. The AN and AW tank designs are nearly identical, and they have approximately a decade remaining in their design service lives at the time of publication. These characteristics make the AN and AW tanks the best candidates for increasing the allowable waste level. The structural analysis investigating the potential for increased liquid level waste in AN-AW tanks was recently completed by Johnson et al. (2021) and provided the basis for nearly a million-gallon increased total storage capacity in AN-AW tank farms. The AN-AW analysis followed evaluation methodologies established by prior DST analysis (AP tank) and considered approaches recommended during the SST AOR.

#### 6.2.2.1 Static Thermal and Operating Load Analysis

The model for analysis of thermal and operating loads includes the soil, reinforced concrete tank, steel primary tank, steel secondary liner, and the contact conditions between each of these structural features (Figure 6.6). The concrete elements simulate the cracking and load redistribution of reinforced concrete under increasing loads. The fractional volumes of the steel rebar reinforcements are defined in the meridional and hoop directions at the tank dome, haunch, wall, and footing locations consistent with specifications in the design drawings. The steel primary tank and secondary liner were modeled using shell elements.

The soil was modeled using the Drucker-Prager pressure-dependent yield criterion as the strength of granular materials increases with confinement, and the Drucker-Prager material model captures such behavior. The at-rest (static) soil pressure on the tank surface resulting from the backfill and soil compaction process was achieved by adjusting the soil-to-tank contact element options until the expected (theoretical) at-rest pressures were developed under gravity loading.

The thermal analysis used the same finite element mesh as the structural model to simulate the transient thermal operating history of the tanks. After the tank reached the historical peak waste temperature of 66°C (150°F), the degraded concrete stiffness and strength properties were applied to the concrete elements. Additional thermal cycles were simulated to capture the effects of thermal ratcheting and concrete creep on load redistribution.

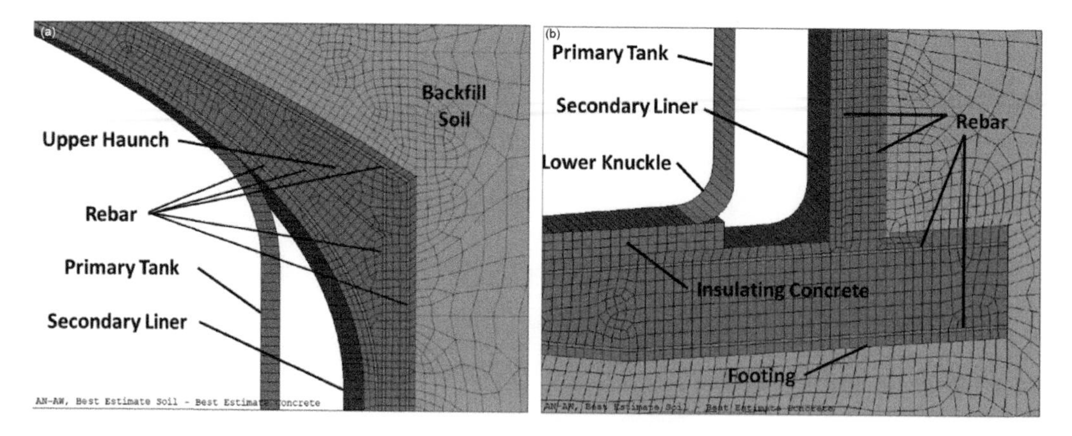


**FIGURE 6.6** Details of the finite element analysis model of DST (a) haunch region and (b) footing showing primary and secondary steel liners. The liners are not visible at the bottom as they were represented with shell elements, which overlapped with solid elements representing insulating concrete. Reprinted with permission from Johnson et al. (2021).

The primary liner is anchored to the cylindrical shell and dome at regular intervals. The anchors in the full model were represented as springs at the dome anchor spacing with the spring stiffness determined from an independent detailed analysis model. The L-bolt threaded anchors (Figure 6.7, left) that attach the steel primary tank to the reinforced concrete dome were modeled in detail to determine the spring stiffness and shear deflection that corresponded to the anchor manufacturer's load-deflection test data. The analysis simulated the plastic deformation of the welded stud and anchor plus crushing of the surrounding concrete (Figure 6.7, right).

#### 6.2.2.2 Dynamic Seismic Analysis

A seismic model of the AN and AW DSTs (Figure 6.8) was developed to capture the soil-structure interaction loads in the reinforced concrete and the low-frequency waste sloshing loads that concentrate in the steel primary tank and the anchor bolts. The latest seismic analysis incorporates three significant updates compared to previous tank analyses. First, the ground motions were

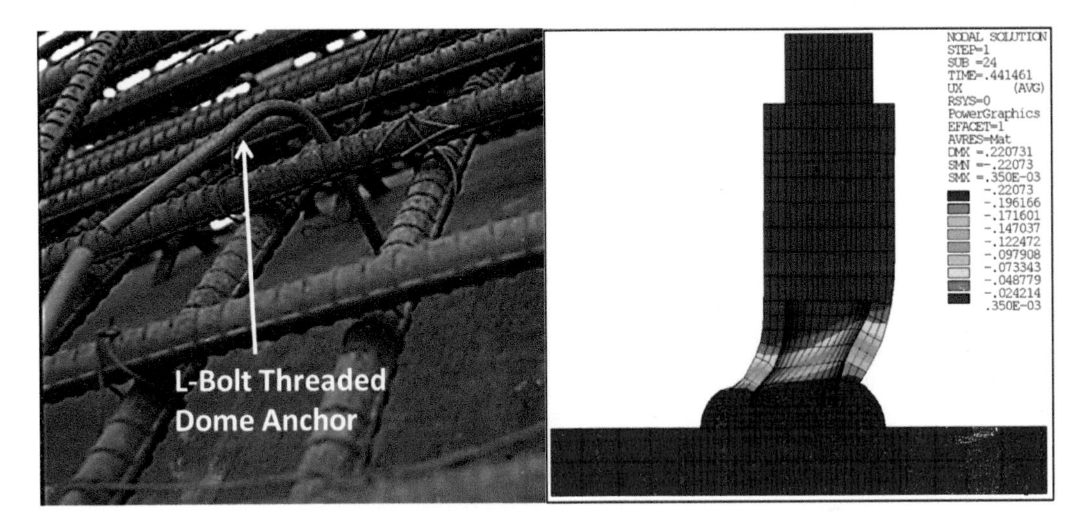


**FIGURE 6.7** L-shaped anchor bolts in the dome of DSTs (left) and detailed bolt model for estimating stiffness of the springs used to represent anchor bolts in the FEA model right. Reprinted with permission from Johnson et al. (2021).

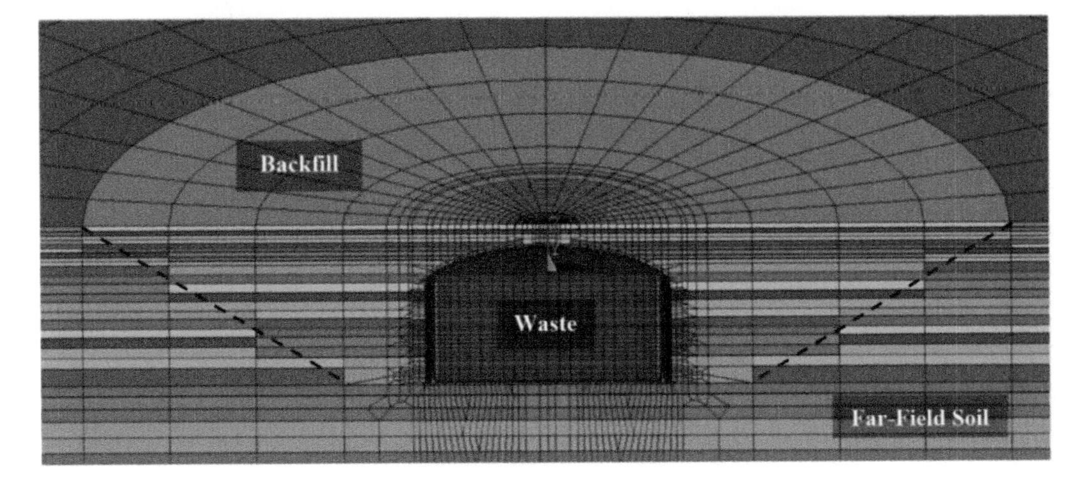


**FIGURE 6.8** Seismic FEA model of AN-AW DST. Reprinted with permission from Johnson et al. (2021).

developed from the latest 2014 probabilistic seismic hazard analysis (PSHA) for the Hanford Site (Coppersmith et al. 2014). PSHA quantifies the probability of exceeding various ground-motion levels at a site considering all possible earthquakes and is used to establish a design basis earthquake and seismic hazard for a given site/region. Next, the low-frequency ground motions were developed separately (from the high-frequency ground motions) for the waste sloshing analysis. Such separation ensured that the bounding spectral accelerations were captured at both high and low frequencies. Considering accurate low-frequency accelerations is essential for DSTs containing liquid waste because of sloshing.

The waste sloshing model also incorporated finite elements formulated specifically to simulate hydrostatic pressures, fluid/solid interactions, and free surface effects such as sloshing in a transient analysis. Submodel sensitivity studies were conducted to confirm that these finite element techniques reproduced known theoretical sloshing solutions in an open-top, straight-walled cylinder.

The fluid interaction with the tank dome in the DST model causes surface wave reflections at a shorter radius than the tank wall, giving rise to additional waves at shorter periods. The contour plot in Figure 6.9 shows an example of the strong convective (free surface) response of waste sloshing in the DST model. The free surface shows superimposed out-of-phase waves caused by reflections off the edge of the dome.

## 6.3 CONCLUDING REMARKS

Comprehensive, state-of-the-art analyses were performed for both SSTs and DSTs at the Hanford Site to assess their structural integrity. While the SST AOR evaluated the integrity of various SSTs, the AN and AW level rise analysis of record evaluated many aspects of a double-shell tank structural integrity.

The reinforced concrete tank shell was evaluated to the force-based design criteria of the American Concrete Institute code for nuclear safety–related concrete structures. Static and seismic load combinations were evaluated, and the force and moment demands were found to be less than the ACI section capacities. A limit load analysis also showed that the current soil and equipment loads above the tank domes are less than one-third of the collapse loads. Concrete shell buckling analysis further confirmed the structural stability of the tanks under current loads.

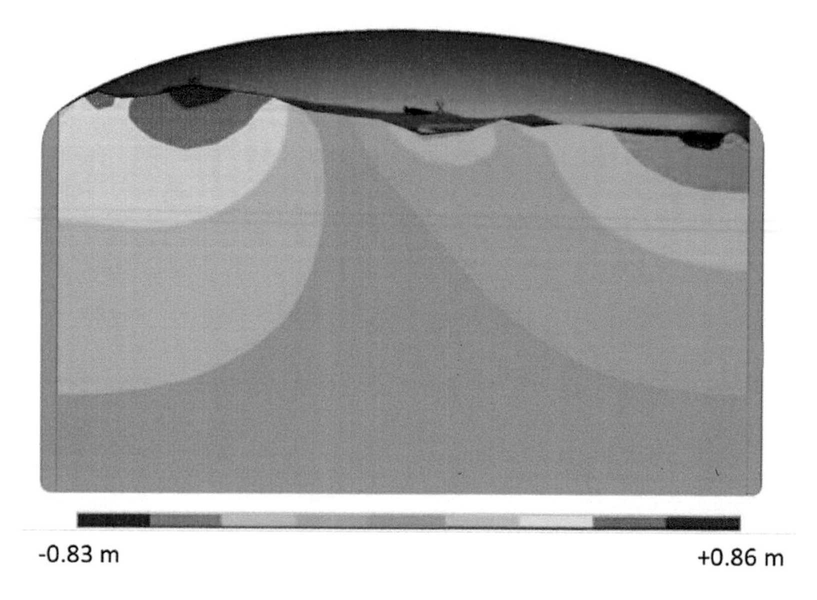


**FIGURE 6.9** Contour plot showing the convective response of waste sloshing in the DST model. Reprinted with permission from Johnson et al. (2021).

The steel primary tank was evaluated to the stress requirements of the ASME Boiler and Pressure Vessel Code. The maximum primary tank stress intensities from static plus seismic waste sloshing loads were only 53% of the ASME allowable stress intensity. Stress corrosion cracking of the steel primary tank wall was also determined to be very unlikely based on the root cause analysis of the AY-102 tank leak and fracture mechanics analysis described by Deibler et al. (2008). The primary tank buckling analysis also showed that the partial vacuum exerted by the tank ventilation system is less than the vacuum limit established from large-displacement, finite element buckling analysis.

The steel secondary liner was evaluated to the strain limits for steel backed by concrete specified in the ASME Boiler and Pressure Vessel Code. The secondary liner strains were less than 20% of the ASME allowable strains.

Finally, the DST anchor bolt analysis showed that at the maximum waste temperature condition, the maximum shear load in the outer ring of anchors was only 63% of the anchor capacity.

The conclusion from these analyses was that the structural capacity of AN and AW tanks exceeds the structural demands (loads) experienced by the tanks. Therefore, it was determined that adequate structural margin remains in the AN and AW tanks with the waste level increased from 10.7 to 11.7 m to provide nearly a million-gallon extra storage capacity in those farms.

More broadly, the Hanford Site's SSTs and DSTs demonstrate the following lessons:

- Structures at legacy industrial sites may be required to function in some capacity well beyond their design service lives as overall site remediation proceeds.
- Periodic evaluation and continuous monitoring of the integrity of such structures is important to ensure the mission of site remediation proceeds in a controlled manner.
- Appropriate steps must be taken to resolve structural issues and avoid any catastrophic failures.

## REFERENCES

Campbell, S. T., J. S. Garfield, C. L. Girardot, J. R. Gunter, J. D. Larson, J. S. Page, and G. E. Soon. 2021. *Double Shell Tank Integrity Program Plan*. RPP-7574, revision 8. Washington River Protection Solutions, Richland, WA.

Coppersmith, K., J. Boomer, K. Hanson, J. Unruh, R. Coppersmith, L. Wolf, R. Youngs, A. Rodriguez Marek, L. Al Atik, and G. Toro. 2014. *Hanford Sitewide Probabilistic Seismic Hazard Analysis*. Report PNNL-23361. Pacific Northwest National Laboratory, Richland, WA.

Deibler, J. E., M. W. Rinker, K. I. Johnson, N. K. Karri, S. P. Pilli, F. G. Abatt, and K. L. Stoops. 2008. *Hanford Double-Shell Tank Thermal and Seismic Project - Summary of Combined Thermal and Operating Loads with Seismic Analysis*. RPP RPT 28968, Rev. 1. PNNL-15721. Pacific Northwest National Laboratory, Richland, WA.

Field, J. G., C. L. Girardot, D. G. Harlow, D. S. Washenfelder, and J. S. Schofield. 2014. *Hanford Single Shell Tanks Leaks Causes, Locations and Rates: Summary Report*. RPP-RPT-54909. Washington River Protection Solutions, Richland, WA.

Hay, B. D. 2023. *Waste Tank Summary Report for Month Ending January 31, 2023*. HNF-EP-0182, revision 142. Washington River Protection Solutions, Inc., Richland, WA.

Johnson, K. I., J. E. Deibler, N. K. Karri, F. G. Abatt, K. L. Stoops, and L. J. Julyk. 2021. *AN-AW Level Rise Structural Integrity Analysis of Record*. RPP-RPT-60175, Rev. 0; PNNL-30097. Pacific Northwest National Laboratory, Richland, WA.

Johnson, K. I., J. E. Deibler, N. K. Karri, S. E. Sanborn, F. G. Abatt, K. L. Stoops, L. J. Julyk, and B. M. Larsen. 2015. A Summary of the Hanford Single-Shell Tank Structural Analysis of Record. *WM2015 Conference*, Phoenix, AZ.

Misiak, T., T. Venetz, M. Gardner, and M. Brown. 2015. Hanford Single-Shell Tank Sidewall Coring Project. *WM2015 Conference*, Phoenix, AZ.

Parker, D. L. and W. B. Barton. 2007. *Retrieval Data Report for Single-Shell Tank 241-S-112*. RPP-RPT-35112. CH2M HILL Hanford Group, Inc., Richland, WA.

Rast, R., M. W. Rinker, D. J. Washenfelder, and J. M. Johnson. 2012. Overview of Hanford Single Shell Tank (SST) Structural Integrity. *Waste Management 2012 Conference*, February 26–March 1, Phoenix, AZ.

# Part III

## Remedy Screening, Evaluation, and Testing

Christian D. Johnson, Katherine A. Muller, and Hilary P. Emerson

In Part II, we described how the nature and extent of contamination at a complex waste site are assessed and characterized. Once the contamination is sufficiently understood and remediation objectives are defined, remedial technologies and strategies must be identified, investigated, and considered for their potential to meet site- or application-specific remediation objectives. Part III focuses on remedy selection with relevant case studies from the Hanford Site.

Remedy selection can include both technologies and engineering or institutional controls (Section III.1). Identifying relevant technologies and down-selecting to those methods with the highest likelihood of being cost-effective and meeting regulatory requirements requires detailed analysis (Section III.2). Then, if there is uncertainty remaining about whether a promising technology is effective and suitable for the site or application in question, the technology must be tested and demonstrated generally beginning with laboratory experiments and scaling up to the field (Section III.3). Throughout this step, new data and analysis can inform updates to the conceptual site model (CSM) and may require revisiting previous remediation steps.

In this context, *site* refers to a specific area or zone being remediated. A site may be part of a larger region, depending on how the remedial objectives are defined. For example, the vadose zone and groundwater may be separately targeted for remediation with different technologies applied at different depths at a specific site. The *application* may include the remediation of environmental media (e.g., subsurface sediments or surface water sources), waste processing, or deactivation and decommissioning of legacy facilities. Once a preferred remedy is selected and approved, the next step is the implementation of the remedy (Part IV; see Section 1.2 of Chapter 1 for context on how these steps fit into the broader remediation process).

DOI: 10.1201/9781003329213-9

## III.1 OVERVIEW OF TECHNOLOGIES AND CONTROLS

Remediation technologies, engineering controls, and institutional (i.e., administrative and legal) controls can all play a role in the overall remediation strategy. Even within a specific area, a multi-pronged approach may be required with a combination of these elements to meet the objectives for the protection of human health and the environment.

*Engineering and institutional controls* are implemented to decrease exposure and risk to human health and the environment. *Engineering controls* reduce risk by containing or reducing exposure to contamination, limiting access to property through physical barriers, or providing alternate sources of uncontaminated water to the site and nearby affected communities. *Institutional controls* limit property use or access to reduce exposure risk through work practices and scheduling, site access controls, building or excavation permits, well drilling prohibitions, zoning restrictions, easements, covenants, deed notices, etc.

*Remediation technologies* are used to reduce the toxicity, mobility, volume, mass, or concentration of contaminants. There are many different types of remediation technologies that may physically contain (e.g., a grout wall or clay cover), remove, treat, or degrade a contaminant of interest based on physical, chemical, biological, or thermal principles (Table I.1 of the Part I introduction includes examples). An appropriate remediation strategy will depend on the contaminants involved, site setting, exposure pathways, receptor locations, and remedial action objectives (RAOs).

Broadly, remediation strategies can be active or passive, and in situ or ex situ. An active remediation strategy requires certain actions to implement. For example, an amendment may be injected to create reducing conditions to precipitate or degrade a contaminant. Passive strategies, alternately, do not require activity besides monitoring and evaluation. Monitored natural attenuation is an example of a passive strategy that allows the natural processes within the subsurface to decrease the contaminant mobility (e.g., adsorption to sediments).

Remediation technologies for the subsurface may treat a contaminant either within the subsurface (in situ) or via a treatment facility or process after removal from the subsurface (ex situ). An example of in situ treatment is contaminant sequestration on or within phosphate minerals emplaced through the injection of liquid amendments. For example, a polyphosphate chemical solution was injected at the 300 Area of the Hanford Site for the enhanced attenuation of uranium via a combination of geochemical mechanisms including adsorption, coprecipitation, and coating (Chapter 11, "Polyphosphate for Enhanced Attenuation of Uranium"). The treatment portion of the pump-and-treat operations at the Hanford Site represents an ex situ treatment method for multiple contaminants historically released to the subsurface (Chapter 10, "Groundwater Remediation with the Pump-and-Treat Technology"). Likewise, the facilities described in Chapter 13 to treat and immobilize the Hanford tank waste are ex situ methods, which occur following retrieval of the waste from holding tanks.

A wide range of technologies have been developed since the 1980s in the U.S. and are still being developed by U.S. Department of Energy (DOE) national laboratories (https://www.energy.gov/national-laboratories), Department of Defense environmental programs (e.g., through the Strategic Environmental Research and Development Program, Environmental Security Technology Certification Program, Naval Facilities Engineering Systems Command, Air Force Civil Engineer Center, and U.S. Army Corps of Engineers), academia, and the environmental industry. Resources such as the Federal Remediation Technologies Roundtable and the Contaminated Site Clean-up Information provided by the Environmental Protection Agency provide a great starting point with technology descriptions, examples, and references. The International Atomic Energy Agency also has resources available through Environet (https://nucleus.iaea.org/sites/conect/ENVIRONETpublic/SitePages/Home.aspx). Additional resources include peer-reviewed journal publications, reports from other complex waste sites, and textbooks.

Because of the many unknowns associated with the cleanup of complex sites, the important factors that impact the performance of a remedy may be identified during different steps of the process, requiring the CSM or other previous steps to be revisited as new information is obtained. This is part of the adaptive site management (ASM) process, which allows for some iteration within the steps (refer to Section 1.4 of Chapter 1 for additional information). In some cases, remedial technologies that were originally identified may not be as effective under site- or application-specific conditions as indicated by earlier evaluations. The performance assessment and ASM approach are used to identify remedy refinements and include alternative technologies or interim objectives. If additional knowledge gaps or site-specific technical challenges are identified, the site assessment or site-specific testing may also be revisited to gather additional data to refine the CSM.

## III.2 REMEDY SCREENING AND EVALUATION

In the remedy screening and evaluation step, remedial technologies are identified, screened, and evaluated to determine whether they can meet the site- or application-specific remediation objectives. After compiling a list of all possible remediation technologies, the list is reduced to the technologies that are potentially relevant. This initial step removes technologies that are not applicable because they do not treat the target contaminants (e.g., soil vapor extraction is not relevant for nonvolatile contaminants) and those that cannot be used for the site- or application-specific conditions (e.g., excavation is not relevant to contamination located hundreds of feet below the ground surface). The culled list of relevant technologies is then screened based on effectiveness, implementability, and relative cost. The CSM provides the context to help define the contaminants and zones targeted for remediation, providing information that may constrain what technologies are feasible and effective and may help assess the implementability and relative cost of a technology. The screening process may consider multiple scenarios for the treatment of different zones or different broad strategies (e.g., the treatment of a source of a groundwater contamination plume versus the treatment of the entire dissolved phase plume). Effectiveness pertains to how well the technology treats the contaminants to meet the RAOs given the contamination context (concentrations, distribution, migration, media, etc.). Implementability relates to the maturity of a technology, the availability of materials, the administrative constraints, and the technical or engineering challenges to putting a remedy into action. Relative cost reflects a rough estimate of the cost for a nominal implementation, typically as a categorical value (e.g., low, medium, or high). The outcome of the screening process is a ranking of potential remediation technologies with respect to the suitability for a remedy. It is important that unbiased decisions are made during the first reduction in the list of potential technologies and screening so that a technology is not summarily dismissed (even if it is believed to be unlikely to be effective).

After potential remediation technologies are identified and ranked during the remedy screening process, "remedial alternatives" are defined (including a baseline "no action" alternative), quantified, and compared for the purpose of determining a preferred remedy. Remedial alternatives can consist of multiple technologies considered together, such as a sequential treatment train after water is pumped to the surface (e.g., to address multiple contaminant types or secondary waste streams), different treatments at different subsurface locations to address different zones of contamination (e.g., the source area versus dilute groundwater contamination), or different treatments at different time points to optimize contaminant immobilization (e.g., an initial reduction step to precipitate a contaminant, followed by a sequestration step to keep the contaminant from re-oxidizing and to immobilize it in the subsurface). A detailed analysis is performed to identify the most effective remedial alternative that best satisfies the statutory or regulatory mandates under site- or application-specific conditions. Preliminary models are created for each remedial alternative to evaluate the potential to achieve RAOs, the short- and long-term effectiveness, the implementability, and the cost. The process should consider risk-informed remedy selection and related decision-making (e.g., IAEA, 2023). The outcome is a preferred remedial alternative and, potentially, a list of recommendations for closing data gaps and minimizing the uncertainties in the CSM or remedial alternative through additional characterization or technology testing (at the laboratory to field scale). An example of remediation technology screening and evaluation is the one carried out for the Hanford Site 200-DV-1 Operable Unit (OU) (DOE/RL 2017, 2019).

## III.3 REMEDY TESTING AND DEMONSTRATION

In the remedy testing and demonstration step, technologies are evaluated, if necessary, at the pilot scale. Where a treatment technology has uncertainties, treatability testing or technology demonstrations (laboratory or field scale) may be used to gather additional data and experience to reduce those uncertainties. Such work can be important for innovative technologies or for demonstrating the effectiveness of a technology under site- or application-specific conditions. Test results provide additional performance, implementability, and cost information for a remedial alternative and support remedy selection decisions.

Depending on the maturity level of a technology, treatability testing may use a phased approach that moves from laboratory- to field-scale demonstrations. Pilot-scale tests (in a laboratory or in the field) help to answer questions related to larger-scale site- or application-specific remedial performance, design, and the overall cost of implementation. For example, a laboratory-scale test of pump-and-treat may be conducted in a high-bay facility with a prototype of the series of treatment columns (often referred to as a train) at a scaled-down size for testing with actual or simulated groundwater or wastewater. Potentially viable treatment technologies may need to undergo bench-scale laboratory tests with site-specific sediments, groundwater, and amendments to quantify site-specific effectiveness and provide information for scale-up. While treatability testing provides insight into the performance of an implemented remedy, such testing is often completed in the laboratory under ideal conditions. Thus, when moving a technology from the laboratory to an actual application, there are often uncertainties about how the laboratory-observed effectiveness will translate to the full-scale application. There may be a reduction in performance; therefore, site- and application-specific conditions (e.g., waste impacts on the subsurface sediments and subsurface heterogeneities) must be evaluated for their impacts on the implementability and efficiency of a treatment. These unknown risks are mitigated via additional testing at the pilot scale or modeling approaches.

The treatability testing underway for the 200-DV-1 OU located in the Central Plateau of the Hanford Site is a good example of the phased process to evaluate the uncertainties using laboratory-scale testing and modeling approaches (DOE/RL 2017, 2019). Based on the knowledge gaps identified in site-specific applications of remediation technologies, the following key unknowns are targeted with the treatability testing:

- Effectiveness of remediation technologies for site-specific perched water locations
- Injectability of remediation technologies at the interface between the vadose zone and the water table
- Effectiveness of remediation technologies for site-specific, deep vadose zone locations
- Required surface barrier size and emplacement depth suitable for mitigating the deep vadose zone contaminant flux toward groundwater.

Depending on the site or application, additional pilot-scale tests under actual conditions may be conducted following the initial laboratory testing. In addition, modeling approaches may be employed to predict the potential effectiveness of technologies at the field scale. Although challenging, these models strive to include complexities like field-scale heterogeneity, remedial amendment delivery, and the complex geochemical conditions created from waste interactions with the subsurface. For example, tests of glass and cementitious waste forms via intermediate-scale field lysimeters are ongoing at the Hanford Site to support the risk assessment for the Integrated Disposal Facility (IDF), which will dispose of immobilized low-activity waste from the Waste Treatment and Immobilization Plant on-site (Meyer et al. 2001, 2020). This facility is a prime example of intermediate-scale testing that can help to answer site-specific and scaling questions as technologies are moved from laboratory- to full-scale implementation. Additional information about the IDF is included in Chapter 14, "Legacy Waste Disposal in the Hanford Mission."

## III.4 CASE STUDIES: THE HANFORD SITE

Part III of this book includes case studies focused on remedy screening, evaluation, and testing from different areas and the historical processes across the Hanford Site, including Chapter 7, "Soil Desiccation Treatability Testing at BC Cribs," and Chapter 8, "Glass Wasteform Testing and Qualification for Long-Term Tank Waste Disposal."

Chapter 7's discussion of the remedy evaluation and treatability testing conducted at the BC Cribs and Trenches Area of the 200-BC-1 OU includes the process for screening and evaluating remediation technologies and the results from the pilot-scale tests. Numerical methods were used to bridge the gap from the pilot-scale tests of desiccation. For the testing phase, monitoring methods are presented to help demonstrate their importance for evaluating the success of a test at the field scale.

Chapter 8 presents the methodologies employed for formulating, characterizing, and selecting a waste form based on the performance of glasses for the future disposal of both low-activity and high-level radioactive waste from Hanford Site tank farms. This chapter discusses specific regulations within the Tri-Party Agreement that require glass waste forms for the Hanford Site. This chapter also considers the development of the process equipment needed to produce the glass waste forms.

## REFERENCES

DOE/RL. 2017. *Technology Evaluation and Treatability Studies Assessment for the Hanford Central Plateau Deep Vadose Zone.* Department of Energy Richland Office, Richland, WA.

DOE/RL. 2019. 200-*DV*-1 *Operable Unit Laboratory Treatability Study Test Plan.* Department of Energy Richland Office, Richland, WA.

IAEA. 2023. *Network of Environmental Management and Remediation - ENVIRONET.* International Atomic Energy Agency, Vienna, Austria. https://nucleus.iaea.org/sites/connect/ENVIRONETpublic/SitePages/Home.aspx.

Meyer, Philip D, R Matthew Asmussen, James J Neeway, Jonathon N Thomle, Ryan Ekre, Diana H Bacon, Gary L Smith, Ridha B Mabrouki, and David J Swanberg. 2020. *Field-Scale Lysimeter Studies of Glass and Cementitious Waste Forms at the Hanford Site-20392.*

Meyer, Philip D, B Peter McGrail, and Diana H Bacon. 2001. *Test Plan for Field Experiments to Support the Immobilized Low-Activity Waste Disposal Performance Assessment at the Hanford Site.* PNNL-13670. Pacific Northwest National Lab. (PNNL), Richland, WA.

# 7 Soil Desiccation Treatability Testing at BC Waste Disposal Cribs

*Adam R. Mangel and Christopher E. Strickland*

## 7.1 INTRODUCTION

The purpose of this chapter is to discuss the treatability testing mechanism of the RCRA/CERCLA process within the context of the closure process of the Hanford Site, as an example of complex waste sites. To illustrate the steps, value, and outcome of this process, this chapter focuses on the BC Cribs and Trenches waste sites of the 200-BC-1 Operable Unit of the Hanford Site. We focus on this site because it is representative of the complex nature of environmental management with commingled technetium-99 (Tc-99) and uranium (U) contamination. In this chapter, we follow the events at the BC Cribs and Trenches waste sites, from early waste disposal through characterization and treatability testing, and close with a perspective on how subsurface technology is evolving to more holistically support vadose zone remediation.

The *BC Cribs* and *Trenches* waste sites are located just south of the 200 East Area of the Hanford Central Plateau within the 200-BC-1 Operable Unit, commonly referred to as the BC Cribs and Trenches Area (Figure 7.1). These 6 cribs and 20 trenches received aqueous waste from chemical processing of uranium fuel rods with high nitrate and radionuclide concentrations in the mid-1950s. Figure 7.2 shows the vertical stratigraphy, Tc-99, and moisture distribution at a well in the Area.

Porous media grain-size variations at the Area range from sands to loamy sands with interbedded zones of silty sand and silt. Previous characterization of the eastern region of the Area, that is, the cribs, showed a plume of mobile contamination beneath the cribs (Serne et al., 2009). At the test location, significant concentrations of Tc-99 and nitrate contamination were observed at multiple depths, from approximately 12.2 m (40 ft) below ground surface (bgs) to approximately 76.2 m (250 ft) bgs. High concentrations of other contaminants were also observed; for example, peak concentrations of Cs-137 were observed near the bottom of the as-built crib excavation and extended several feet deeper. A schematic of the cribs and trenches disposal mechanism is shown in Figure 7.3. These unlined infiltration structures held volumes of liquid waste while it seeped into the ground, with the understanding that the 100-m (330-ft)-thick soil below the Area would effectively capture the radionuclides and other contaminants of concern (COCs) and prevent groundwater impacts.

## 7.2 RISKS POSED TO GROUNDWATER FROM TECHNETIUM-99 AND URANIUM

During Hanford's plutonium production period, low-level liquid waste products generated from chemical processing of uranium fuel rods were discharged directly to the ground through a system of cribs and trenches located in the 200-BC-1 Operable Unit (OU). The BC Cribs and Trenches received more than 117,000 $m^3$ of radioactive liquid waste discharged to the soil.

DOI: 10.1201/9781003329213-10

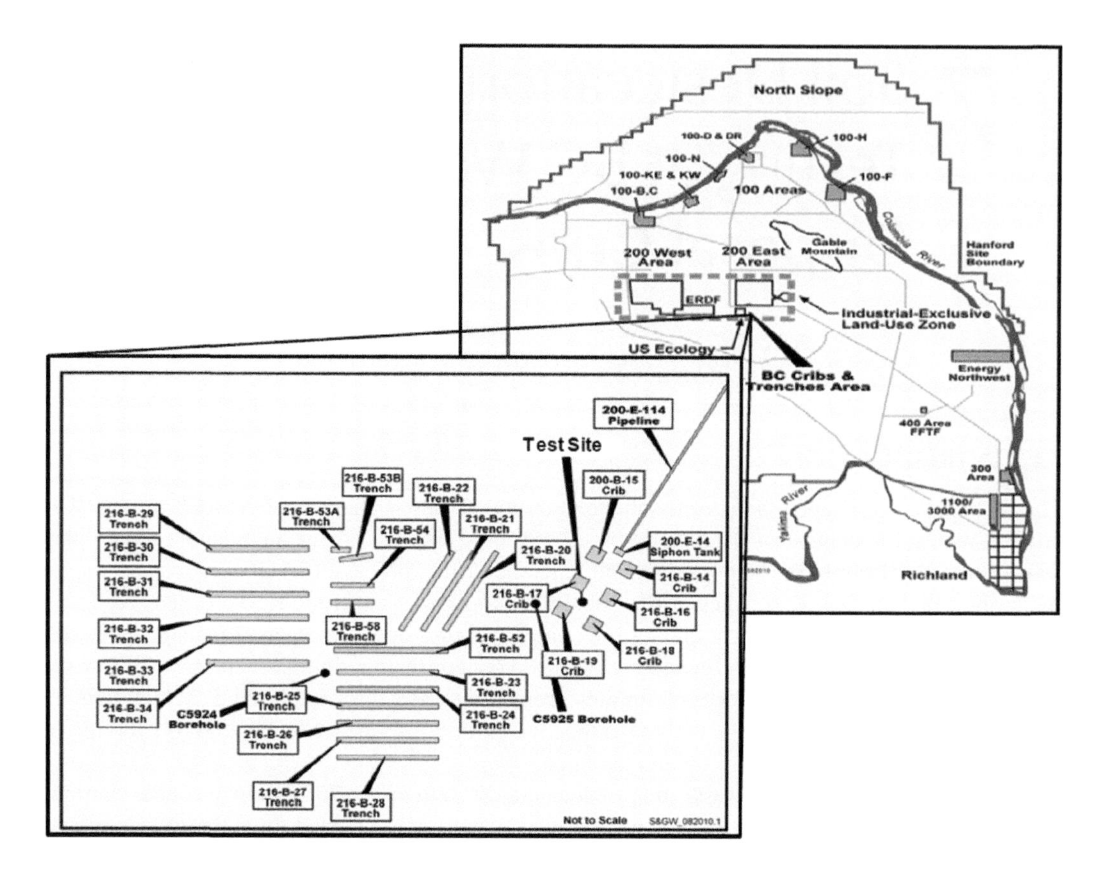


**FIGURE 7.1** Location of the BC Cribs and Trenches Area on the Hanford Site and the arrangement of the cribs and trenches.

Wastewater effluent from the 26 cribs and trenches contains approximately 410 curies of Tc-99 and is primarily located between 30 and 70 m (98 and 230 ft) depth (Corbin et al., 2005; Ward et al., 2004). Although Tc-99 is the primary contaminant of concern, approximately 3,740 kg of U was released in addition to other contaminants of concern (e.g., tritium, strontium-90, iodine-129, cesium-137, and chromium). Despite no evidence indicating that the contamination has reached the groundwater at BC Cribs and Trenches, Tc-99 and U represent the greatest risk to groundwater due to the amounts released, their mobility in the vadose zone, and the relatively long half-lives (211,000 years for Tc-99, and 250,000 years, 705 million years, and 4.5 billion years for U-234, U-235, and U-238, respectively) (Icenhower et al., 2008; Zachara et al., 2007). For additional information on the fate and transport of Tc-99 and U in the subsurface, reviews by Icenhower et al. and Zachara et al., respectively, summarize the geochemistry of these contaminants. Using data from numerical models, laboratory analyses, field investigations, and information on historical discharges, the EPA and Ecology identified risks associated with Tc-99 and U contamination. These risks were communicated to the U.S. DOE in a letter requesting development of a strategy for improved methods to understand the nature and extent of vadose zone contamination, specifically Tc-99, and to develop remedial options for addressing such contamination (Appendix A, DOE-RL, 2007). To develop the appropriate technology for characterizing, remediating, and monitoring the deep vadose zone contamination, the U.S. DOE worked with the EPA and Ecology to create a Treatability Test Plan (DOE-RL, 2007) under a Remedial Investigation/Feasibility Study (RI/FS) for the Hanford 200 Area (DOE-RL, 1999). Under this RI/FS, it was determined that a treatability test for soil desiccation (or drying) should be carried out as it was identified as a promising in situ treatment technology for mitigating risks posed by vadose zone contaminants to the groundwater

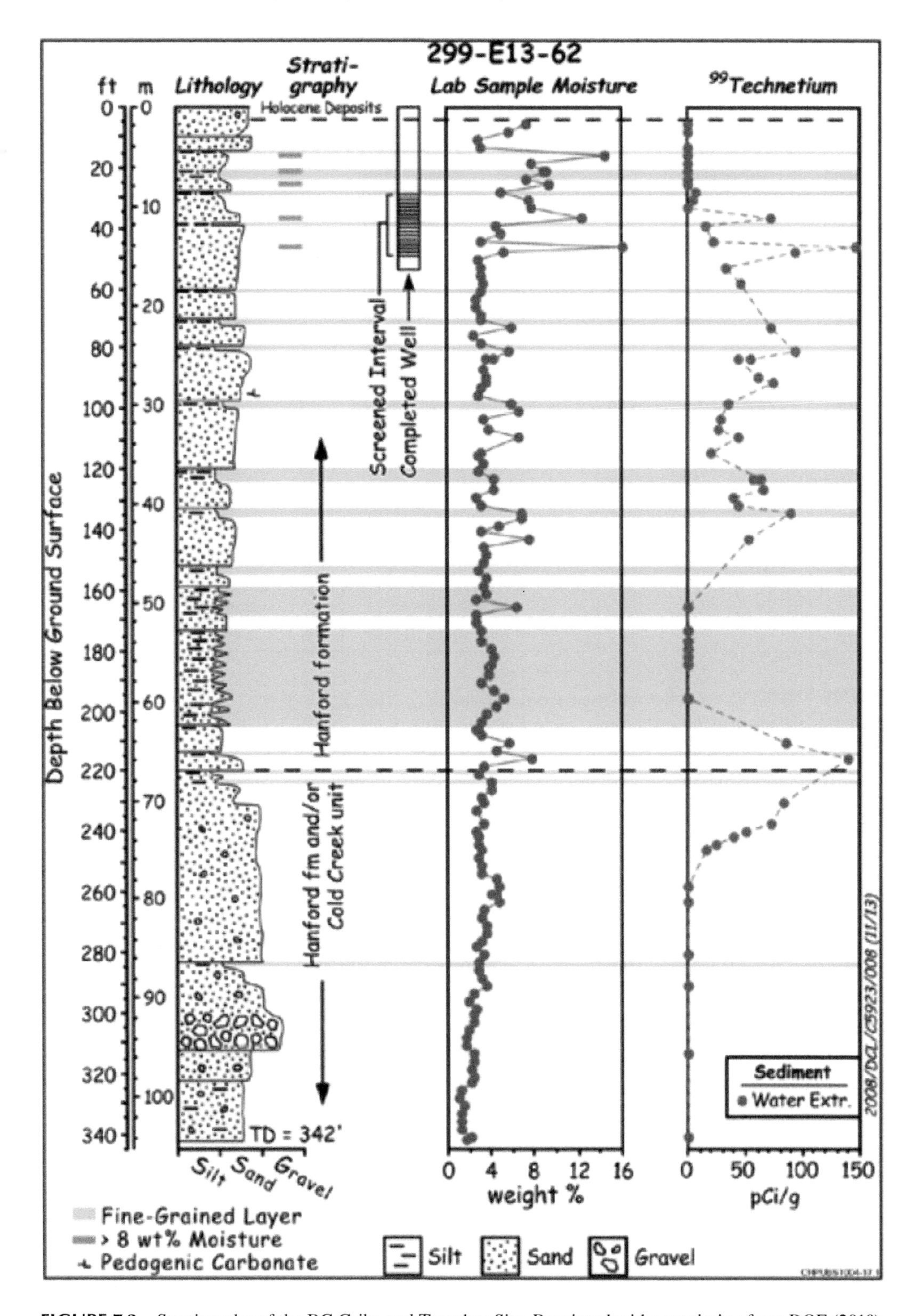


**FIGURE 7.2** Stratigraphy of the BC Cribs and Trenches Site. Reprinted with permission from DOE (2010).

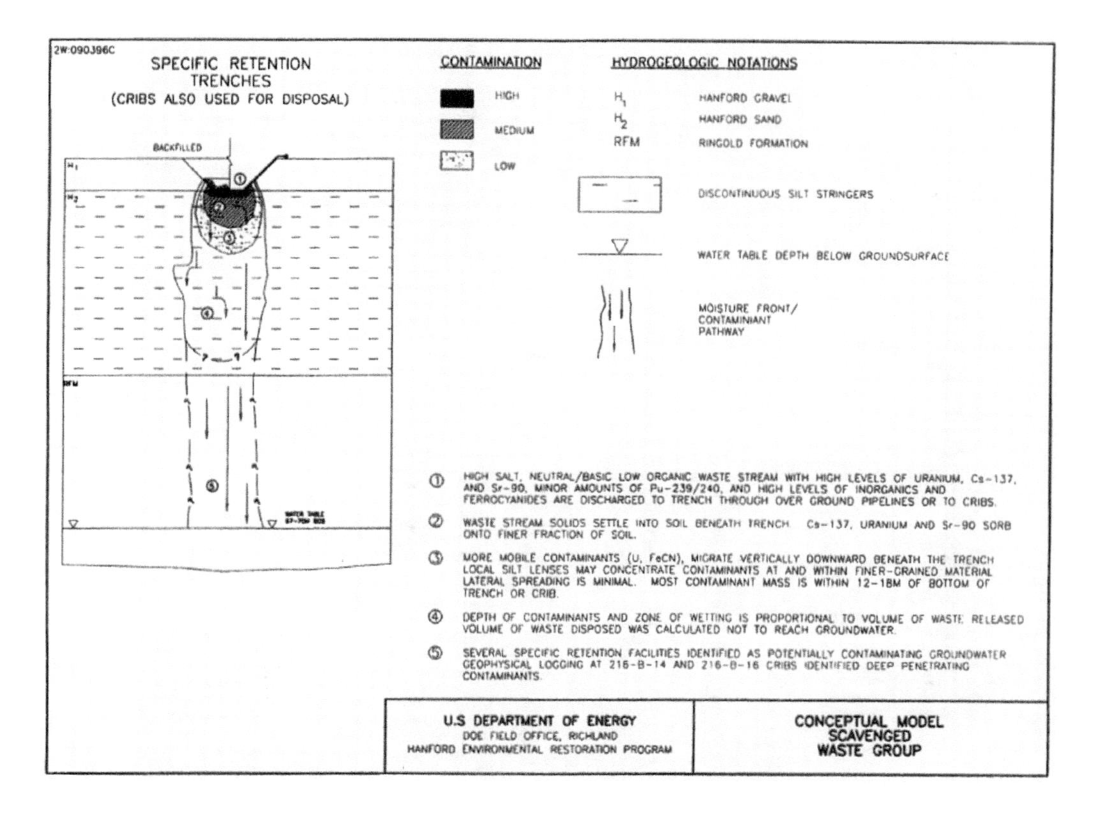


**FIGURE 7.3** Cross-sectional and plan view schematic of a Hanford crib/trench showing a contaminant plume and the depth to the groundwater table. Reprinted with permission from DOE-RL (1997).

table (FHI, 2006). The BC Cribs and Trenches Area was identified as a representative location for Tc-99 and U contamination across other waste sites within the Central Plateau of the Hanford Site and selected for the soil desiccation treatability test.

In this chapter, we summarize the overlying regulatory framework of RI/FS and treatability tests and illustrate how development and experimentation supported the evaluation of selected remedies. We briefly discuss the RI/FS for the 200 Area of the Hanford Site and focus on the soil desiccation treatability testing performed at the BC Cribs and Trenches Area under the Deep Vadose Zone Treatability Test Plan for the Hanford Central Plateau (DVZ-TT). The DVZ-TT is one component of the RI/FS for the Hanford 200 Area and represents the underlying regulatory framework that drives operations toward records of decision and waste site closure.

## 7.3 REMEDY INVESTIGATION, FEASIBILITY, IMPRACTICABILITY, AND TREATABILITY

As discussed in Chapters 1 and 2 of this book, the Tri-Party Agreement (TPA) is the governing document comprising a legal agreement and an action plan, developed by the Washington State Department of Ecology (Ecology), the U.S. Environmental Protection Agency (EPA), and the U.S. Department of Energy (DOE). The action plan portion of the document guides cleanup of the Hanford Site through designation of responsibilities and establishment of a timeline, with distinct milestones that govern progress toward site closure. Regulations stipulated by the RCRA, CERCLA, and the TPA are combined into a common regulatory framework under the RI/FS for the 200 Area of the Hanford Site (DOE-RL, 1999). As part of an adaptive site management

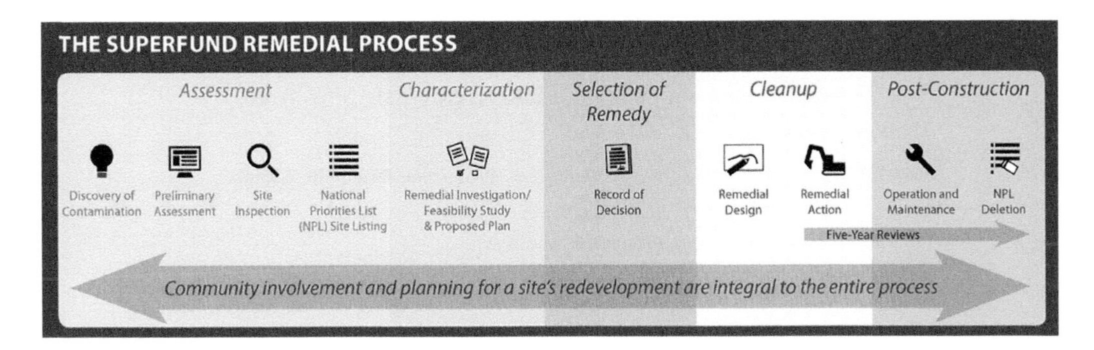


**FIGURE 7.4** Treatability testing in the RCRA/CERCLA process.

strategy, these decisions and milestones are revisited as additional data and/or testing become available and may be adjusted over time.

An RI/FS was issued by the U.S. DOE for the Hanford 200 Area in April 1999 (DOE-RL, 1999). The plan addressed roughly 700 waste sites and associated infrastructure but did not include environmental remediation planning for waste storage tank farms or facilities/buildings. The integrated regulatory approach for the 200 Area, which uses the CERCLA process (Figure 7.4) as the basis for assessment, dictates the grouping of waste sites under the RI/FS to support a consistent characterization strategy and an analogous site approach. The analogous site approach selects a representative site from a group and uses data to support remedial action decisions on other sites in that group. Ultimately, individual waste sites must be sampled and characterized to confirm that remedial decisions are appropriate and to provide important operational data for application of the remedy.

For the RI/FS, sites were grouped according to their association with different operations occurring at the Hanford Site. As described in Section 1.2, the 200 Area was the epicenter of isotopic separations and plutonium purification, which resulted in the generation of substantial quantities of waste by-products, which needed to be disposed of (DOE-RL, 1999). In general, processes included fuel dissolution, plutonium isolation, uranium recovery, cesium/strontium recovery, and finally waste treatment and storage. Using these processes as guiding information, a waste site grouping study determined that grouping the waste sites of the 200 Area by the process that was responsible for waste generation provided advantages for the cleanup process due to the similarities in waste constituents and geographic location (DOE-RL, 1997). The final waste site groups are listed in Table 7.1.

The use of the analogous site approach is critical due to the large number of waste sites that exist in the 200 Area. Grouping of the waste sites based on geography and production processes creates advantages in applying remediation strategies for the sites, as many of the sites within a given group can be treated with the same approach.

The BC Cribs and Trenches Area was determined to fall under the Scavenged Waste group, as they received waste products primarily from the U Plant's Uranium Recovery Project (URP). The URP was an effort to recover additional uranium from the bismuth phosphate's process waste that had been stored in the tank farms. It was expected that the waste could be used to supply reactors with uranium while reducing the total waste volume in the tank farms, which were approaching capacity. After it was discovered that the process generated more waste than it removed, the URP ceased, and the waste was treated with ferrocyanide to precipitate, or scavenge, cesium-137 (Cs-137) and strontium-90 (Sr-90) (DOE-RL, 1997). The waste from this process was deemed acceptable for ground disposal in the BC Cribs. The disposal of this waste at the BC Cribs site resulted in a contaminated depth interval between 30 and 70 m (98 and 230 ft) of the vadose zone where elevated concentrations of U, Tc-99, Cs-137, Sr-90, and ferrocyanide are observed (Corbin et al., 2005; Ward et al., 2004).

**TABLE 7.1**
**The 23 Waste Site Groups for the 200 Area Waste Sites from DOE-RL (1997)**

| Priority Ranking (in 1997) | Processing-Based Site Groups |
|---|---|
| 1 | Scavenged Waste Group |
| 2 | Chemical Sewer Group |
| 3 | Plutonium/Organic-Rich Process Waste Group |
| 4 | Gable Mountain/B-Pond and Ditches Cooling Water Group |
| 5 | S-Pond/Ditches Cooling Water Group |
| 6 | 200 North Cooling Water Group |
| 7 | 300 Area Chemical Laboratory Waste Group |
| 8 | T-Ponds/Ditches Cooling Water Group |
| 9 | Miscellaneous Waste Group |
| 10 | U-Ponds/Z-Ditches Cooling Water Group |
| 11 | Uranium-Rich Process Waste Group |
| 12 | Organic-Rich Process Waste Group |
| 13 | Tank Waste Group |
| 14 | Nonradioactive Landfills and Dumps Group |
| 15 | Steam Condensate Group |
| 16 | 200-Area Chemical Laboratory Waste Group |
| 17 | Radioactive Landfills and Dumps Group |
| 18 | General Process Waste Group |
| 19 | Fission Product-Rich Processes Waste Group |
| 20 | Plutonium Process Waste Group |
| 21 | Septic Tanks and Drain Fields Group |
| 22 | Tanks/Lines/Pits/Boxes Group |
| 23 | Unplanned Releases Group |

## 7.4 THE HANFORD CENTRAL PLATEAU TREATABILITY TEST FOR THE DEEP VADOSE ZONE

Knowing the risks posed to groundwater from Tc-99 and U in the deep vadose zone, the DOE began negotiations with Ecology and the EPA for addressing Tc-99 and U contamination in the 200 Area nontank farm operable units. In selecting a remediation strategy, section 121(b) of CERCLA specifies that remedies "utilize permanent solutions and alternative treatment technologies or resource recovery technologies to the maximum extent practicable." Preferred remedial actions under CERCLA include those that "permanently and significantly reduce the volume, toxicity, or mobility of hazardous substances, pollutants, and contaminants."

Deep excavation and removal of contamination may seem like the most effective method, but the method is not a panacea. For the BC Cribs and Trenches Area, the contamination resides at depths of 30–70 m (98–230 ft), exposure of workers to radiation, and the impact on groundwater in other areas render excavation impractical (DOE-RL, 2010). Remediation of the deep vadose zone therefore requires in situ treatment and surface barrier technologies to achieve remediation and closure of the Area's waste sites. In the early 2010s, these in situ technologies had not been sufficiently tested to enable an adequate evaluation as a remedial alternative, and, therefore, required substantial investigation through a scientific testing and development component under a treatability test.

### 7.4.1 Issuance of the Treatability Test Plan

Waste site characterization and treatability investigations are two of the main components of the RI/FS process. As waste site characterization and technology information is collected and reviewed, additional treatability studies may be required to fill some of these data gaps. In this manner, the process can be iterative, circling back to the characterization process to fill knowledge gaps and then into treatability options until sufficient information is gathered to make decisions. In general, treatability studies are performed to provide valuable waste site-specific data necessary to support remedial action decisions. Ultimately, information gleaned from treatability testing is used in selecting a viable remedy for a given waste site and used to inform operational components and implementation of the selected remedy. Remedies that illustrate potential to be effective in meeting cleanup goals for a given waste site can be fine-tuned through additional design and technology performance evaluations to implement the final remedy.

A TPA change package (M-15-06-02) was submitted on February 13, 2007, which established Milestone M-015-50. This milestone required the DOE to submit a Treatability Test Work Plan for cleanup of deep vadose zone Tc-99 and U to Ecology and EPA. The establishment of this milestone within the TPA not only provided the legal and regulatory framework to govern the remediation, but also ensured the appropriate focus of attention and resources toward remediation of the deep vadose zone. The multiple components of the Treatability Test Work Plan that address Tc-99 and U contamination are shown in Figure 7.5.

The Deep Vadose Zone Treatability Test Plan for the Hanford Central Plateau (DVZ-TT) was published in March 2008 by the U.S. DOE (DOE-RL, 2007). The primary objective of the DVZ-TT was to document and describe required testing and investigations of the identified remedial technologies. Once the investigations are completed, technical performance data can be used to provide a technical basis to evaluate the technology as part of a remedy in subsequent remedial alternative assessments. The technologies identified under the DVZ-TT were all selected as promising technologies for reducing the migration and flux of contamination from the vadose zone after thorough discussion and selection across multiple agencies. At the time, the development of improved methods for understanding vadose zone contamination was being conducted through the individual waste site programs. Investigations of potential remediation approaches were addressed by evaluating treatment technologies (e.g., CH2M, 2007) and conducting two technical workshops comprising panels of outside experts with input solicited from corresponding regulatory agencies. These panels evaluated the value of specific technologies, like electrical resistivity measurements, to characterize stratigraphy and contamination in the deep vadose zone (FHI, 2007) and evaluated multiple potential treatment technologies for immobilizing Tc-99 in the deep vadose zone (FHI, 2006). At the time, significant efforts were also being made in laboratory studies to address the behavior of Tc-99 and the effect of specific remediation techniques on Tc-99 mobility. Table 7.2 shows a summary of all the technologies evaluated under the DVZ-TT (DOE-RL, 2007), which were selected as a result of this effort. Table 7.3 shows the specific meetings where these ideas were developed into a strategy.

After an initial excavation treatability test, soil desiccation technology was identified as the primary objective of the DVZ-TT due to the emphasis on the highly mobile Tc-99 and U contamination at several waste sites, and the similarities in operational execution between desiccation and reactive gas injection. For the remainder of this chapter, we will focus on the development, testing, and implementation of the soil desiccation remedy.

### 7.4.2 Impracticability of Excavation Demonstrated through Treatability

An excavation-based treatability test was conducted at selected BC Cribs and Trenches Area waste sites (Figure 7.6) in 2008 to support the Deep Vadose Zone Treatability Test Plan (DOE-RL, 2007), the RI/FS for the 200 Area (DOE-RL, 1999), and the Focused Feasibility Study for the BC Cribs

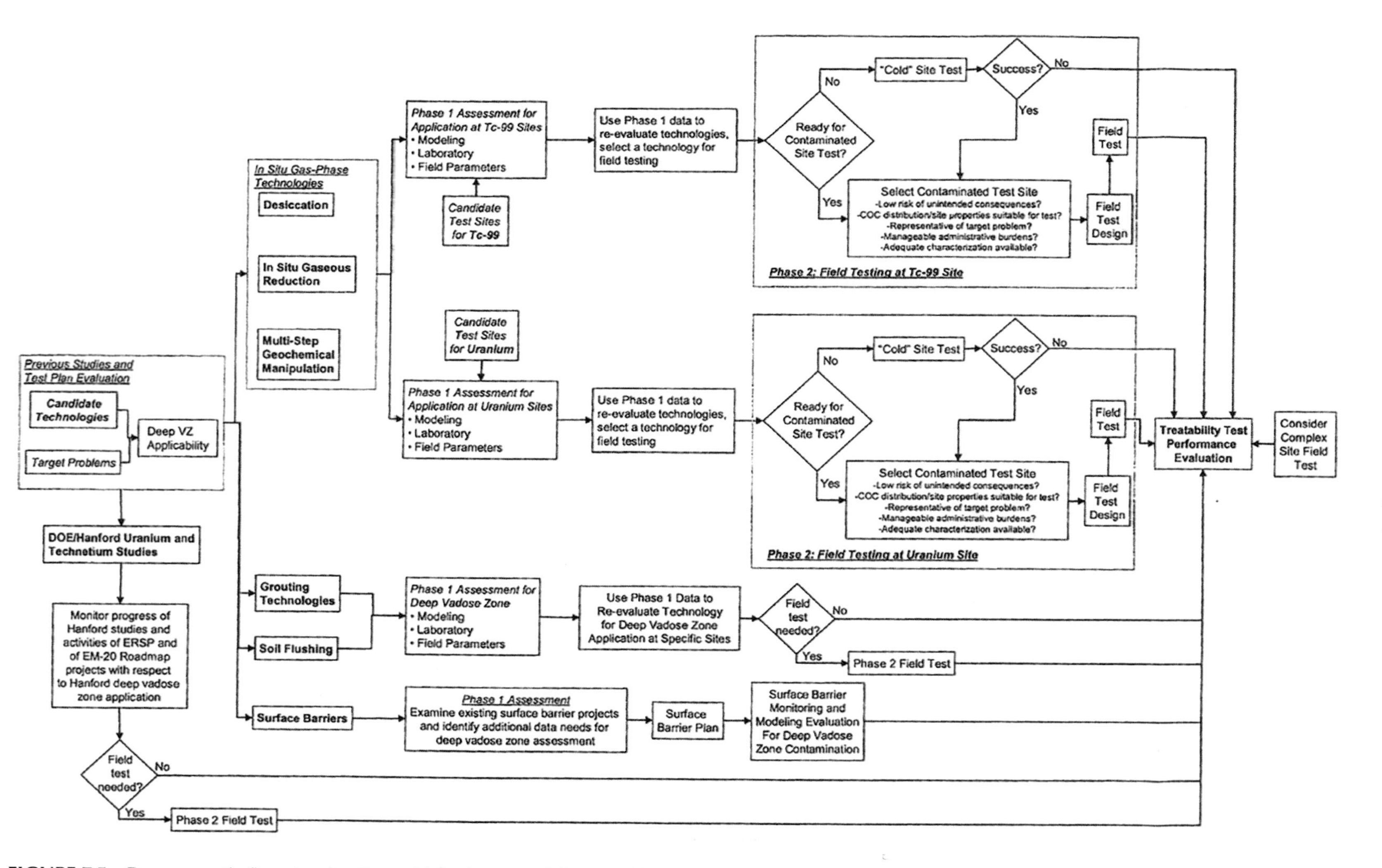


**FIGURE 7.5** Programmatic flow showing the multiple elements of the treatability testing approach. Reprinted with permission from DOE/RL-2007-56.

**TABLE 7.2**
**Remediation Technologies Evaluated under the DVZ-TT**

| Technology Evaluated | Technology Description | Selection for Inclusion in Treatability Test Plan |
|---|---|---|
| Desiccation | Desiccation involves drying a targeted portion of the vadose zone by injecting dry air and extracting soil moisture. Because desiccation removes water already in the vadose zone, it reduces the amount of pore fluid that could transport contaminants into the deep vadose zone, impedes water movement, and augments the impact of surface water infiltration control | YES – Removing water from the vadose zone via desiccation is promising |
| In situ gaseous reduction | A reducing gas (e.g., hydrogen sulfide) is used to directly reduce some contaminants and render them less soluble while they remain reduced or can reduce sediment-associated iron, which can subsequently reduce contaminants | YES – Because in situ gaseous reduction has the potential to immobilize technetium-99 and uranium and has been demonstrated at the field scale for similar applications, it is included for further study in the Treatability Test Plan |
| Multi-step geochemical manipulation | Geochemical manipulation is in the developmental stage. The technique involves introducing gases into the vadose zone that change Eh and/or pH and creates conditions for precipitation of minerals with coprecipitation of contaminants | YES – While this multi-step process is still conceptual, it builds on the successful development and demonstration of in situ gaseous reduction and provides a potential for more effective immobilization of contaminants such as technetium-99 and uranium |
| Grout injection | Grout injection is a means of treating subsurface contaminants by injecting grout of a binding agent into the subsurface to physically or chemically bind or encapsulate contaminants. There are multiple types of ground/binding materials. Grouting technologies have the potential for use as part of a remedy for the deep vadose zone | YES – Grouting technologies have the potential for use as part of a remedy for the deep vadose zone |
| Soil flushing | Soil flushing operates by adding water and an appropriate mobilizing agent, if necessary, to move contaminants and flush them from the vadose zone and into the groundwater where they are subsequently captured by a pump-and-treat system | YES – Soil flushing provides a potential mechanism to remove contaminants from the subsurface; however, efforts need to determine whether it is feasible to implement soil flushing in a way that minimizes uncertainties for applications in the deep vadose zone |
| Surface barriers | Reduction of water infiltration by surface barriers diminishes the hydraulic driving force for contaminant migration downward through the vadose zone to the water table | YES – Surface barriers are a baseline technology for near-surface contamination, and previous technology screening studies identified surface barriers as a promising technology for the deep vadose zone |

*Source:* From DOE-RL (2007).

and Trenches Area waste sites (DOE/RL-2004-66, 2005). The objective of this treatability test was to refine the estimates of worker dose and cost for partial excavation of near-surface soil at BC Cribs and Trenches. Worker dose and cost estimates prior to the test were based on limited characterization data and assumptions regarding excavation details. Using the analogous waste site approach of the RI/FS, three waste sites were identified as being representative of the different

**TABLE 7.3**
**Meetings in Which the Deep Vadose Treatability Test Plan Was Discussed (DOE-RL, 2007)**

| Date | Deep Vadose Zone–Integrated Project Team[a] | Deep Vadose Zone Strategy Working Group[a] | Monthly Groundwater Meetings Held at Washington Department of Ecology Facilities[b] | Deep Vadose Zone Treatability Test Plan Workshop |
|---|---|---|---|---|
| Nov. 13, 2006 | X | | | |
| Jan. 16, 2007 | X | | | |
| Feb. 14, 2007 | X | | | |
| Mar. 14, 2007 | X | | | |
| Apr. 24, 2007 | X | | | |
| May 15, 2007 | | | X | |
| May 22, 2007 | X | | | |
| Jul. 25, 2007 | | X | | |
| Aug. 15, 2007 | | | X | |
| Aug. 23, 2007 | | X | | |
| Oct. 4, 2007 | | X | | |
| Oct. 17, 2007 | | | X | |
| Feb. 7, 2008 | | | X | |
| Feb. 19, 2008 | | | | X |

[a] Team includes representatives from DOE-RL, DOE-ORP, regulatory agencies, and contractors. Several different integrated project teams have been formed to address various cross-cutting issues at the Hanford Site.

[b] Team includes representatives from DOE-RL, DOE-ORP, Tribal Nations, regulatory agencies, contractors, and the State of Oregon.

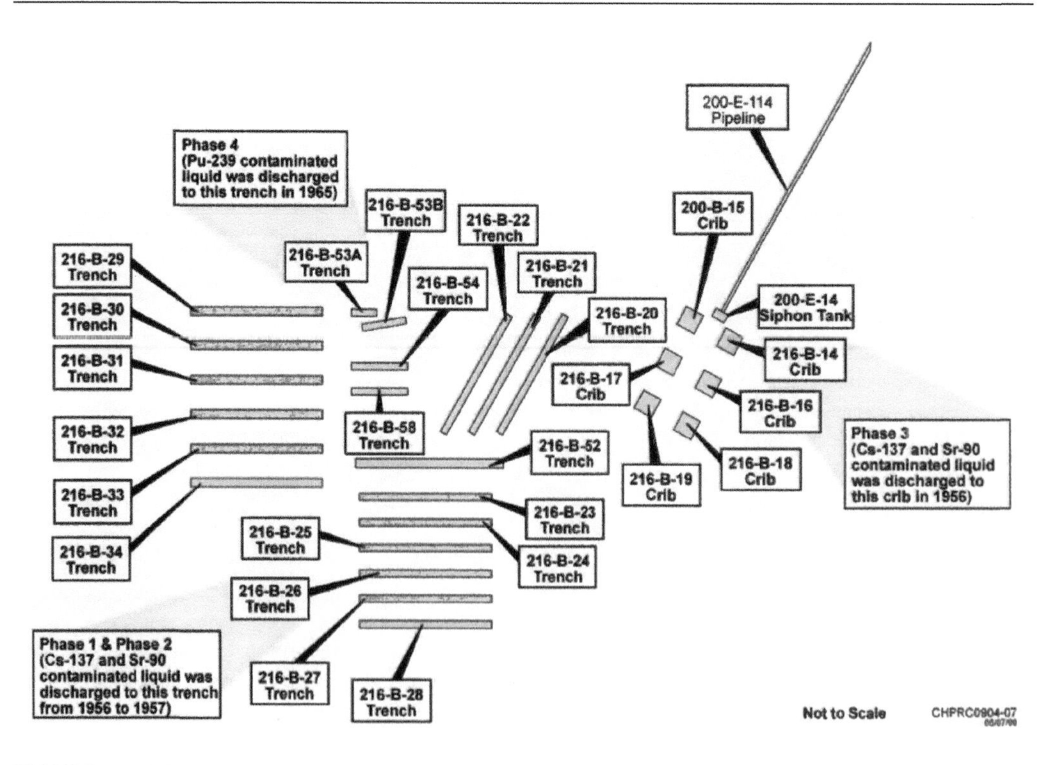


**FIGURE 7.6** Plan view of the BC Cribs and Trenches Area showing the layout of the waste sites and those sampled during the excavation-based treatability test. Reprinted with permission from DOE-RL (2010a).

construction, waste discharge methods, and waste characteristics that were carried out at the BC Cribs and Trenches Area. Two important pieces of information were gleaned from this excavation-based treatability test:

1. Financial expenses and dose rates observed during the excavation of the 216-B-26 crib collected during this test were recommended as important data points in evaluating the practicability of additional excavations.
2. Subsurface characterization data that delineated the distribution of radioactivity beneath the 216-B-26 and 216-B-53A trenches, and 216-B-14 crib were recommended as valuable data for refining conceptual models of the BC Cribs and Trenches Area.

A final report for the test was published in February 2010 (DOE-RL, 2010) and concluded that excavation could be practical for some of the waste sites in the BC Cribs and Trenches Area, but most of the waste sites would produce extremely high dose rates for workers and result in excessive risk associated with excavation. This solidified the justification behind the planned investigation of in situ treatments for vadose zone contamination at analogous waste sites within the Hanford 200 Area.

### 7.4.3 The Treatability Test Plan for the Soil Desiccation Remedy at BC Cribs

Soil desiccation is described as the injection of dried air in tandem with extraction of an equal volume of wet air from a network of wells. Desiccation removes moisture from the vadose zone through evaporation, which effectively reduces the downward movement of contaminants (Figure 7.7). Thisin situ technology has significant promise in reducing contaminant movement by decreasing

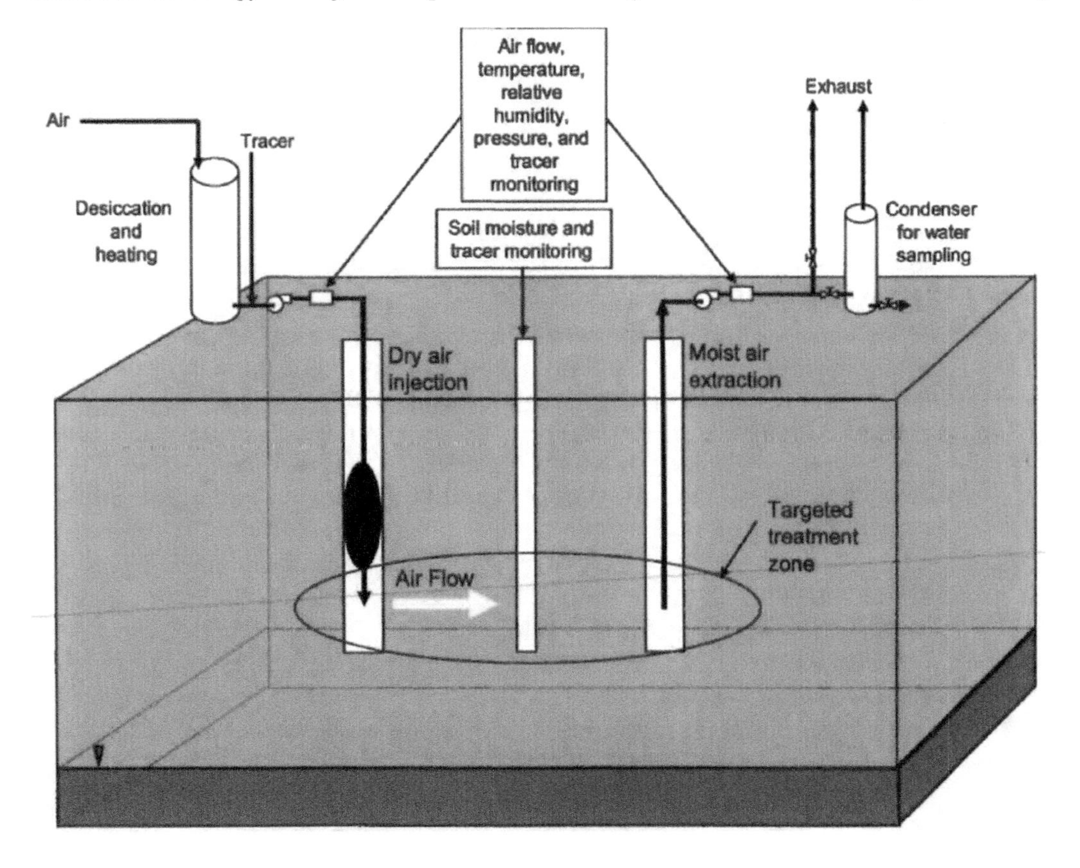


**FIGURE 7.7** Conceptual design for the field-scale soil desiccation test. Reprinted with permission from DOE-RL (2007).

water content and was identified as the leading remediation technology for investigation under the treatability test. In addition, soil desiccation could be combined with additional technologies, which increase sequestration of contaminants like Tc-99 and U by chemical processes like reduction and precipitation.

Prior to the treatability test, some screening-level testing was completed at the Hanford Site under the Hanford Subsurface Air Flow and Extraction (SAFE) Project (Truex, 2004). Results included details on the extraction of roughly 1,000 gallons of water during a 2-week test, but the volume of water extracted included water present both in the vadose zone and in the injected air. This screening-level investigation did not provide details on how the technology could be applied, potential strengths and weaknesses, or how the geology might affect the technology performance, both short- and long-term. As a result of this evaluation, it was recommended by Truex (2004) that treatability testing of in situ soil desiccation be carried out, because it had the least uncertainty of technologies evaluated in the screening-level investigation.

Additionally, in 2005, DOE-RL and the Hanford Site cleanup contractor convened an independent technical panel to review alternative remediation technologies for treating the Tc-99 contamination at a workshop in Richland, Washington. The panel was composed of experts from academia, national laboratories, and industry in vadose zone transport, infiltration control, hydrology, geochemistry, environmental engineering, and geology. The panel reviewed the findings of documents like Truex (2004) and Ward et al. (2004) and concluded that an evaluation of the effectiveness of desiccation would be beneficial at several Hanford vadose zone locations.

Specific activities recommended by the panel included numerical modeling of the injection/extraction of desiccated air, laboratory testing to determine fluid transport properties and compare model predictions, and informed field testing that would further advance the understanding of soil desiccation technology under waste site–specific conditions. The treatability test for soil desiccation incorporated these recommendations and includes design analysis activities, conceptual and numerical modeling, and laboratory work to study fundamental controls on the desiccation process (Phase 1). Subsequent field testing (Phase 2) included evaluation of the desiccation process, operational parameters, and the long-term performance for mitigation of contaminant transport. The overall objectives of the soil desiccation treatability test were as follows:

- To determine the design parameters for applying soil desiccation:
    - Operational parameters such as air flow rate and injected air properties (e.g., temperature and humidity).
    - Soil moisture reduction targets to achieve acceptable reduction of contaminant transport in the vadose zone.
- To demonstrate field-scale desiccation for targeted areas within the vadose zone.
    - Quantify the air flow, water extraction rate, and other operational parameters to evaluate implementability of the process on a large scale.
    - Determine the extent of soil moisture reduction in the targeted treatment zone to evaluate the short-term effectiveness of the process.
    - After desiccation is completed, determine the rate of change in soil moisture for the desiccated zone.
- To determine the best types of instrumentation for monitoring key subsurface and operational parameters to provide feedback to operations and for evaluating long-term effectiveness.
- To determine the number of injection and extraction wells, screened intervals, type of equipment and instrumentation, and operational strategy such that costs for full-scale application can be effectively estimated.

Laboratory-scale components of this treatability test were focused on providing a technical basis for implementing the remedy and determining the performance target that satisfies regulatory thresholds for contaminant transport and impact on groundwater. Numerical modeling of the desiccation process supported laboratory-scale experimentation, as numerical models of laboratory-scale tests can be well constrained using experimental controls. Numerical models were then used to bridge the gap in scale from the laboratory to the field, as typically less information is available at the field scale. In the following section, selected modeling, laboratory, and field-scale components of the soil desiccation treatability test will be discussed. We will also illustrate, using examples, how they integrated to support the goals of the treatability test, and, ultimately, the RI/FS for the Hanford 200 Area.

## 7.5 TECHNICAL EXAMPLES SUPPORTING THE SOIL DESICCATION TREATABILITY STUDY

The vadose zone is the soil layer above the groundwater table. Soils in the vadose zone are wet, but the water does not fully occupy the void space between the soil grains, that is, pore space. Fluid flow, that is, flux, occurs in the vadose zone at a very slow rate, for example, centimeters per year, and is controlled by the amount of water in the soil. For example, water deposited at the surface during a precipitation event will infiltrate into the ground and slowly move downward. For waste sites, this water may encounter contaminants as it moves through the vadose zone soil, which become entrained or dissolve into the pore fluid. This water is eventually transmitted to the groundwater table, where it can flow below ground to environmental receptors, for example, surface water, humans, animals, plants, and cause impacts on their health and wellbeing.

The movement of water through the vadose zone is predominantly downward and can be thought of as a balancing act between gravity and capillary action. Gravity drives the vadose zone water and solutes toward the groundwater table, whereas capillary forces retain water in the soil pore spaces. A real-world example of capillary action is easily demonstrated by lowering a paper towel into a bowl of water; the water in the paper towel will rise above the level of water in the bowl since the paper towel has a high capillary strength. Essentially, the pressure in the paper towel is lower than the pressure in the water, which drives water into the paper towel.

For the vadose zone at Hanford, soil desiccation technology was explored as a solution to limit the amount of contamination entering the groundwater table from the vadose zone. Essentially, soil desiccation seeks to remove the pore water in the vadose zone through evaporation, which removes the gravity component of the fluid transport, increases the capillary force, and reduces the overall flow of water to the groundwater table.

### 7.5.1 COMPUTATIONAL, LABORATORY, AND FIELD-SCALE ADVANCES IN SUPPORT OF TREATABILITY TESTING

To evaluate soil desiccation as a remediation technology and achieve the objectives outlined in Section 7.4, several key components of understanding were identified that had the potential to significantly impact the outcome of any applied remedy. Key components identified in the DVZ-TT included the following:

1. The impact of subsurface geology, soil moisture, and geochemistry;
2. the short-term effectiveness of remediation;
3. the long-term effectiveness of remediation; and
4. the large-scale implementation of the technology.

The following studies supported the treatability test to provide confidence that the technology could be reliably evaluated at the field scale. These studies involved the advancement of computational, laboratory, and field-scale capabilities and are discussed within the context of the treatability test for soil desiccation.

### 7.5.2 Sensor Evaluation for Desiccation and Rewetting

A critical component of subsurface remedy implementation is the ability to accurately measure and predict subsurface conditions during application of the remedy. Information derived from observational and predictive tools has profound impacts on the value of a given remedy, seeing how these data are ultimately used to evaluate and verify remedy performance.

Intermediate-scale laboratory experiments performed by Oostrom et al. (2009, 2012) investigated uncertainties in physical mechanisms and the performance of commercial off-the-shelf sensors and numerical models in capturing and predicting the soil desiccation response (Figure 7.8). Uncertainties based on energy limitations, osmotic effects, the effects of heterogeneity, and contaminant remobilization were addressed in these studies. Experimental instrumentation included thermistors, thermocouple psychrometers, dual-probe heat pulse sensors, heat dissipation units, and humidity probes. All instrument types used for this experiment observed the desiccation front indicating that uncertainty in subsurface moisture conditions could be reduced using a network of these sensors. Limitations in the sampling volume of the sensors were noted; however, given that a response to the changing conditions was recorded only when the drying front was very close to a sensor. Rewetting signatures were observed only in the heat dissipation unit and humidity probe data, indicating that these sensors would be integral in evaluating the long-term effectiveness of the remediation technology.

In addition to the laboratory desiccation experiments, data were used to verify the performance of a numerical model, which is a key predictive tool for understanding the field-scale response of the soil desiccation remedy (Ward et al., 2005; White and Oostrom, 2006). Independently obtained

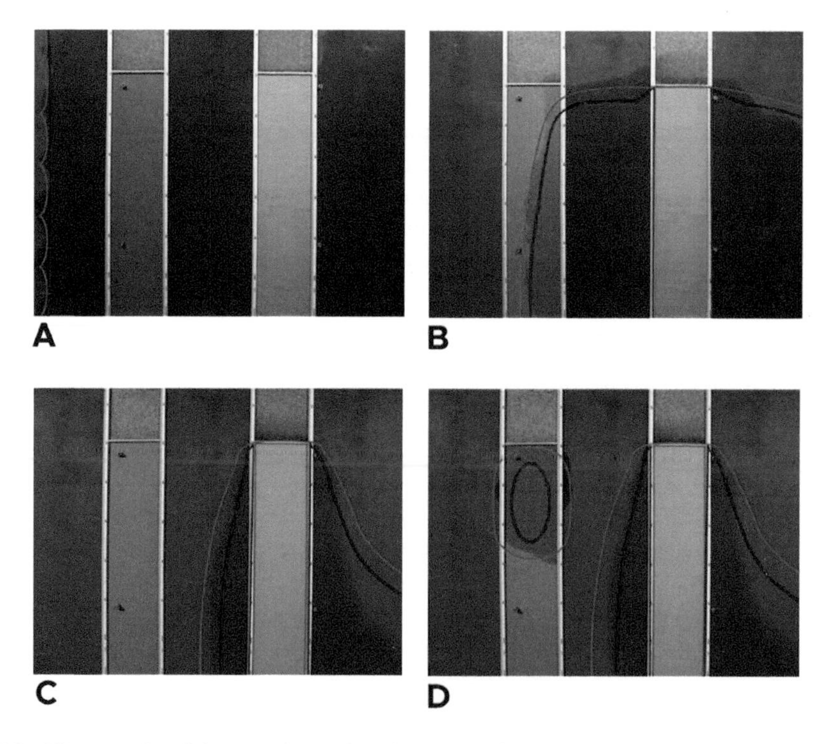


**FIGURE 7.8** Photographs of the experimental tanks used in Oostrom et al. (2009) showing the impacts of heterogeneities on the desiccation of porous media. Blue and red lines indicate the simulated 1% and 10% moisture content contours, respectively. (a) One day of desiccation. (b) One week of desiccation. (c) Four weeks of desiccation; the flow cell has been desiccated except for the 200-mesh sand wellbore and its immediate surroundings. (d) At the end of the redistribution period after injection of 500 mL of water in the upper part of the 100-mesh sand wellbore.

hydraulic and thermal porous medium properties were used to parameterize the numerical modeling. The numerical simulation results reasonably matched the observed experimental data, indicating that the simulator captures the pertinent gas flow and transport processes related to desiccation and rewetting and may be useful in the design and analysis of field tests.

### 7.5.3 Updating Theoretical Models to Interpret Soil Desiccation

Desiccation creates moisture conditions that are not commonly observed in many hydrologic studies. Therefore, the understanding of how water moves through desiccated soils needed to be refined within the DVZ-TT to understand the field-scale impacts of the remedy. To accurately predict subsurface moisture across the full range of moisture conditions, Zhang (2011) developed additional features of common soil water retention models that enabled numerical models to capture the extremely dry conditions created in the subsurface.

Traditional water retention models that describe how capillary forces in soils capture pore fluid, like the Brooks-Corey (Brooks and Corey, 1966) and the Mualem-van Genuchten (van Genuchten, 1980) models, cannot replicate extremely dry conditions observed during desiccation, as they were not developed using this consideration. By extending the lower bound of existing water retention functions and conductivity models from residual water content to oven-dry conditions (i.e., zero water content), these models could be applied to the range of expected conditions for soil desiccation remediation.

A state-dependent residual water content for a soil drier than a critical value was defined by Zhang (2011), which did not require refitting of the retention parameters from the unextended model and reduce to the unextended form when the soil is wetter than the critical value. This allowed for extension of existing water retention data to conditions observed in soils where desiccation was applied, reducing uncertainty in models used to support the laboratory and field experiments of the desiccation treatability study. Zhang (2012) also modified the model for hydraulic conductivity at low moisture conditions by introducing a correction factor to describe the film flow–induced hydraulic conductivity for natural porous media. By accounting for desiccated conditions in these refined conceptual models, this reduces uncertainty in the estimation of contaminant flux in the vadose zone in the higher level numerical model used to predict contaminant fluxes at field scales (Figure 7.9).

### 7.5.4 Monitoring Field-Scale Soil Desiccation

A clustered monitoring borehole network was constructed within the 200-BC-1 Operable Unit of the Hanford Central Plateau in support of the field-scale soil desiccation component of the soil desiccation treatability test. The clustered monitoring network consists of several sensor boreholes containing thermistors, gas-sampling ports, heat dissipation units, and electrodes for cross-hole electrical resistivity monitoring. Each sensor borehole has a logging well replicate placed nominally adjacent comprising an unscreened PVC-lined borehole designed to accommodate the use of wireline and borehole geophysical tools, for example, borehole radar and neutron probe.

Dry nitrogen ($N_2$) gas was used at a controlled temperature of 20°C to provide a constant inlet condition with a relative humidity of zero. Injection occurred at a stable flow rate of 510 $m^3$/h for 164 days except during a 13-day interval from April 21, 2011, through May 4, 2011, when there was no injection. Extraction of soil gas occurred for the duration of the test at a flow rate of 170 $m^3$/h. Figure 7.7 depicts the lateral layout of injection, extraction, and the monitoring locations. A gas-impermeable membrane barrier was installed at the surface centered over the well network to inhibit soil gas flow at the ground surface. Temperature, borehole neutron probe, electrical resistivity, and borehole ground-penetrating radar data were collected at regular intervals throughout the test. A full discussion of the field test can be found in DOE-RL (2012).

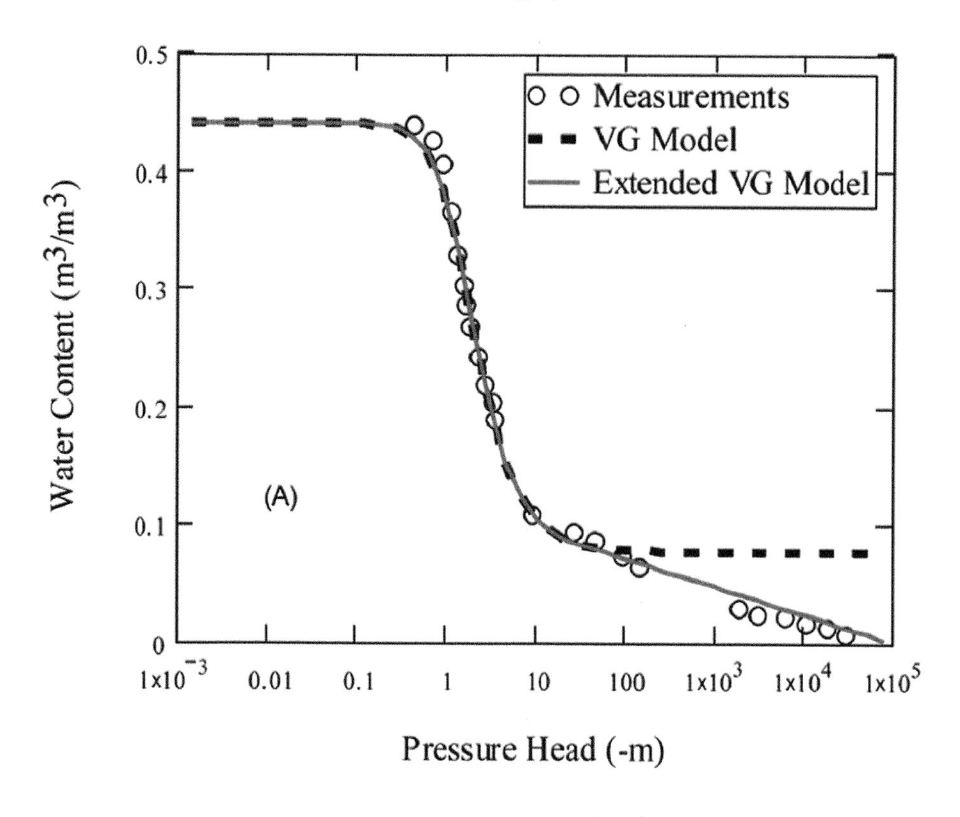


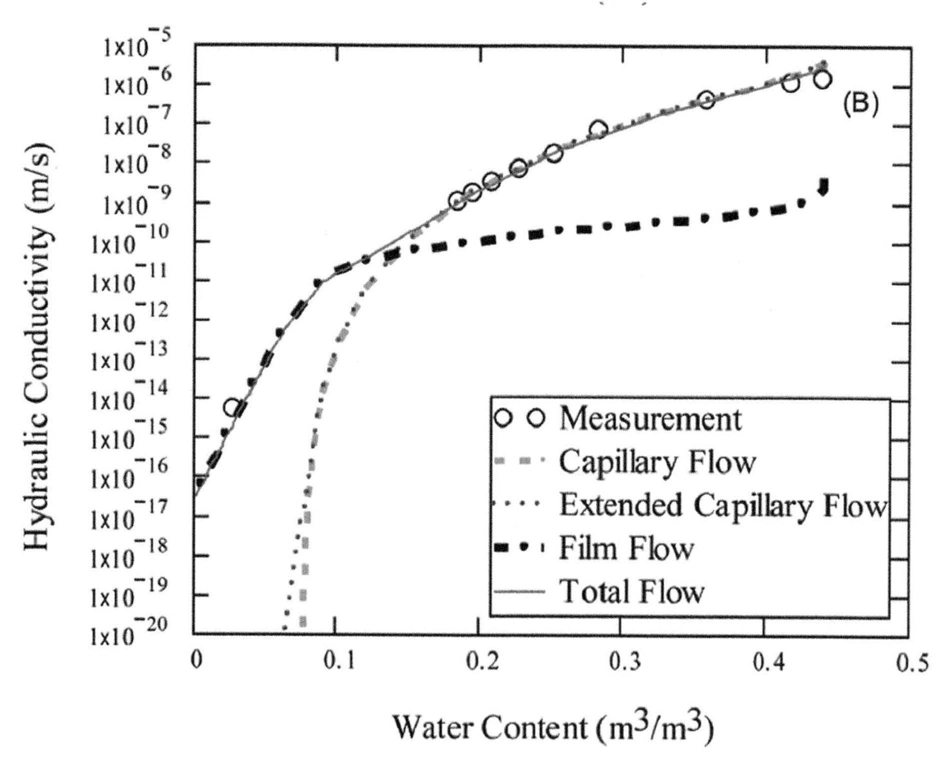


**FIGURE 7.9** Plots showing the extension of the water retention and unsaturated hydraulic conductivity models to values less than the residual moisture content parameter of the Mualem-van Genuchten soil model. Reprinted with permission from Zhang (2012).

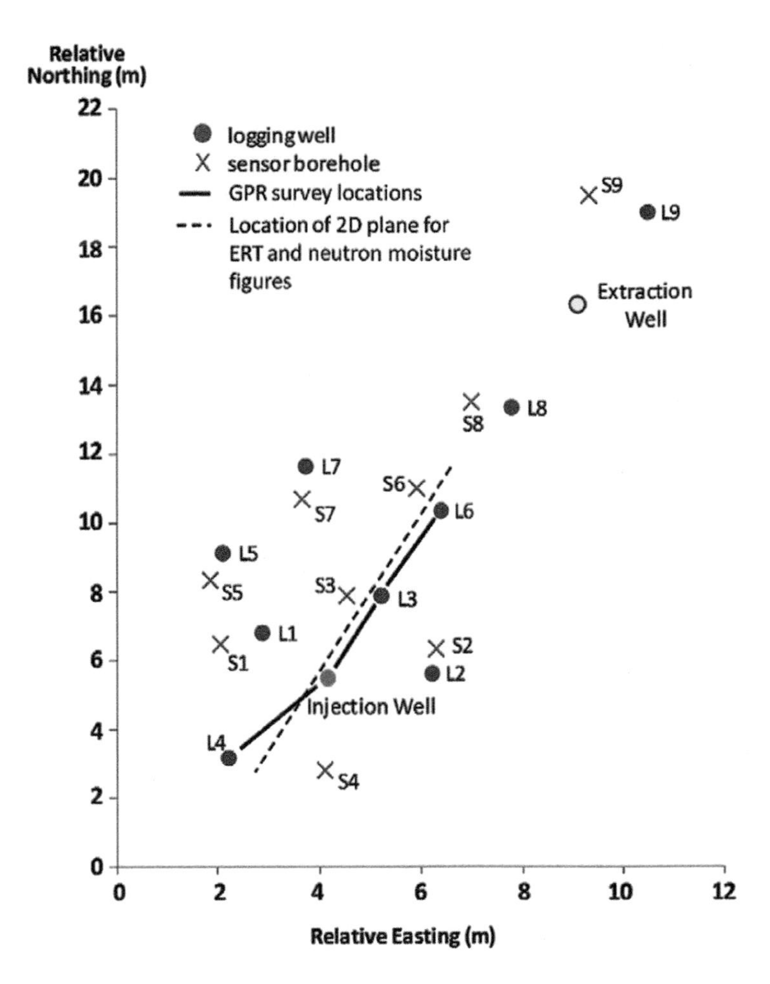


**FIGURE 7.10** Relative location of the test waste site wells. Boreholes with an S-designation contained distributed sensors (e.g., temperature) and electrical resistivity electrodes. Boreholes with an L-designation were open boreholes completed with sealed PVC casing for neutron logging and ground-penetrating radar measurements. Reprinted with permission from Truex et al. (2013).

Thermistors were used to monitor subsurface temperature at the S-designated boreholes shown in Figure 7.10. At each location, thermistors were installed every 4 ft from depths of 10–70 ft below ground surface. Temperatures were logged at 10-minute intervals, and data were used to estimate a 3D temperature field using interpolation (Figure 7.11). Cold zones are observed shortly after the onset of desiccation due to the evaporative cooling effect of the dry nitrogen. Once water is removed from an area, the zone warms toward the inlet gas temperature.

Similarly, borehole neutron probe data were collected at 7.5-cm-depth intervals at the L-designated boreholes in Figure 7.12. Data were interpolated to create a 3D volumetric moisture content (VMC) field. A two-dimensional slice through the well field is shown in Figures 7.13 and 7.14 with two different types of geophysical surveys, which show similar spatial and temporal trends for volumetric moisture content to those observed in the temperature data. The distribution of moisture content with time can be used to identify where desiccation has reached a specified threshold moisture content, as observed in the 13- to 17-m-depth interval out to a radial distance of about 3 m from the injection well (red zone). Interpretation of the two-dimensional moisture content representation should consider that interpolation does not incorporate subsurface conditions that can impact the distribution of desiccation away from the measurement point.

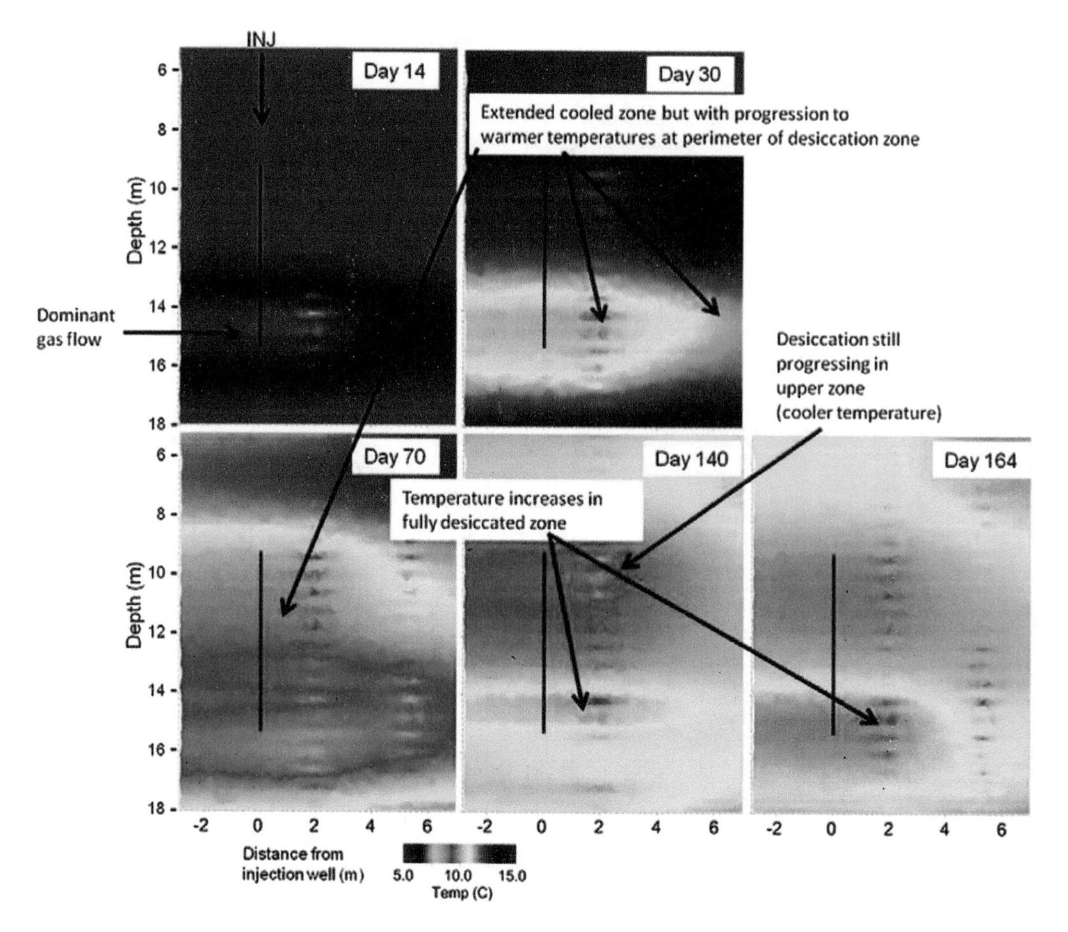


**FIGURE 7.11** Data from the temperature probe network collected during the field-scale soil desiccation test showing the evaporative cooling effect as moisture is removed. Data are interpolated between borehole locations. Reprinted with permission from Truex et al. (2013).

Electrical resistivity tomography (ERT) and ground-penetrating radar (GPR) were also evaluated as part of a field-scale soil desiccation treatability test. Trends in moisture redistribution over a broad zone in the vicinity of the test waste site were observed in the ERT and GPR data that aligned with observed trends in temperature and neutron probe data. The estimation of volumetric moisture from GPR data was adversely impacted by high electrical conductivity in the subsurface, which limited the interpretation of the GPR data. Regardless of the limitation on the GPR data, the geophysical data were paramount in illustrating the spatial extent of the dry conditions created during the injection, whereas sensor-based and wireline tools (e.g., neutron probe) data required interpolation, which fully ignores the physics governing fluid flow in unsaturated media. Despite this, the ancillary soil moisture information from borehole neutron data directly showed that volumetric moisture fell below 1% at boreholes within a 2-m radius of the injection well, creating a significant dry zone that could limit the flux of contamination to groundwater. Reliable moisture information at a waste site is particularly valuable within the context of a remediation strategy, as moisture ultimately governs the relative permeability of the liquid phase, which has the potential to carry vadose zone contamination to the groundwater table.

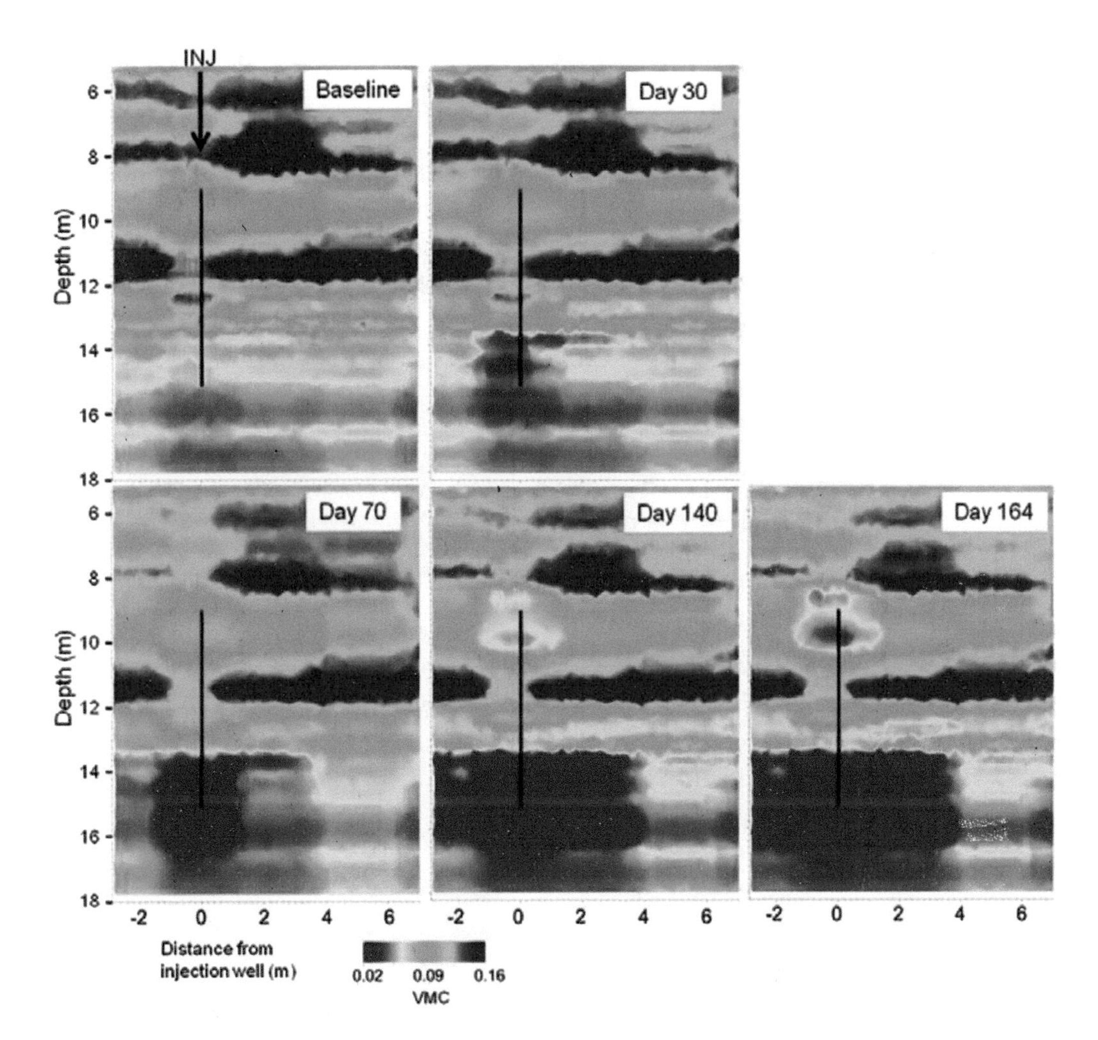


FIGURE 7.12 Data from time-lapse neutron probe measurements taken during the field-scale soil desiccation test showing the reduction in volumetric moisture content. Data are interpolated between borehole locations showing changes in volumetric water content as estimated from surveys. Reprinted with permission from Truex et al. (2013).

## 7.6 EVOLUTION OF TECHNOLOGY TO SUPPORT REMEDIATION OF COMPLEX WASTE SITES

Monitoring tools and predictive models are used to ensure remedial strategies are effective at removing or immobilizing subsurface contamination. Similar to the laboratory work performed under the treatability test to understand the ability of sensors to measure desiccation signals and the ability of models to replicate desiccated conditions, monitoring and predictive technologies should continually be advanced.

Contaminant migration in the vadose zone at the Hanford Site is a slow process taking hundreds to thousands of years for contamination to reach the groundwater table under natural conditions. However, notably, the volumes of waste released in different areas significantly impacted the flow conditions and water table so that some areas have had contaminants travel through the vadose zone to groundwater. The use of groundwater monitoring alone may be unable to provide timely

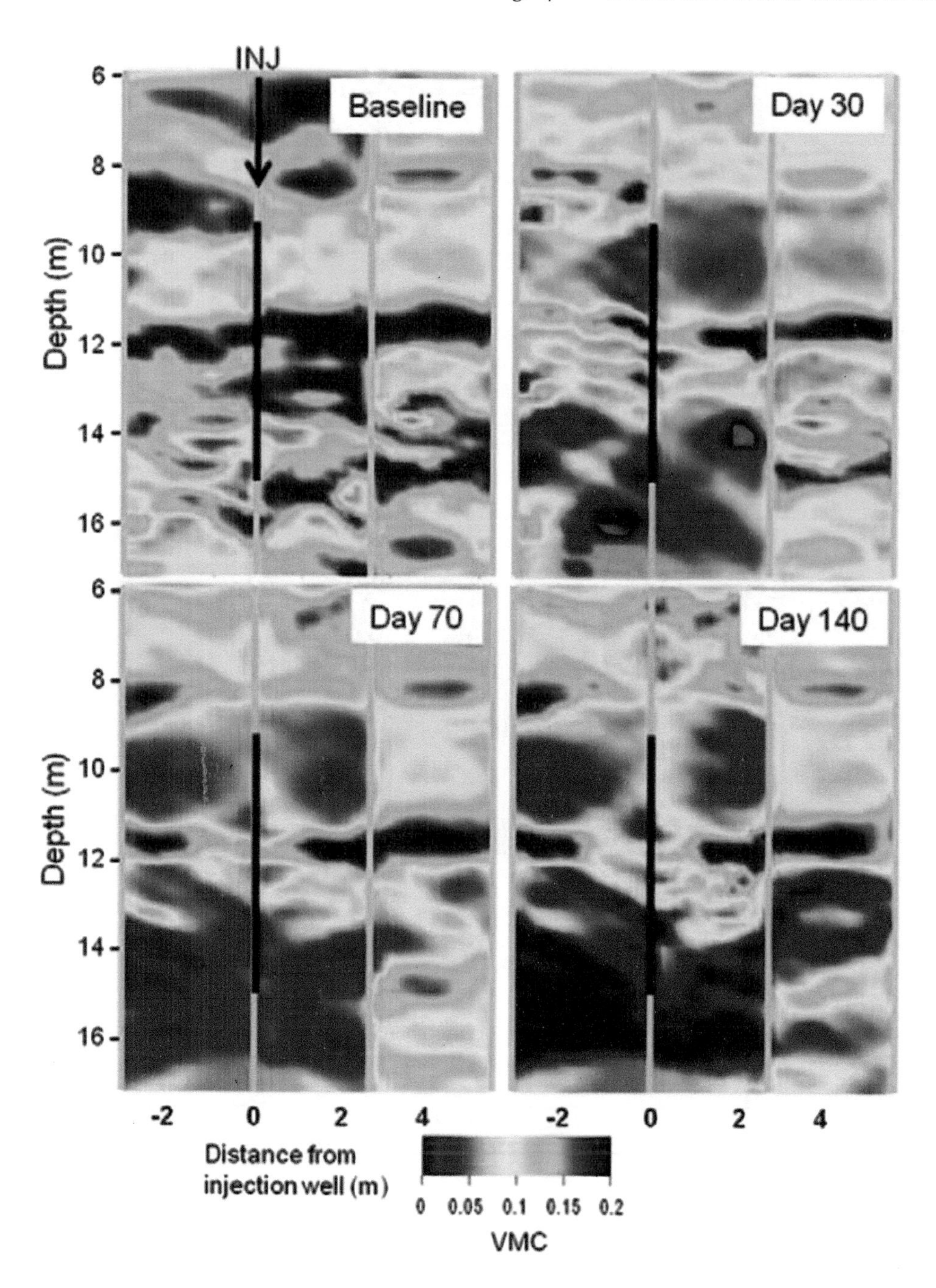


**FIGURE 7.13** Data from time-lapse ground-penetrating surveys collected during the field-scale soil desiccation test showing changes in volumetric water content as estimated from surveys. Reprinted with permission from Truex et al. (2013).

information to verify long-term remedy performance; therefore, monitoring the long-term behavior of remediated subsurface systems and the performance of remedial actions will be critical to treatability testing and waste site closure. Monitoring components for the deep vadose zone include methods and technologies for directly and indirectly measuring moisture conditions and contaminant flux to groundwater, providing early-warning monitoring of unexpected or unacceptable deep vadose zone behaviors. There is a clear need to identify appropriate technologies and monitoring

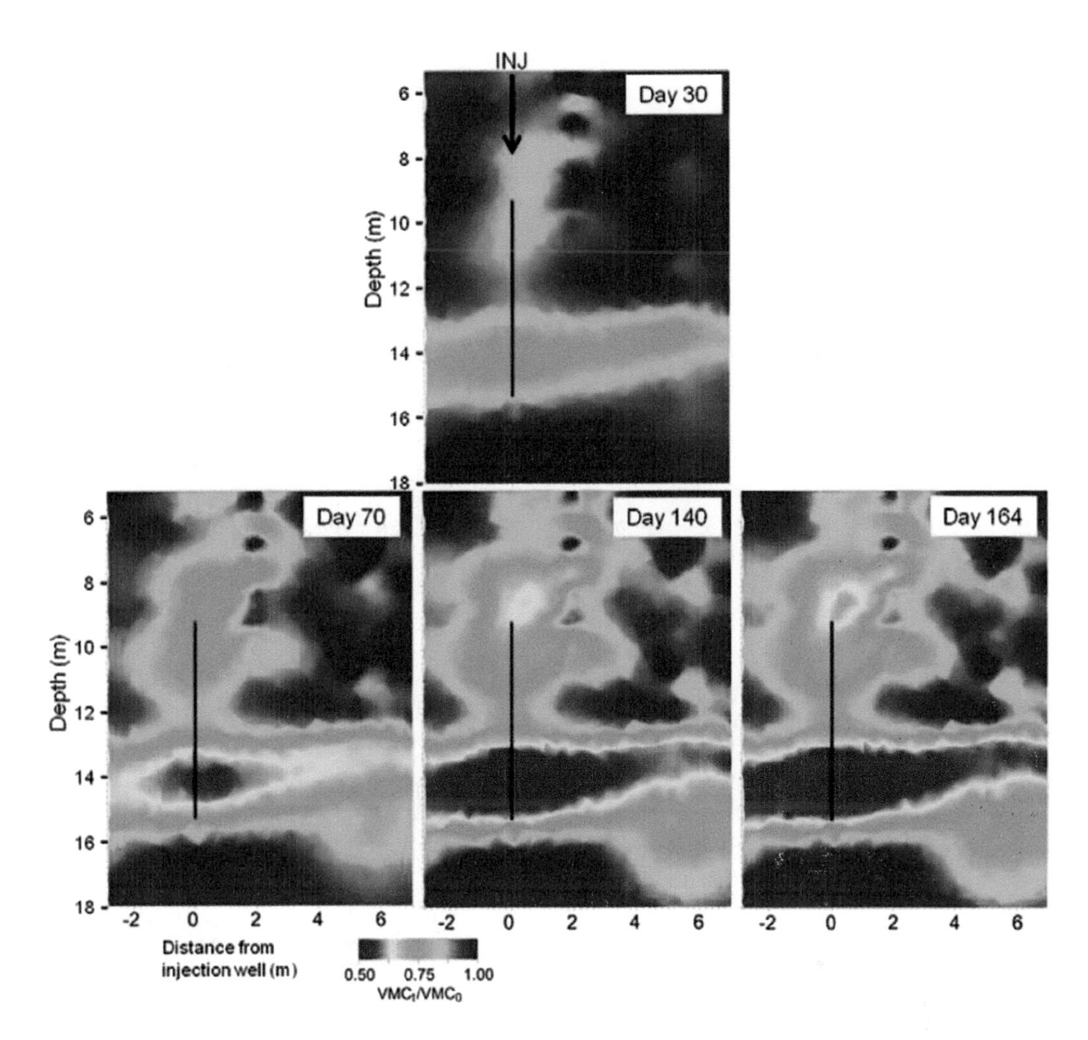


**FIGURE 7.14** Data from time-lapse borehole electrical resistivity measurements taken during the field-scale soil desiccation test showing changes in volumetric water content calculated based on electrical resistivity measurements. Reprinted with permission from Truex et al. (2013).

configurations for a comprehensive vadose zone monitoring strategy applicable to the Hanford Central Plateau, especially considering that monitoring technologies will continue to develop over time and recommendations for monitoring approaches may change.

Important parameters of the subsurface are required to ensure that subsurface conditions are such that contaminant flux to the groundwater is below regulations or guidelines. The flux rate of contaminants depends on multiple factors, including the mobility of the contaminant, the heterogeneity of the sediment, and the driving hydrologic forces. While there are no current technologies that can directly measure flux to the groundwater from the vadose zone, there are several technologies that can be deployed to provide hydrologic state data at different scales in time and space (Figure 7.15).

Monitoring technologies have seen many recent advancements and have been successfully demonstrated for a wide range of groundwater and vadose zone applications, including soil desiccation, surface barriers, and amendment infiltration/injections (Table 7.4). Geophysical methods, for example, have seen increased application in environmental characterization and monitoring given the ability of the methods to remotely sense geologic structure and hydrologic state of the subsurface. Multi-physics approaches are particularly attractive for subsurface monitoring as the independent

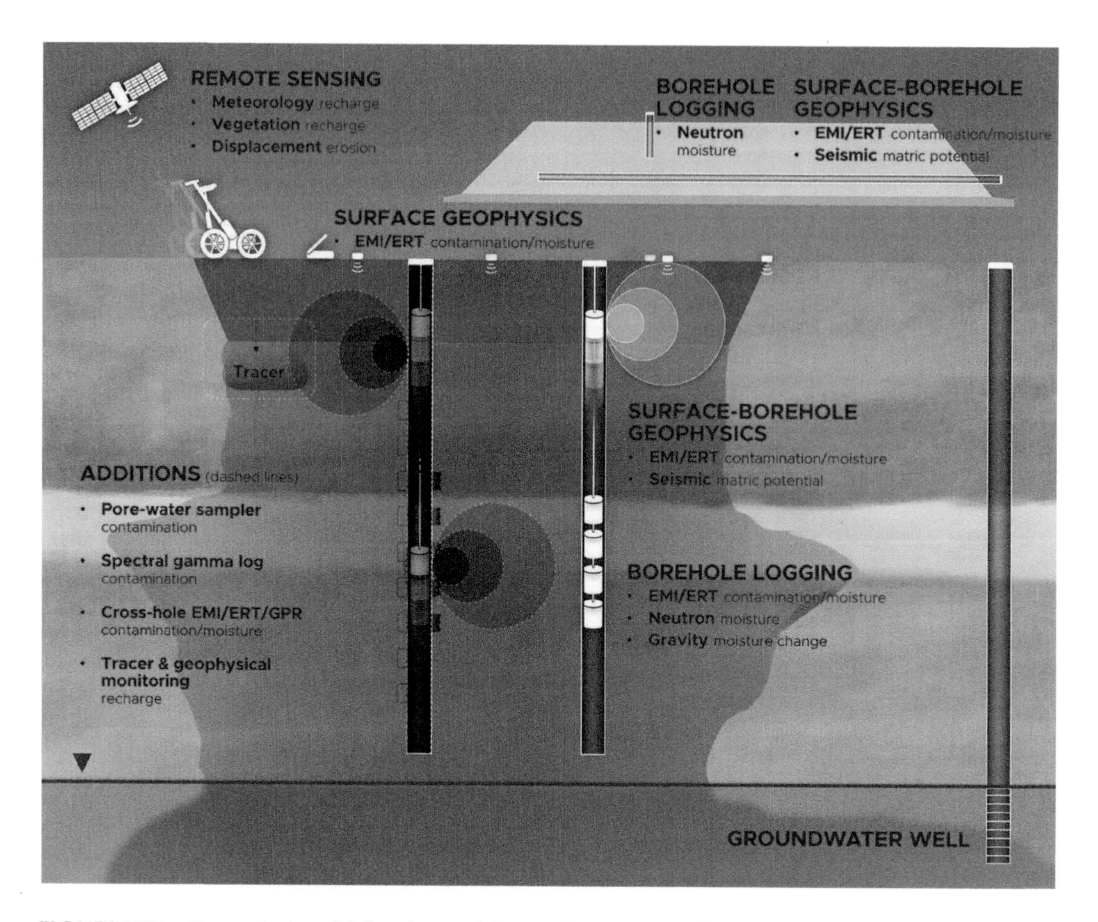


**FIGURE 7.15** Conceptual model for advanced deep vadose zone monitoring methods that includes remediation and monitoring objective-driven sampling of air, water, and soil phases at multiple scales in time and space. Technologies in figure correspond to technologies listed in Table 7.4. Reprinted with permission from Strickland et al. (2018).

data help to reduce issues with nonuniqueness in geophysical inverse problems by jointly constraining subsurface parameters through unique constitutive relationships (Linde et al., 2006; Doetsch et al., 2010).

Ultimately, selection and implementation of remediation technologies will require vadose zone characterization/monitoring programs that are designed to support effective remedy design and implementation (e.g., refine the conceptual site model), verify the effectiveness of remedial actions, and provide an early indication of future contaminant flux from vadose zone sources to groundwater. Additionally, the intensity of monitoring needed should be considered over the lifetime of the applied remedy. The evolution of monitoring intensity and the types of monitoring technologies used over time must be consistent with the stage of remediation and be suitable for transition to long-term monitoring. It is anticipated that the monitoring configuration will evolve over time, where more intense methods/frequency of monitoring will progress to less intensive methods/frequency of monitoring.

Due to the very long-time scales involved, predictive numerical modeling will be needed to make remedy decisions and will require accurate waste site-specific information. Monitoring data must be integrated with modeling analyses to facilitate feedback between predictive model results, verification monitoring, and refinement of the conceptual site model.

In some cases, a technology selected for a monitoring configuration would require additional development and testing to demonstrate that it has sufficient maturity for the designated use (e.g.,

**TABLE 7.4**
**Deep Vadose Zone Monitoring Technologies Depicted in Figure 7.15**

| Monitoring Method | Types of Technologies | Summary of Information Provided |
|---|---|---|
| 1. Remote sensing of surface conditions | InSAR, LiDAR, hyperspectral, meteorological | Spatial distribution of surface boundary conditions important to vadose zone moisture and contaminant transport (e.g., recharge, erosion, intrusion) |
| 2. Surface-based geophysics for the subsurface | EMI, GPR, ERT, Seismic (EMI, GPR, and ERT can be coupled with a tracer) | Subsurface 3D distribution of moisture content/matric potential, subsurface infrastructure indirect indicators of subsurface hydrogeologic properties: autonomous (e.g., ERT) can be applied to provide continuous temporal data |
| 3. Single borehole geophysical logging | EMI, GPR, ERT, Seismic, gravity, spectral gamma, neutron probe | Subsurface vertical profile of moisture content/matric potential, radionuclide concentration (e.g., spectral gamma). Subsurface infrastructure indirect measures of subsurface hydrogeologic properties; autonomous (e.g., ERT) can be applied to provide continuous temporal data |
| 4. Cross-hole or surface-borehole geophysical logging | EMI, GPR, ERT, Seismic (EMI, GPR, and ERT can be coupled with a tracer) | Subsurface 2D (and 3D) cross-sections of moisture content/ matric potential subsurface infrastructure indirect measures of subsurface hydrogeologic properties; autonomous techniques (e.g., ERT) can be applied to provide continuous temporal data |
| 5. Point-location sensors | In situ sensors or sensors deployed at a fixed measurement location | Temperature, moisture content, matric potential, humidity, and contaminant chemical data at a specific location can be arranged as a vertical set of sensors in a borehole or at the surface |
| 6. Point-location pore water/soil gas samplers | Sampling ports to collect water or gas for aboveground analysis by a field instrument or in the laboratory | Water or gas chemical constituent analyses |
| 7. Groundwater monitoring well | Sample collection or in situ probes | Groundwater contaminant constituent concentration flux and water level information to evaluate vadose zone impact to groundwater |

electromagnetic imaging needs additional development of data interpretation for subsurface imaging). Each of the identified technologies has strengths and weaknesses, such that no single technology is the best choice, and a combination of technologies is recommended. For all of the geophysical and remote sensing techniques, improved data interpretation and data management will also improve their use for meeting monitoring objectives.

## REFERENCES

Brooks, R.H., and A.T. Corey. 1966. "Hydraulic properties of porous media affecting fluid flow." *Proceedings of the ASCE Journal of the Irrigation and Drainage Division* 92:61–88.

CH2 M. 2007. *Central Plateau Vadose Zone Remediation Technology Screening Evaluation.* RPP-ENV-34028 Rev 0, CH2M HILL Plateau Remediation Company, Richland, WA.

Corbin, R.A., B.C. Simpson, M.J. Anderson, W.F. Danielson III, J.G. Field, T.E. Jones, and C.T. Kincaid. 2005. *Hanford Soil Inventory Model Rev. 1.* RPP-26744, Rev. 0, CH2M HILL Hanford Group, Inc., Richland, WA.

DOE. 2010. *Field Test Plan for the Soil Desiccation Pilot Test.* DOE/RL-2010-04, Rev. 0, U.S. Department of Energy Richland Office, Richland, WA.

DOE-RL. 1997. *Waste Site Grouping for 200 Areas Soil Investigations.* DOE/RL-96-81 Rev. 0, U.S. Department of Energy, Richland Operations Office, Richland, WA.

DOE-RL. 1999. *200 Areas Remedial Investigation/Feasibility Study Implementation Plan - Environmental Restoration Program.* DOE/RL-98-28 Rev. 0, U.S. Department of Energy, Richland Operations Office, Richland, WA.

DOE-RL. 2007. *Deep Vadose Zone Treatability Test Plan for the Hanford Central Plateau*. DOE/RL-2007-56 Rev. 0, U.S. Department of Energy, Richland Operations Office, Richland, WA.

DOE-RL. 2010. *BC Cribs and Trenches Excavation-Based Treatability Test Report*. DOE/RL-2009-36 Rev. 1 Reissue, U.S. Department of Energy, Richland Operations Office, Richland, WA.

DOE-RL. 2012. *Deep Vadose Zone Treatability Test for the Hanford Central Plateau: Soil Desiccation Pilot Test Results*. DOE/RL-2012-34 Revision 0. U.S. Department of Energy, Richland Operations Office, Richland, WA.

DOE/RL-2004-66. 2005. *Focused Feasibility Study for the BC Cribs and Trenches Area Waste Sites, Draft A*. U.S. Department of Energy, Richland Operations Office, Richland, WA.

Doetsch, J., N. Linde, and A. Binley. 2010. "Structural joint inversion of time-lapse crosshole ERT and GPR traveltime data." *Geophysical Research Letters* 37(24), 1–6.

FHI. 2006. *Evaluation of Vadose Zone Treatment Technologies to Immobilize Technetium-99*. WMP-27397, Rev. 1, Fluor Hanford, Inc., Richland, WA.

FHI. 2007. *Technical Evaluation of Electrical Resistivity Methods*. SGW-34795, Fluor Hanford, Inc., Richland, WA.

Icenhower, J.P., N. Qafoku, W.J. Martin, and J.M. Zachara. 2008. *The geochemistry of technetium: A summary of the behavior of an artificial element in the environment. PNNL-18139*. Pacific Northwest National Laboratory, Richland, WA.

Linde, N., A. Binley, A. Tryggvason, L.B. Pedersen, and A. Revil. 2006. "Improved hydrogeophysical characterization using joint inversion of cross-hole electrical resistance and ground-penetrating radar traveltime data." *Water Resources Research* 42(12): 1–16.

Oostrom, M., T.W. Wietsma, J.H. Dane, M.J. Truex, and A.L. Ward. 2009. "Desiccation of unsaturated porous media: Intermediate-scale experiments and numerical simulation." *Vadose Zone Journal* 8: 643–650.

Oostrom, M., T.W. Wietsma, C.E. Strickland, V.L. Freedman, and M.J. Truex. 2012. "Sensor and numerical simulator evaluation for porous medium desiccation and rewetting at the intermediate laboratory scale." *Vadose Zone Journal* 11(1): 1–10.

Serne, R.J., A.L. Ward, W. Um, B.N. Bjornstad, D.F. Rucker, D.C. Lanigan, and M.W. Benecke. 2009. *Electrical Resistivity Correlation to Vadose Zone Sediment and Pore-Water Composition for the BC Cribs and Trenches Area*. PNNL-17821, Pacific Northwest National Laboratory, Richland, WA.

Strickland, C.E., M.J. Truex, R.D. Mackley, and T.C. Johnson. 2018. *Deep Vadose Zone Monitoring Strategy for the Hanford Central Plateau*. Report PNNL-28031, Pacific Northwest National Laboratory, Richland, WA.

Truex, M.J. 2004, *Feasibility Study Evaluation of In Situ Technologies for Immobilization of Technetium beneath the BC Cribs*. Letter Report, Pacific Northwest National Laboratory, Richland, WA.

Truex, M.J., T.C. Johnson, C.E. Strickland, J.E. Peterson, and S.S. Hubbard. 2013. "Monitoring vadose zone desiccation with geophysical methods." *Vadose Zone Journal* 12(2): 1–14.

van Genuchten, M.T. 1980. "A closed-form equation for predicting the hydraulic conductivity of unsaturated soils." *Soil Science Society of America Journal* 44(5):892–898.

Ward, A.L., G.W. Gee, Z.F. Zhang, and J.M. Keller. 2004. *Vadose Zone Contaminant Fate and Transport Analysis for the 216-B-26 Trench*. PNNL-14907, Pacific Northwest National Laboratory, Richland, WA.

Ward, A.L., M.D. White, E.J. Freeman, and Z.F. Zhang. 2005. *STOMP: Subsurface Transport Over Multiple Phases, Version 1.0. Addendum: Sparse Vegetation Evapotranspiration Model for the Water-Air-Energy Operational Mode*. PNNL-15465, Pacific Northwest National Laboratory, Richland, WA.

White, M.D., and M. Oostrom. 2006. *STOMP: Subsurface Transport Over Multiple Phases, Version 4.0. User's Guide*. PNNL-15782, Pacific Northwest National Laboratory, Richland, WA.

Zachara, J.M., C. Brown, J. Christensen, J.A. Davis, E. Dresel, C. Liu, S. Kelly, J. McKinley, J. Serne, and W. Um. 2007. *A Site-wide Perspective on Uranium Geochemistry at the Hanford Site*. PNNL-17031. Pacific Northwest National Laboratory, Richland, WA.

Zhang, Z.F., C.E. Strickland, J.G. Field, D.L. Parker, and R.E. Clayton. 2012. "Evaluating the performance of a surface barrier for reducing soil-water flow." *Vadose Zone Journal* 11(3). https://doi.org/10.2136/vzj2011.0117

Zhang, Z.F. 2011. "Soil water retention and relative permeability for conditions from oven-dry to full saturation." *Vadose Zone Journal* 10(4):1299–1308.

# 8 Wasteform Development and Qualification for Tank Waste Vitrification and Disposal

*Jose Marcial and Derek Dixon*

This chapter overviews the methodology of decision-making about wasteforms for the safe and efficient immobilization and storage of radioactive wastes resulting from nuclear material production. The process of developing radioactive waste glasses that are acceptable with respect to operational limitations and requirements of the selected melter technology while also satisfying performance criteria for storage and disposal affects the decision-making and is detailed here. The Hanford Site, located in southeastern Washington State, USA, will be used as a case study. The historical context of tank waste management at the Hanford Site is discussed in detail in Chapters 1 and 5, while the history relevant to wasteform selection and the technological developments that ensued will be provided in this chapter. Chapter 14 describes the Integrated Disposal Facility at Hanford into which the waste glasses will be dispositioned.

## 8.1 HANFORD RADIOACTIVE TANK WASTE

One hundred and seventy-seven underground tanks (149 single-shell and 28 double-shell) were originally used at the Hanford Site to store the highly radioactive wastes resulting from plutonium production operations between 1944 and 1988. The waste from some of these tanks has since been retrieved. There remains a combined volume of about 54 million gallons (205,000 $m^3$) of highly radioactive waste (Rodgers 2023). Technologies used to separate plutonium from irradiated nuclear fuel advanced over four decades of Hanford operations from the early bismuth phosphate carrier precipitation method, to the REDOX solvent extraction method, and finally to the PUREX solvent extraction methods, which resulted in a wide range of chemical and radiochemical compositions in the waste tanks (Agnew et al. 1997). In addition, subsequent tank farm operations by Hanford Site contractors that included evaporation, waste transfers and blending, separation processes, and corrosion control additions introduced changes to tank waste compositions. The forms of waste within tanks include saltcake, supernatant, and sludge, with heterogeneous distributions of waste components within these forms (Hay 2022).

With respect to the Hanford vitrification process, low-activity waste (LAW)[1] refers to the liquid supernatant and water-soluble components of dissolved saltcakes and sludges after cesium and transuranic elements are removed, while high-level waste (HLW) refers to the sludge with a minor fraction of liquid and the soluble cesium and transuranic elements that were removed from LAW (Deng et al. 2016). These HLW and LAW designations may not apply to other countries or other U.S. sites depending on time frame of definition. Overall, LAW accounts for a majority of the waste volume, while HLW accounts for the majority of the waste radioactivity (Kruger 2013, Bernards et al. 2017). Hanford is somewhat unique because not only will the HLW be immobilized in borosilicate glass, but LAW will also be immobilized in a borosilicate glass. In contrast, the West Valley Demonstration Project (WVDP, in western New York) and the Savannah River

DOI: 10.1201/9781003329213-11

Site (SRS, in South Carolina) immobilized HLW in a borosilicate glass, but the low-level waste fractions of their tank wastes were/are immobilized in a cementitious wasteform commonly referred to as saltstone or grout.

## 8.2 HANFORD TANK WASTE COMPARISON TO WASTE SITES

The compositional and phase complexity of Hanford tank waste is unique compared to nuclear fuel reprocessing wastes at other U.S. and international sites. For example, when comparing Hanford to the SRS, there are differences in separations where Hanford used several separation processes (e.g., Bi-P) that were not used at the SRS, and as a result, Hanford tank waste can contain a significant fraction of Bi and P. For many other countries, HLW is the result of reprocessing spent nuclear fuel from energy production. These wastes typically do not have the compositional complexity found in Hanford's tank wastes since they have gone through fewer separation processes and fewer years of storage (Lutze and Ewing 1988). In addition to the significant compositional differences, the volume of tank waste at Hanford is greater than wastes stored at other sites (Lutze and Ewing 1988).

For many sites, the simplicity of the waste chemistry provides an opportunity to simplify the vitrification flowsheet while satisfying process, disposal, and regulatory requirements. For example, WVDP had a small volume of a relatively consistent tank waste composition, which allowed the use of an essentially fixed glass composition that satisfied all requirements for the entire inventory of wastes. Meanwhile, the compositional complexity at Hanford coupled with the pace of waste retrieval and processing expectations will require near real-time adjustments to the additives as dramatic waste composition swings occur.

## 8.3 HANFORD WASTEFORM SELECTION

The selection of the borosilicate wasteform for the Hanford legacy waste (and other U.S. sites) began with the technical review of wasteforms by an independent panel of experts who met annually from 1979 to 1981 (Alternative Waste Form Peer Review Panel 1981) and identified borosilicate glass as the preferred wasteform to immobilize HLW stored throughout the nation (e.g., sites in the states of Washington, Idaho, New York, and South Carolina). The initial driver for the wasteform decision was the West Valley site that served as the leading edge for HLW vitrification operations in the U.S. (DesCamp and McMahon 1996, Petkus et al. 2003). The expert panel considered a list of 15 waste forms including SYNROC, grout/cementitious materials, porous glass matrix, tailored ceramic, coated sol-gel spheres, FUETAP concrete, lead-iron phosphate glass, glass ceramic, clay ceramic, titanate ion exchanger, and borosilicate glass (Alternative Waste Form Peer Review Panel 1980). Major considerations in the downselection process included the following: (i) high chemical durability, (ii) no crystal structure requirements (i.e., wasteform could accept a broad chemical composition with a reasonably high loading of waste into the final wasteform), and (iii) could be produced using a fast and continuous process. The panel placed emphasis on "scientific merit over engineering practicality" (Alternative Waste Form Peer Review Panel 1979), and the breadth of international research on the borosilicate wasteform weighed heavily on this metric.

The Department of Energy (DOE) issued separate records of decision (ROD) listing borosilicate glass as the desired wasteform for HLW first for the WVDP, the Defense Waste Processing Facility (DWPF) at the SRS, and eventually for Hanford (U.S. Department of Energy 1982a, 1982b, 1988, 1995). A subsequent ROD established requirements for the performance of the release of potentially hazardous substances (U.S. Department of Energy 1997, 2004), which helped define durability, or wasteform performance requirements. As a result of these defined requirements, the Hanford LAW will be vitrified and stored on the Hanford Site at the Integrated Disposal Facility (IDF) on account of the wastes' hazardous constituents (U.S. Department of Energy 2000). The reader is referred to Chapter 1, which further discusses the site assessment and the corresponding cleanup program for Hanford.

## 8.4 HANFORD MELTER TECHNOLOGY DEVELOPMENT

The Hanford Site serves as a useful case study for waste immobilization technology development because the large quantity and compositional variability of Hanford waste introduce significant technical challenges.

Contemporary melter technologies include the Joule-heated liquid-fed ceramic melter (LFCM), the cold crucible induction melter (CCIM), the hot walled induction melter (HWIM), the hot walled resistance melter (HWRM), the in-can melter (ICM), and the GeoMelt™ in-container vitrification (ICV). The LFCM and ICV are Joule-heated by passing an electric current through the glass melt. The CCIM is heated by coupling an induction field to the glass melt. The glass melt is heated in the HWRM, HWIM, and ICM by heating the metal wall of the container using either resistance (HWRM and ICM) or induction (HWIM and ICM). In each case, the waste is fed to the melter with glass-forming chemicals (GFCs) onto a heated melt pool. The melt pool heats and, in case of slurry melter feed, dries the melter feed, which forms an insulating pile or cold cap on top of the melt pool. The GFCs are often premelted into a frit. Also, to accelerate the melter feed-to-glass conversion process, the waste is calcined prior to charging into the melter. Calcining is a thermal treatment of the waste to remove gas-evolving components (such as nitrates, carbonates, and sulfates) and to dewater. Without calcining, these components will evolve gases inside of the melter, and without appropriate measures, the evolved gases may become trapped and lead to foaming, which in turn can prevent heat transfer into the cold cap with concomitant deleterious impacts on the processing rate. In contrast, the LFCM has been developed to mitigate foaming and eliminate the need for a calciner.

Today, various countries, including Russia and South Korea, utilize CCIM technology for vitrifying low-level waste, while France is the only country to use a CCIM to process HLW from commercial spent nuclear fuel reprocessing operations. The benefit of a CCIM or ICV is that high temperatures can be achieved, and therefore, high waste loading feeds can be practicably processed especially if the waste is enriched in high atomic number elements (which can greatly increase the melting temperature). The ICV and ICM have the advantage that glass is disposed of in the melter itself, so crystal constraints used to ensure the melt will pour into a disposal container are not needed. One limitation of HWIM, HWRM, and ICM is their size because heat must be supplied from the outside of the melter to the bulk melt, and thus limiting the size of the melters (Ojovan and Lee 2005). The CCIM has a similar challenge where the energy field can only penetrate a certain depth into the glass to supply heat. The high temperatures of CCIM can also lead to volatility of elements that are desired in the glass, causing a greater need for offgas recycling. On the contrary, the LFCM and ICV can be made relatively larger since they rely on an internal heat source and so can process more waste per melter due to their large melt surface area. Therefore, the LFCM has been used at three sites holding significant quantities of radioactive waste from legacy operations, including Hanford. The remainder of this chapter focuses on the LFCM technology.

However, it was the LFCM technology that was initially developed at Pacific Northwest Laboratory (PNL) for the WVDP and development continued for the Hanford Site (Buelt and Chapman 1978, Chapman et al. 1979, Barnes and Larson 1981). Preliminary testing that led to proof-of-concept design culminated with the development of the LFCM with testing as early as 1972 (Chapman and McElroy 1989). Duratek Inc. and its successors commercialized the LFCM and especially pursued the bubbler technology described later that significantly increased its production rate. The LFCM technology was the basis for the melters used at the WVDP, the DWPF, and the Hanford Waste Treatment and Immobilization Plant (WTP) (Callow et al. 2004, Matlack et al. 2004, Chapman et al. 1986).

The WTP melters can process large volumes of glass of varying composition. Processing rates are enhanced as a result of technological advancements such as bubbler technology (Chapman 2004). For the WTP melters, a slurry melter feed (in the case of both LAW and HLW melter feeds) is continuously charged into the melter on top of the pool of molten glass (Buelt and Chapman 1978,

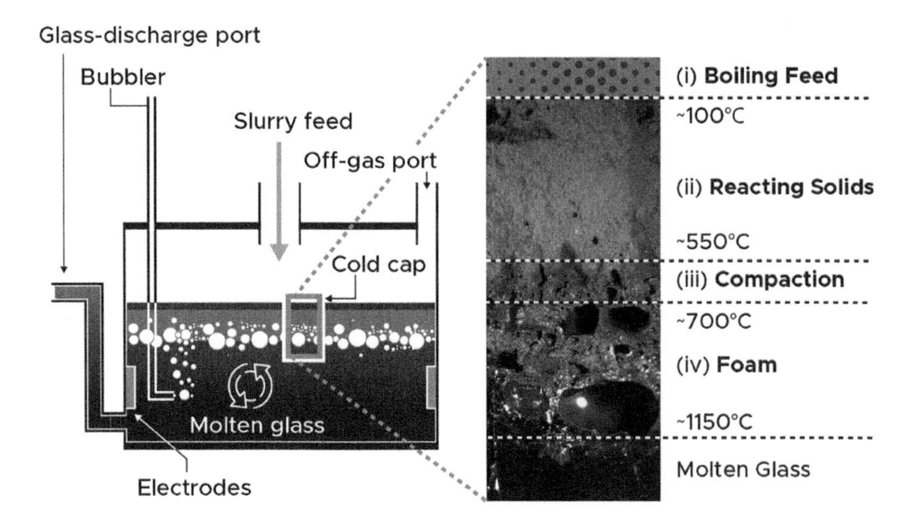


**FIGURE 8.1** Schematic of the WTP melter including an inlay showing the microstructure of a reacting simulated cold cap sample prepared using laboratory-scale experiments. Reprinted with permission from McCarthy et al. (2018).

Barnes and Larson 1981, Chapman and McElroy 1989, Hrma et al. 2019b, McCarthy et al. 2018, Dixon et al. 2022a, Jin et al. 2015, Luksic et al. 2018, George et al. 2020, Xu et al. 2015, Pokorny and Hrma 2012, Abboud et al. 2020, Guillen et al. 2018, Zamecnik et al. 2002, Matlack et al. 2011). The nominal melt pool operating temperature of 1,150°C will be maintained using electrodes immersed in the melt. An electrical current running through the electrodes will travel through the melt causing Joule-heating of the molten glass pool. As melter feed slurry is charged on top of the pool, the cold cap (see the schematic in Figure 8.1) is formed. At the top of the cold cap exists boiling slurry with the bulk of the cold cap being comprised primarily of dried feed at temperatures between 100°C and 450°C with a thin (centimeter-scale) reaction zone at the interface between the cold cap bottom and the molten glass pool. There can exist a primary foam, caused by chemical decomposition reactions of gas-evolving melter feed components such as nitrates, carbonates, and sulfates, and in the underlying glass-forming melt, there can exist a secondary foam, which is caused by oxidation-reduction reactions of transition metals in the glass-forming melt (Hrma et al. 2019a, b). To enhance heat transfer from the melt through the cold cap, air is injected into the melter using a series of bubblers, which force the transfer of heat by convection. With respect to impacts on operational efficiency, the enhanced heat transfer reduces the time to convert the incoming melter feed to a glass and therefore provides a basis for higher glass production rates over nonbubbled systems. The increased production rates ultimately translate into a potential reduction of overall mission life.

## 8.5 WASTEFORM PACKAGING AND CONSIDERATIONS FOR CHEMICAL DURABILITY

During the continuous processing of Hanford radioactive waste in the WTP melter, the final glass product will be discharged into stainless-steel vessels. The dimensions of the LAW containers are approximately 2.3 m tall and 1.2 m in diameter, while the HLW canisters are approximately 4.4 m tall and 0.61 m in diameter (Figure 8.2; Guillen 2014). The different sizes were selected based on the anticipated dimensional constraints of their respective final disposal sites.

However, the dimensions of the canisters/containers can play a role in the resulting microstructure of the final wasteform, which can have implications on long-term performance or durability. As a silicate melt cools in the canister/container from the nominal pour temperature, it will form a solid glass (which is a metastable state) if the cooling rate is sufficiently fast to prevent atomic ordering and hence crystallization. However, if the cooling rate is sufficiently slow, then

**FIGURE 8.2** (a) Image demonstrating the HLW (left) and LAW (right) stainless-steel canisters/containers for storage of vitrified radioactive waste. (b) Image of an HLW canister containing simulated radioactive glass. (c) Image showing an example of the appearance of a borosilicate radioactive waste glass. (Courtesy of A. Kruger).

crystallization or devitrification can occur, which, depending on the type and extent, may impact durability (Jantzen and Bickford 1985). In the large-scale stainless-steel canisters/containers, the slowest cooling will be observed in the center of the canisters/containers, based on measurements of temperature evolution and complementary computational fluid dynamic simulations (Jantzen and Bickford 1985, Amoroso 2011). To bound the potential thermal impact on the glass microstructure, laboratory glasses are subjected to two thermal heat treatments: (i) quenched where molten glass produced in the laboratory is poured directly onto a stainless-steel plate and (ii) canister centerline cooling (CCC) representing the thermal profile along the centerline of the canisters/containers. Both quenched and CCC glasses are then subjected to various physical characterization techniques including assessments of durability to understand the difference, if any, between the two bounding thermal profiles. Based on the knowledge gained from multiple studies, Hanford waste glasses are formulated to avoid known compositional regions where crystallization can have detrimental impacts on durability. One such compositional area is the crystallization of aluminosilicate phases (see Figure 8.3), which can remove highly chemically durable components from the glass ($Al_2O_3$ and $SiO_2$) to form the crystalline phase resulting in a residual glass that is more susceptible to the release of hazardous components (Kim et al. 1995, Riley et al. 2001, Marcial et al. 2016, McCloy et al. 2010, Goel et al. 2019).

## 8.6 WASTEFORM TESTING AND QUALIFICATION

As previously mentioned, one of the most significant challenges faced by Hanford's vitrification process is the compositional complexity of the waste not only within a tank but among tanks. As such, the option of combining waste with glass frit is not optimal for Hanford waste. Recognizing decisions on GFC additions need to be made in advance of processing in the melter, glass formulation requires analysis of the incoming tank waste. Samples of the radioactive waste are compositionally analyzed and that information is used as the technical basis to determine an optimal set of GFC additions to ensure the melter feed can be processed through

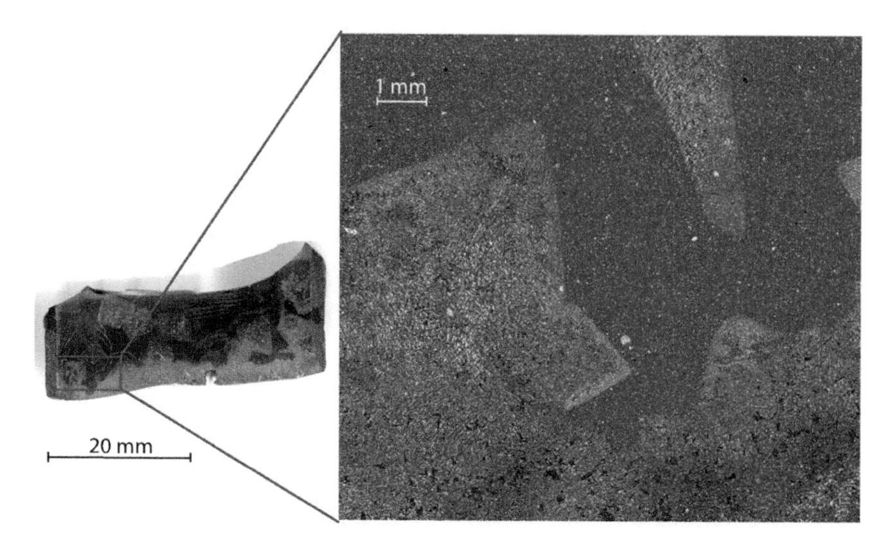


**FIGURE 8.3** (Left) Optical microscope image along the cross-section of a CCC heat-treated HLW simulant glass sample and (right) scanning electron microscope–backscattered electron (SEM-BSE) image of crystals in the microstructure of the sample, the large triangular features are nepheline, sodium aluminosilicate crystal, while the small dots are transition metal spinel. (Courtesy of M. J. Schweiger and J. S. McCloy.)

the melter and result in a glass wasteform that satisfies durability and regulatory requirements. The process of determining the type and amount of GFCs (mostly minerals plus boric acid, lithium carbonate, and/or sodium carbonate) is based on a computer algorithm for both LAW (Kim et al. 2012) and HLW (Vienna and Kim 2023). The computational output of the type and amount is based on a delicate balance of the impact each GFC has on the various processes and durability properties of interest. To complicate matters, GFCs will have competing effects driving specific properties in opposite directions that require a fundamental understanding of the glass composition – property relationships that are developed through laboratory studies in advance of radioactive operations (Marcial et al. 2024). For example, silica is added to enhance chemical durability but will increase viscosity; boron and sodium enhance processibility but can reduce durability depending on concentration; lithium enhances thermal conductivity but can also reduce durability; and vanadium improves sulfur solubility. The competing effects of not only the GFCs but also the composition of the waste must be accounted for in the algorithms to target an acceptable melter feed.

With respect to chemical durability, several laboratory-scale tests are available to assess this critical property (Thorpe et al. 2021). For Hanford HLW and LAW glasses, the seven-day product consistency test (PCT) (ASTM 2021) and the toxicity characteristic leaching procedure (TCLP, EPA 1992) are used to evaluate the acceptability of the glass. The vapor hydration test (VHT) (ASTM 2018) response is also required to satisfy constraints for Hanford LAW glass. In general, for the PCT procedure, quenched or heat-treated glass is crushed, washed, then sieved to obtain a specific particle size range (75–150 μm). Figure 8.4 provides an example SEM image of glass powder before alteration (ASTM 2021). The powdered glass is placed in deionized water to achieve a desired solid surface area-to-liquid volume ratio of $2000\,m^{-1}$ and then transferred into a 90°C oven for seven days.[2] After the seven-day period, aliquots from the leachate solution are taken, acidified (to prevent precipitation), and analyzed, often by inductively coupled plasma-optical emission spectroscopy (ICP-OES). The analytical results are used to determine the elemental release rates of key glass components, which are then compared to the release rates of a glass standard to determine acceptability. The normalized PCT responses of B, Na, and Li must be below those of the DWPF Environmental Assessment glass for HLW glasses, and the

**FIGURE 8.4** (a) Optical image of VHT sample before and after alteration at 200°C and (b) SEM-BSE image of glass powder prior to seven-day PCT. (Courtesy of A. Kruger.)

normalized releases of B and Na must be below 2 g/m$^2$ for the Hanford LAW glass (Jantzen et al. 1992, U.S. Department of Energy 2000).

In the TCLP, the waste form is crushed and passed through a screen with a 9.5-mm aperture. The material is added to a bottle full of a slightly acidic extraction fluid in a 1:20 mass ratio and rotated end over end for 18 hours at room temperature. The resulting solution is filtered, acidified, and analyzed for hazardous components using ICP. The concentrations of hazardous components must satisfy regulatory requirements to delist the HLW glass or satisfy land disposal restrictions for the LAW glass (cite delisting and LDR petitions). In the VHT procedure, quenched or heat-treated glass is cut into a coupon of dimensions roughly 10 mm × 10 mm × 1.5 mm, polished to 600 grit (30 µm surface finish), then suspended in a stainless-steel vessel using platinum wire. The test is then performed at 200°C in a water saturated vessel with sufficient excess water to corrode the glass, yet not enough to cause reflux. VHT experiments are often performed for between 7 and 24 days. After termination of the VHT alteration experiment, samples are bisected, and the alteration layer thickness is measured. For LAW glass, the PCT limit is 2 g/m$^2$ and the VHT limit is 50 g/m$^2$/d (U.S. Department of Energy 2000).

## 8.7 PROCESS TESTING

Although the development of composition-property models and an integrated algorithm have provided the technical basis for targeting specific glasses or the acceptable glass composition region, the means of predicting melt rate or cold cap behavior are relatively immature. The recent work described by Lee et al. (2021) provides an empirical correlation for melt rate based on results from scaled melter tests using simulated Hanford tank waste feed. The correlation helps with understanding the chemistry of the cold cap and the parameters affecting melt rate, but the chemical and physical complexities mean a theoretical understanding is not yet complete, and there remains a need for process testing.

Melting rate (or the time to convert the incoming melter feed to glass) and the dynamic behavior of the cold cap (e.g., formation of a foam layer) can have significant negative impacts on production rates, as well as radionuclide retention. Therefore, since the development of the LFCM in the 1970s and the selection of borosilicate as the primary wasteform for Hanford waste in the early 1980s, the need to assess the impact of melter feed composition on key factors influencing full-scale processing and component retention has been a major focus. Research in this area seeks to provide a fundamental understanding of the complex reactions that occur within the cold cap, and with that knowledge, operators may be able to avoid critical processing issues (e.g., foam formation, killer scum, melter feed mounding), which can have detrimental impacts on full-scale operations. Crucible testing can be useful for investigating the chemical reactions of vitrification, particularly in tracing the pathways of incorporation in the glass matrix for radionuclides of interest and their surrogates (Matlack et al. 2010, 2011, Jin et al. 2014, 2015, 2020, Luksic et al. 2016, 2018, George et al. 2019, 2020, Xu et al. 2015). Collaborating researchers from PNNL, Idaho National Laboratory, and University of Chemistry and Technology Prague have developed crucible-scale experimental methods and computational models to correlate the chemical behavior data gained with crucible-scale methods to known average processing rates in large-scale melters using the "melting rate correlation equation," which was utilized to build computational fluid dynamic models of the melter cold cap (Pokorny and Hrma 2012, 2014, Abboud et al. 2020, 2021, Guillen et al. 2018, 2020). Limitations arise when crucible testing is used to directly assess the dynamic process of vitrification, and as such, scaled melters are helpful for the continued testing of melter feeds.

Test melters have been used to evaluate Hanford melter feed processing since the turn of the century (Matlack et al. 1997, 2000, 2001), and they range in size from 1/3 scale (the DuraMelter 3300 (DM3300) system operated by Duratek at their Maryland facility (Matlack 2004)) to around 1/1,000 scale like both the Large C system previously operated by Savannah River Technology Center (Zamecnik et al. 2002) and the continuous laboratory-scale melter (CLSM) operated by PNNL (Riley et al. 2009, Dixon et al. 2020a, 2022a). Due to the advantage of the smaller melter size, the Large C system and the CLSM have processed actual Hanford tank waste (Zamecnik et al. 2002, Dixon et al. 2018, 2019, 2020b, 2022b, 2022c) while achieving the appropriate chemical steady state via turnovers of glass inventory in the systems, thus generating valuable data about the retention of radionuclides, like $^{99}$Tc, contained in low quantities in the actual LAW.

The various scaled melter systems relative to the WTP melters are shown in Figure 8.5. These systems have been valuable for generating dynamic production data for melter feeds, determining the retention factors for radionuclides (or surrogates) in the type of environments that will be experienced during waste vitrification, and assessing the impact of offgas condensate recycling on the retention factors. When the scaled melters are connected to offgas systems prototypic of the WTP vitrification facilities, decontamination factors around different units in the melter or offgas systems can be calculated from the data collected during processing (Matlack et al. 2011, 2012, 2018).

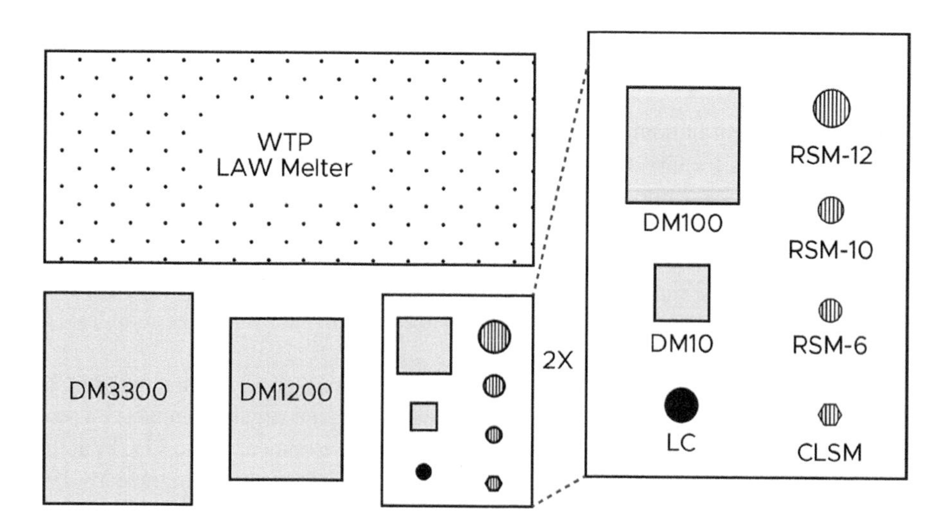


**FIGURE 8.5** Relative size of the Waste Treatment and Immobilization Plant low-activity waste (WTP LAW) melter and scaled melter vessels, defined as the DuraMelter (DM), Large C (LC), research-scale melter (RSM), and continuous laboratory-scale melter (CLSM). Reprinted with permission from Dixon et al. (2022a).

## 8.8 FUTURE DEVELOPMENTS FOR WASTE FORM QUALIFICATION AND TESTING

Based on current trends in research and design, future glass formulation efforts seek to expand the processing flexibility and improve the efficiency of the WTP using developments in modern materials science tools including machine learning. These growing methods could be used to enhance the statistical design of glass wasteforms, assess gaps in the available composition-property models, and develop new glass compositions capable of enhancing glass properties such as chemical durability and waste loading. Along with modeling efforts, benchtop-scale tests are being performed with the aim of developing correlations capable of estimating the production rates within large-scale facilities. Due to the quantity and complexity of LAW and HLW at the Hanford Site, it is anticipated that computer modeling, benchtop-scale tests, and scaled melter work will continue in the effort to enhance waste loading, enhance the compositional operating window, and address emerging issues as the WTP moves through start-up and operations. The effect of such research can be difficult to quantify, but is seen in budget and schedule improvements, like those proposed in the WTP system plan (Bernards et al. 2017). Additionally, enhanced operational windows may translate into a broader range of potential tank waste processing flowsheet options, higher waste loading per unit of glass, and increased operational efficiency.

## NOTES

1 At the time of writing, the terms LAW and HLW at Hanford carry regulatory significance and the waste is not formally classified from a regulatory perspective until it is fully treated or immobilized.
2 Note: The parameters listed (90°C for seven days) are specific to the product consistency test method A (PCT-A) under ASTM C1285-21.

## REFERENCES


Abboud, A.W., G.P. Guillen, W.C. Eaton, D.R. Dixon, M. Hall, and A.A. Kruger. (2021). An artificial neural network with computational fluid dynamics for estimating cold-cap coverage in a laboratory-scale melter, *WM2021 Conference, Paper 21174*, Phoenix, AZ.

Abboud, A.W., D.P. Guillen, and R. Pokorny. (2020). Effect of cold cap coverage and emissivity on the plenum temperature in a pilot-scale waste vitrification melter. *Int. J. Appl. Glass Sci.* 11, 357–368.

Agnew, S.F., J. Boyer, R.A. Corbin, T.B. Duran, J.R. FitzPatrick, K.A. Jurgesen, T.P. Ortiz, and B.L. Young. (1997). *Hanford Tank Chemical and Radionuclide Inventories HDW Model.* LA-UR-96-3860, Rev 4. Los Alamos National Laboratory, Los Alamos, NM.

Alternative Waste Form Peer Review Panel. (1979). *The Evaluation and Review of Alternate Waste Forms for Immboliziation of High-Level Radioactive Wastes.* DOE/TIC-10228. U.S. Department of Energy, Washington, DC.

Alternative Waste Form Peer Review Panel. (1980). *The Evaluation and Review of Alternate Waste Forms for Immboliziation of High-Level Radioactive Wastes.* Report number 2 DOE/TIC-11219. U.S. Department of Energy, Washington, DC.

Alternative Waste Form Peer Review Panel. (1981). *The Evaluation and Review of Alternate Waste Forms for Immboliziation of High-Level Radioactive Wastes.* Report Number 3 DOE/TIC-11472. U.S. Department of Energy, Washington, DC.

American Society for Testing and Materials. (2018). *ASTM C1663-18 Standard Test Method for Measuring Glass or Ceramic Durability by Vapor Hydration Test.* ASTM International, West Conshohocken, PA.

American Society for Testing and Materials. (2021). *ASTM C1285-21, Standard Test Methods for Determining Chemical Durability of Nuclear, Hazardous, and Mixed Waste Glasses and Multiphase Glass Ceramics: The Product Consistency Test (PCT).* ASTM International, West Conshohocken, PA.

Amoroso, J. (2011). *Computer Modeling of High-level Waste Glass Temperatures within DWPF Canisters during Pouring and Cool Down.* SRNL-STI-2011-00546. Savannah River National Laboratory, Aiken, SC.

Barnes, S.M., and D.E. Larson. (1981). *Materials and Design Experience in a Slurry-Fed Electric Glass Melter.* PNL-3959. Pacific Northwest Laboratory, Richland, WA.

Bernards, J.K., T.M. Hohl, R.T. Jasper, S.L. Orcutt, S.D. Reaksecker, C.S. Smalley, A.J. Schubick, T.L. Waldo II, E.B. West, L.M. Bergmann, R.O. Lokken, A.N. Praga, S.N. Tilanus, and M.N. Wells. (2017). *River Protection Project System Plan*. ORP-11242, Rev. 8. U.S. Department of Energy Office of River Protection, Richland, WA.

Buelt, J.L., and C.C. Chapman. (1978). *Liquid-Fed Ceramic Melter: A General Description Report*. PNL-2735. Pacific Northwest Laboratory, Richland, WA.

Callow, R.A., K.S. Matlack, and I.L. Pegg. (2004). *Comparison of Off-Gas Emissions from Tests with LAW Simulants on the DM100, DM1200, and DM3300 Melters*. VSL-04S4850-1. Vitreous State Laboratory, The Catholic University of America, Washington, DC.

Chapman, C.C. (2004). *Investigation of Glass Bubbling and Increased Production Rate*. REP-RPP-069, Revision 0. Duratek Proprietary Data, Washington, DC.

Chapman, C.C., Y.B. Katayama, J.L. Buelt, L.R. Bunnell, and S.C. Slate. (1979). *Vitrification of Hanford Wastes in a Joule-Heated Ceramic Melter and Evaluation of Resultant Canisterized Product*. PNL-2904. Pacific Northwest Laboratory, Richland, WA.

Chapman, C.C., and J.L. McElroy. (1989). *Slurry-Fed Ceramic Melter-A Broadly Accepted System to Vitrify High-Level Waste*. PNL-SA-16209. Pacific Northwest National Laboratory, Richland, WA.

Chapman, C.C., J.M. Pope, and S.M. Barnes. (1986). Electric melting of nuclear waste glasses state of the art. *J. Non-Cryst.* Solids 84, 226–240.

Deng, Y., R.C. Chen, R. Gimpel, B. Slettene, M.R. Gross, K. Jun, and R. Fundak. (2016). *Flowsheet Bases, Assumptions, and Requirements*. 24590-WTP-RPT-PT-02-005, Rev. 8. Waste Treatment Plant, Richland, WA.

DesCamp, V.A. and C.L. McMahon. (1996). *Vitrification Facility at the West Valley Demonstration Project*. DOE/NE/44139-77. West Valley Nuclear Services Co., Inc., West Valley, NY.

Dixon, D.R., M.A. Hall, J.B. Lang, D.A. Cutforth, C.M. Stewart, and W.C. Eaton. (2020a). *Continuous Laboratory-Scale Melter Runs for System Evaluation*. PNNL-30073, Rev. 0 (EWG-RPT-024, Rev. 0). Pacific Northwest National Laboratory, Richland, WA.

Dixon, D.R., M.A. Hall, J.B. Lang, D.A. Cutforth, C.M. Stewart, and W.C. Eaton. (2022a). Retention analysis from vitrified low-activity waste and simulants in a laboratory-scale melter. *Ceramics Int.* 48, 5955–5964. DOI:10.1016/j.ceramint.2021.08.102.

Dixon, D.R., J.B. Lang, M.A. Hall, C.M. Stewart, D.A. Cutforth, W.C. Eaton, J. Marcial, A.M. Westesen, and R.A. Peterson. (2022b). *Vitrification of Hanford Tank 241-AP-101 Waste and Simulant*. PNNL-33600, Rev. 0 (RPT-DFTP-035, Rev. 0). Pacific Northwest National Laboratory, Richland, WA.

Dixon, D.R., C.M. Stewart, J.J. Venarsky, J.A. Peterson, G.B. Hall, T.G. Levitskaia, J.R. Allred, W.C. Eaton, J.B. Lang, M.A. Hall, D.A. Cutforth, A.M. Rovira, and R.A. Peterson. (2018). *Vitrification of Hanford Tank Waste 241-AP-105 in a Continuous Laboratory-Scale Melter*. PNNL-27775 (RPT-DFTP-010, Rev. 0). Pacific Northwest National Laboratory, Richland, WA.

Dixon, D.R., C.M. Stewart, J.J. Venarsky, J.A. Peterson, G.B. Hall, T.G. Levitskaia, J.R. Allred, W.C. Eaton, J.B. Lang, M.A. Hall, D.A. Cutforth, A.M. Rovira, and R.A. Peterson. (2019). *Vitrification of Hanford Tank Waste 241-AP-107 in a Continuous Laboratory-Scale Melter*. PNNL-28361, Rev. 0 (RPT-DFTP-014, Rev. 0). Pacific Northwest National Laboratory, Richland, WA.

Dixon, D.R., A.M. Westesen, M.A. Hall, C.M. Stewart, J.B. Lang, D.A. Cutforth, W.C. Eaton, and R.A. Peterson. (2020b). *Vitrification of Hanford Tank 241-AP-107 with Recycled Condensate*. PNNL-30189, Rev. 0 (RPT-DFTP-024, Rev. 0). Pacific Northwest National Laboratory, Richland, WA.

Dixon, D.R., A.M. Westesen, M.A. Hall, C.M. Stewart, J.B. Lang, D.A. Cutforth, W.C. Eaton, and R.A. Peterson. (2022c). *Vitrification of Hanford Tank Wastes for Condensate Recycle and Feed Composition Changeover Testing*. PNNL-32344, Rev. 1 (RPT-DFTP-033, Rev. 1). Pacific Northwest National Laboratory, Richland, WA.

George, J.L., P. Cholsaipant, D.-S. Kim, T.G. Levitskaia, M.S. Fujimoto, I.E. Johnson, and A.A. Kruger. (2019). Effect of sulfate sequestration by Ba-Sn composite material on Re retention in low-activity waste glass. *J. Non-Cryst. Solids* 510, 151–157.

George, J.L., D.-S. Kim, and A.A. Kruger. (2020). Effects of iron oxalate on rhenium incorporation into low-activity waste glass. *J. Non-Cryst. Solids* 545, 120257.

Goel, A., J. McCloy, R. Pokorny, and A.A. Kruger. (2019). Challenges with vitrification of Hanford High-Level Waste (HLW) to borosilicate glass - An overview. *J. Non-Cryst. Solids X* 4, 1–19.

Guillen, D.P. (2014). Thermal predictions of the colling of waste glass canisters. *Proceedings of 2014 American Nuclear Society Winter Meeting, INL/CON-14-32319*, Idaho National Laboratory, Idaho Falls, ID.

Guillen, D.P., A.W. Abboud, R. Pokorny, M. Hall, W.C. Eaton, and A.A. Kruger. (2020). Computationally aided design of a small-scale waste-glass melter, *WM2020 Conference*, Paper 20452, Phoenix, AZ.

Guillen, D.P., J. Cambareri, A.W. Abboud, and I.A. Bolotnov. (2018). Numerical comparison of bubbling in a waste glass melter. *Ann. Nucl. Energy* 113, 380–392.

Hay, B.D. (2022). *Waste Tank Summary Report for Month Ending in April 30, 2022.* HNF-EP-0182 Rev. 412. Washington River Protection Solutions, Richland, WA.

Hrma, P., J. Klouzek, R. Pokorny, S. Lee, and A.A. Kruger. (2019a). Heat transfer from glass melt to cold cap: Gas Evolution and foaming. *J. Am. Ceram. Soc.* 102(10), 5853–5865.

Hrma, P., R. Pokorny, S. Lee, and A.A. Kruger. (2019b). Heat transfer from glass melt to cold cap: Melting rate correlation equation. *Int. J. Appl. Glass Sci.* 10(2), 143–150.

Jantzen, C.M., N.E. Bibler, D.C. Beam, C.L. Crawford, and M.A. Pickett. (1992). *Characterization of the Defense Waste Processing Facility (DWPF) Environmental Assessment (EA) Glass Standard Reference Material (U).* WSRC-TR--92-346-Rev. 1. Westinghouse Savannah River Co., Savannah River Site, Aiken, SC.

Jantzen, C.M., and D.E. Bickford. (1985). Leaching of devitrified glass containing simulated srp nuclear waste, *Materials Research Society Proceedings*, Pittsburgh, PA, 1985, Vol. 44, pp. 135–146.

Jin, T., D.-S. Kim, and M. Schweiger, (2014). Effect of sulfate on rhenium partitioning during melting of low-activity-waste glass feeds, *WM2014 Conference, Paper 14116*, Phoenix, AZ.

Jin, T., D.-S. Kim, A.E. Tucker, M.J. Schweiger, and A.A. Kruger. (2015). Reactions during melting of low-activity waste glasses and their effects on the retention of rhenium as a surrogate for technetium-99. *J. Non-Cryst. Solids* 425, 28–45.

Jin, T., D. Mar, D. Kim, B.L. Weese, and A.A. Kruger. (2020). Reactions during conversion of simplified low-activity waste glass feeds. *Thermochim. Acta* 694, 178783Kim, D.S., J.D. Vienna, and A.A. Kruger. (2012). *Preliminary ILAW Formulation Algorithm Description.* 24590 LAW RPT-RT-04-0003, Rev. 1. ORP-56321, Revision 0. U. S. Department of Energy Office of River Protection, Richland, WA.

Kruger, A.A. (2013). Advances in glass formulations for hanford high-aluminum, high-iron and enhanced sulphate management in HLW streams – 13000, *Proceedings for Waste Management Conference*, Vol. 1, pp. 1–15. Phoenix, AZ.

Lee, S., D.A. Cutforth, D. Mar, J. Klouzek, P. Ferkl, D.R. Dixon, R. Pokorny, M.A. Hall, W.C. Eaton, P. Hrma, and A.A. Kruger. (2021). Melting rate correlation with batch properties and melter operating conditions during conversion of nuclear waste melter feeds to glasses. *Int. J. Appl. Glass Sci.* 2021, 398–414.

Luksic, S.A., D.-S. Kim, W. Um, G. Wang, M.J. Schweiger, C.Z. Soderquist, W. Lukens, and A.A. Kruger. (2018). Effect of Technetium-99 sources on its retention in low activity waste glass. *J. Nucl. Mater.* 503, 235–244. DOI:10.1016/j.jnucmat.2018.02.019.

Luksic, S.A., B.J. Riley, K.E. Parker, and P. Hrma. (2016). Sodalite as a vehicle to increase Re retention in waste glass simulant during vitrification. *J. Nucl. Mater.* 479, 331–337.

Lutze, W. and R.C. Ewing. (1988). *Radioactive Waste Forms for the Future.* North-Holland Physics Publishing, Amsterdam, The Netherlands.

Marcial, J., J. Crum, O. Neill, and J. McCloy. (2016). Nepheline structural and chemical dependence on melt composition. *Amer. Mineral.* 101(2), 266–276.

Marcial, J., B.J. Riley, A.A. Kruger, C.E. Lonergan, and J.D. Vienna. (2024). Hanford low-activity waste vitrification: A review. *J. Haz. Mat.* 461, 132437.

Matlack, K.S., H. Abramowitz, M. Brandys, I.S. Muller, R.A. Callow, N. D'Angelo, R. Cecil, I. Joseph, and I.L. Pegg. (2012). *Technetium Retention in WTP LAW Glass with Recycle Flow-Sheet: DM10 Melter Testing.* VSL-12R2640-1, Rev. 0. Vitreous State Laboratory, The Catholic University of America, Washington, DC.

Matlack, K.S., H. Abramowitz, I.S. Muller, I. Joseph, and I.L. Pegg. (2018). *DFLAW Glass and Feed Qualifications for AP-107 to Support WTP Start-Up and Flow-Sheet Development.* VSL-18R4500-1, Rev. 0. Vitreous State Laboratory, The Catholic University of America, Washington, DC.

Matlack, K.S., W. Gong, T. Bardakci, N. D'Angelo, P.M. Bizot, R.A. Callow, M. Brandys, and I.L. Pegg. (2004). Bubbling Rate and Foaming Tests on the DuraMelter 1200 with LAWC22 and LAWA30 Glasses. VSL-04R4851-1. Vitreous State Laboratory, The Catholic University of America, Washington, DC.

Matlack, K.S., S.P. Morgan, and I.L. Pegg. (2000). *Melter Tests with LAW Envelope B Simulants to Support Enhanced Sulfate Incorporation.* VSL-00R3501-1, Rev. 0. Vitreous State Laboratory, The Catholic University of America, Washington, DC.

Matlack, K.S., S.P. Morgan, and I.L. Pegg. (2001). *Melter Tests with LAW Envelope A and C Simulants to Support Enhanced Sulfate Incorporation.* VSL-01R3501-2, Rev. 0. Vitreous State Laboratory, The Catholic University of America, Washington, DC.

Matlack, K.S., I.S. Muller, R.A. Callow, N. D'Angelo, T. Bardakci, I. Joseph, and Pegg, I.L. (2011). *Improving Technetium Retention in Hanford LAW Glass - Phase 2.* VSL-11R2260-1, Rev. 0, Vitreous State Laboratory, The Catholic University of America, Washington, DC.

Matlack, K.S., I.S. Muller, S.S. Fu, and I.L. Pegg. (1997). *Results of Melter Tests Using TWRS LAW Envelope B Simulants.* Final Report, Rev. 0. Vitreous State Laboratory, The Catholic University of America, Washington, DC.

Matlack, K. S., W. Gong, and P. Glass. (2004). *Formulation testing to increase sulfate volatilization.* Vitreous State Laboratory, Catholic University of America, Washington DC

Matlack, K.S., I.S. Muller, I. Joseph, and I.L. Pegg. (2010). *Improving Technetium Retention in Hanford LAW Glass - Phase 1.* VSL-10R1920-1, Rev. 0. Vitreous State Laboratory, The Catholic University of America, Washington, DC.

McCarthy, B.P., J.L. George, D.R. Dixon, M. Wheeler, D.A. Cutforth, P. Hrma, D. Linn, J. Chun, M. Hujova, A.A. Kruger, and R. Pokorny. (2018). Rheology of simulated radioactive waste slurry and cold cap during vitrification. *J. Am. Ceram. Soc.* 101, 5020–5029.

McCloy, J.S., C. Rodriguez, C. Windisch, C. Leslie, M.J. Schweiger, B.R. Riley, and J.D. Vienna. (2010). Alkali/akaline-earth content effects on properties of high-alumina nuclear waste glasses. In American Ceramic Society (ed.), *Advances in Materials Science for Environmental and Nuclear Technology.* John Wiley & Sons, Inc., Hoboken, NJ, pp. 63–76.

Ojovan, M.I. and W.E. Lee. (2005). *An Introduction to Nuclear Waste Immobilisation.* Elsevier Ltd., Oxford, UK.

Petkus, L.L., J. Paul, P.J. Valenti, H. Houston, and J. May. (2003). A complete history of the high-level waste plant at the west valley demonstration project, *Waste Management Conference* 2013, Abstract #244, Tucson, AZ.

Pokorny, R. and P. Hrma. (2012). Mathematical modeling of cold cap. *J. Nucl. Mater.* 429(1–3), 245–256.

Pokorny, R. and P. Hrma. (2014). Model for the conversion of nuclear waste melter feed to glass. *J. Nucl. Mater.* 445(1–3), 190–199.

Riley, B.J., J.V. Crum, W.C. Buchmiller, B.T. Rieck, M.J. Schweiger, and J.V. Vienna. (2009). *Initial Laboratory-Scale Melter Test Results for Combined Fission Product Waste.* PNNL-18781. Pacific Northwest National Laboratory, Richland, WA.

Riley, B.J., P. Hrma, J. Rosario, and J.D. Vienna. (2001). Effect of crystallization on HLW glass corrosion. *Ceram. Trans.* 132, 257–265.

Rodgers, M.J. 2023. *Waste Tank Summary Report for Month Ending* June 30, *2023.* HNF-EP-0182, Revision 426. Washington River Protection Solutions, Richland, WA.

Thorpe, C.L., J.J. Neeway, C.I. Pearce, R.J. Hand, A.J. Fisher, S.A. Walling, N.C. Hyatt, A.A. Kruger, M. Schweiger, D.S. Kosson, C.L. Arendt, J. Marcial, and C.L. Corkhill. (2021). Forty years of durability assessment of nuclear waste glass by standard methods. *npj Mater. Degrad.* 5(1), 61.

U.S. Department of Energy. (1982a). *Long-Term Management of Liquid High-Level Radioactive Waste at the Western New York Nuclear Service Center.* DOE/EIS-0081. West Valley; Record of Decision, West Valley, NY.

U.S. Department of Energy. (1982b). *Environmental Impact Statement Defense Waste Processing Facility Savannah River Plant.* DOE/EIS-0082. US Department of Energy, Washington DC.

U.S. Department of Energy. (1988). *Disposal of Hanford Defense High-Level, Transuranic, and Tank Wastes.* DOE/EIS-0113. Hanford Site, Richland, WA; Record of Decision (ROD), Richland, WA.

U.S. Department of Energy. (1995). *Record of Decision; Defense Waste Processing Facility at the Savannah River Site.* DOE/EIS-0082 S1. US Department of Energy, Washington DC

U.S. Department of Energy. (1997). *Record of Decision for the Tank Waste Remediation System, Hanford Site.* DOE/EIS-0189. US Department of Energy, Washington DC.

U.S. Department of Energy. (2000). *Design, Construction, and Commissioning of the Hanford Tank Waste Treatment and Immobilization Plant.* Contract DE-AC27-01RV14136, as amended. U.S. Department of Energy, Office of River Protection, Richland, WA.

U.S. Department of Energy. (2004). *Solid Waste Program,* Hanford Site, Richland, WA*: Storage and Treatment of Low-Level Waste and Mixed Low-Level Waste; Disposal of Low-Level Waste and Mixed Low-Level Waste, and Storage, Processing, and Certification of Transuranic Waste for Shipment to the Waste Isolation Pilot Plant.* DOE/EIS-0286. US Department of Energy, Richland, WA.

US Environmental Protection Agency Method 1311, Revision I, 1992, Final Update I to the Third Edition of the Test Methods for Evaluating Solid Waste, Physical/Chemical Methods, EPA publication SW-846.

Vienna, J.D., and D. Kim. (2023). *Preliminary IHLW Formulation Algorithm Description.* 24590-HLW-RPT-RT-05-001, Rev. 1. U. S. Department of Energy Office of River Protection, Richland, WA.

Xu, K., D.A. Pierce, P. Hrma, M.J. Schweiger, and A.A. Kruger. (2015). Rhenium volatilization in waste glasses. *J. Nucl. Mater.* 464, 382–388, DOI:10.1016/j.jnucmat.2015.05.005.

Zamecnik, J.R., C.L. Crawford, and D.C. Koopman. (2002). *Large Scale Vitrification of 241-AN-102 (Envelope C) Sample.* WSRC-TR-2002-00093, Rev. 0. Westinghouse Savannah River Company, Aiken, SC.

# 9 Plutonium Finishing Plant Demolition and Interim Subsurface Disposal Structure Stabilization

*Calvin H. Delegard, Carolyn I. Pearce, Hilary P. Emerson, Andrea M. Hopkins, and Theodore J. Venetz*

## 9.1 PLUTONIUM FINISHING PLANT BUILDING DE-INVENTORY, DEACTIVATION, DECOMMISSIONING, AND DEMOLITION

With the collapse of the Soviet Union in late 1991, U.S. demand for continued defense Pu production disappeared. This eliminated the need for the Plutonium Finishing Plant (PFP), for which operations had already ceased since June 1989 (see Section 4.1 in Chapter 4.1 for further description). The PFP was left in operational standby at the effective end of its processing and production mission in 1989 with significant Pu-bearing scrap, product-quality Pu oxide and metal inventories, and associated process chemical holdings. Without a production mission, the process lines fell into disrepair while radiological control, Pu inventory maintenance, environmental monitoring, and physical security measures remained in place with their attendant costs. Accordingly, in October 1996, the DOE called for PFP's permanent shutdown and, by extension, its decommissioning (Hebdon et al. 2003). With the DOE decision, negotiations occurred among the concerned parties – the DOE, the U.S. Environmental Protection Agency (EPA), and the Washington Department of Ecology (Ecology) – under the Tri-Party Agreement (TPA) described in Chapter 1. However, the first step, disposition and stabilization of the Pu-bearing nuclear materials, a DOE responsibility, led to disputes and an impasse among the TPA parties in 1997. This deadlock resulted in poor publicity, work delays, and further relational erosion. Collaborative negotiations in 2000–2001 among the TPA members built trust, established agreement, and effectively allowed cleanup to move forward (Hebdon et al. 2003).

The PFP was a Hazard Category II nuclear facility, defined by the DOE as a non-reactor nuclear facility with the potential for a nuclear criticality[1] event, because of its extensive inventory of radioactive and fissile plutonium. The plutonium itself also represented a valuable national resource, obtained at great cost. Accordingly, preparations began in the early 1990s to itemize, stabilize, package, and remove the solid and solution Pu inventory held within the PFP. The initial characterization and assessment of the PFP and its contents are described in Chapter 4, Section 1.6. This was followed by listing the hazards and Pu-bearing residues, decontaminating, to the extent possible, the PFP process equipment, and packaging and disposing of it as non-transuranic (non-TRU) waste. It was then necessary to prepare the facility itself for decommissioning by emptying pipelines and ductwork and fixing the Pu-bearing surface contamination. Finally, the process entailed demolishing the structure, removing and burying the debris, and stabilizing the remaining brownfield. This sequence of de-inventory, deactivation, decommissioning, and demolition can be envisioned as a snake eating its tail in that the facility itself was used to effect its own annihilation. The process was often iterative as additional hazards, legacy Pu-bearing scraps, or wastes[2] were identified or regulations changed over time.

DOI: 10.1201/9781003329213-12

To "consume" the Pu-bearing inventory and then decontaminate or fix the radioactive contamination in the process equipment, ductwork, piping, walls, and floors before the building could be demolished, the PFP had to retain, as needed, operational gloveboxes, material processing capabilities, ventilation, water, electrical service, radiation protection, physical security, and qualified personnel who would be working themselves out of their jobs. The PFP and associated site structures were vast and complex, encompassing ~23,000 $m^2$ of Pu-processing floorspace. Their contents included the following: 231 gloveboxes, hoods, and conveyor spaces; 52 long (up to 7.2 m) but small-diameter (~15-cm) critically safe bank tanks (later dubbed "pencil" tanks), largely in the Plutonium Reclamation Facility (PRF), which housed solution storage and solvent extraction columns; four remote process cells (large enclosed spaces, most notably, the PRF canyon, an ~$9.8 \times 16.5 \times 9.8$-m room with a "penthouse" extending a further three glove-boxed stories to accommodate the tall solvent extraction columns); four major chemical storage areas; and ~2.7 km of contaminated exhaust ductwork (Charboneau et al. 2006a; Swartz 2016). Additional historical processing details are included in Section 4.1 of Chapter 4 for reference.

The deactivation and decommissioning of a facility of this nature was unprecedented, and challenges were identified in the following areas:

1. Information management:
   i. Information on material, legacy Pu-bearing scrap, and wastes required documentation spanning 40+ years; facilities were in multiple phases of characterization and assessment, with some facilities still operational and some having been shut down for 30+ years.
   ii. Information about some material, scrap, and wastes was classified and thus not readily available to plan and conduct work.
2. Hazardous material characterization:
   i. Characterization of hazardous material, both chemical and radiological, presented technical challenges, hazards, and risks to facility operators.
   ii. Parts of the PFP were required to remain operational not only for characterization but also for processing and treatment of Pu-bearing materials prior to dispositioning.
3. Regulatory limits:
   i. Regulations were still being established for cleanup and dispositioning of these complex materials, and the science of risk assessment for radionuclides was refining over time.[3]

To achieve these ends, milestones for the complex PFP decommissioning activities were integrated into the wider TPA by collaborative negotiations in 2001–2002. The TPA parties prepared removal action work plans to ensure safe and compliant decommissioning work under applicable, relevant, and appropriate requirements (ARARs) while following budget and schedule (Hopkins 2008). The agreed-upon approach addressed requirements for both DOE decommissioning and the Comprehensive Environmental Response, Compensation, and Liability Act (CERCLA) while monitoring health and safety, data management, cost and schedule, and end-point completion and verification. This early collaboration provided the regulatory and administrative framework for the successful completion of PFP building removal in late 2021, that, though lengthy, was a significant feat as PFP was considered one of the most hazardous and complicated buildings within the DOE Environmental Management Complex (DOE-EM 2022).

### 9.1.1 De-Inventory of Pu

About 8,000 packages of Pu-bearing scrap (impure solids, sludges, solutions, and pure materials) remained in the PFP after its 1989 shutdown, with some of these packages containing numerous smaller packages within them. For example, a 7-gallon can that was accounted for as a single item might hold 12–15 one-quart containers of slag and crucible scrap of varying quantities, highlighting the complexity in documenting these materials. PFP operations applied existing plant facilities, new rudimentary Pu chemical precipitation facilities, and furnaces to run 950°C to convert the higher concentration materials (i.e., containing >30 wt% Pu, dry basis) to stable oxide and metal forms (Charboneau et al. 2006a) and refined analytical measurements methods and specialized Pu packaging capabilities to prepare them according to evolved DOE "3013" standards for stabilization, packaging, and long-term storage (DOE-STD-3013; DOE 2018).

The conversion of Pu-rich solutions to form stable Pu oxide used new simple Pu oxalate precipitation facilities followed by hotplate roasting and then furnace calcination of the oxalate solids to 950°C. The 950°C furnace treatment also was used to stabilize solid Pu oxide scrap under the 3013 strictures. Pu metal pieces had to be greater than 50 g in mass and brushed free of Pu oxide to meet the 3013 criteria; smaller pieces were burned to Pu oxide. Unique Pu-rich scrap from "polycubes" (Pu oxide mixed in polystyrene and cast into orthogonal shapes; Figure 9.1) and chloride salt–bearing pyrochemical scrap required alternative processes. Tests showed that the polycubes could be safely burned in air to totally volatilize and oxidize the aged polystyrene (Jones 1999). However, the sodium and potassium chloride salts (NaCl and KCl) in pyrochemical scrap volatilize appreciably at 950°C, severely damaging downstream equipment, whereas heating magnesium and calcium chloride ($MgCl_2$ and $CaCl_2$) hydrates in other scrap above 600°C converts them to stable oxides but evolves corrosive hydrogen chloride (HCl) gas. Therefore, methods to non-destructively identify such salt-bearing scrap items and certify stabilization at 750°C were developed (Boak et al. 2003) and incorporated into the (DOE-STD-3013; DOE 2018) standard. Though stakeholders challenged the need for purified products (e.g., Pu oxide derived from solutions), fearing further operational mission for PFP, the need for stabilization and retention of Pu-rich materials for beneficial use and safeguard security remained and prevailed (Sections 2 and 16 of Gerber 1997).

A further collection of about 300 small radioactive items having a wide variety of material forms (e.g., sealed capsules, metals, powders, solutions, fuel pins) and isotopes also were processed for disposal (Venetz 2007a). They included radiation and neutron sources, chemical and non-destructive analysis standards, and fuel test articles. Though some could be sent directly to laboratories in the DOE system for beneficial uses, others required unique technical solutions for stabilization prior to discarding. Full process development was not warranted for items of simple chemical form, but more complex and difficult items required hazard assessments and careful handling to ensure safe stabilization. Many difficult items contained reactive metals – sodium-bonded, mixed-nitride fuel capsules; a lithium-bonded, uranium nitride test capsule; and uranium metal samples. Other items

**FIGURE 9.1** Polycube entering muffle furnace for stabilization test. (Courtesy of the Hanford Site photo archive.)

included neptunium and $^{233}U$, natural, and enriched uranium solutions and a $PuO_2$-polystyrene compact source sealed in a welded pipe. Addressing these and other non-standard items was performed throughout de-inventory by personnel in the plant's process support laboratories. Their flexibility and resourcefulness were key not only here but also in devising safe and effective methods to stabilize larger PFP inventory strata such as polycubes and chloride salt–bearing items and in developing the Pu solution stabilization process.

Legacy non-irradiated plutonium/uranium mixed oxide fuel comprised the largest Pu inventory held in the PFP exclusion zone that was not associated with routine PFP operations – 56 intact assemblies and 4,125 fuel pins. These fuels and other special nuclear materials had been stored at PFP to take advantage of its extensive existing material protection infrastructure. The fuels were packaged into Hanford Unirradiated Fuel Packages (HUFPs) that first had to be designed, certified, fabricated, and delivered. The filled HUFPs were dispatched off-site. Slightly irradiated Fast Flux Test Facility and Los Alamos Molten Plutonium Reactor Experiment fuels also present within the PFP exclusion zone were sent to Hanford's Interim Storage Area (Evans 2009). About 800 Pu-bearing vibrationally packed (VIPAC) mixed oxide (~2 wt% Pu in uranium) fuel pins were shipped to the Waste Isolation Pilot Plant (WIPP).

By February 2004, 2,239 (and ultimately 2,257) 3013 containers of stabilized Pu-rich materials had been prepared and 1,998 vessels of lower-concentration residues (e.g., ash, reduction slag and crucible scrap, dilute alloys, oxides, sludges, and dilute solutions) had been packaged in pipe overpack containers (POCs) (Venetz 2007b, 2009). The lower-concentration process residues were processed according to environmental and worker safety, technical risk, and safeguard attractiveness guidelines. PFP operations diluted these materials and packaged them as POCs for disposal at WIPP in New Mexico (Hopkins et al. 2003) with some grouted slag and crucible scrap stored at Hanford's T Plant for future treatment. Both the Pu-rich materials and lower-concentration residues had been retained at PFP for Pu recovery during the Cold War. By September 2009, the last of the 3,013 containers had been shipped to the Savannah River Site and the 13 HUFPs dispatched. The timely de-inventory of the PFP exclusion zone and removal of the Pu saved \$100 million in security maintenance costs and another \$100 million in security upgrades (Venetz 2009).

### 9.1.2 Deactivation and Removal of Process Chemical Inventories

Deactivation of process lines, largely housed in gloveboxes, also included their decontamination to remove surface deposits and pockets of collected Pu. Roughly 100 kg of Pu in various chemical forms was estimated (generally by non-destructive analyses) to have accumulated in the process equipment, ventilation ducts, and piping (Charboneau et al. 2006a). The Pu removal potentially allowed sufficiently decontaminated hardware to be disposed of by on site burial rather than at much greater expense to WIPP as transuranic (TRU) waste. Accordingly, means to scrub the Pu from the process equipment and gloveboxes and dispose safely of the wash rags and the chemically neutralized agents were developed (Ewalt et al. 2005; Scheele et al. 2016). Though, ultimately, none of the chemical decontamination methods allowed gloveboxes to meet low-level waste criteria, these methods did remove significant material to ease subsequent size reduction.

Worker innovations were key to decontamination operations (Charboneau et al. 2006b). Visibility was restored to aged and marred glovebox windows by technologies used for aircraft canopies. Augur-type tools were developed to dislodge Pu-bearing materials from process piping. Other tools dislodged Pu contamination and debris in hard-to-access areas, for example, behind fume hood baffle plates and within ductwork. Workers developed collapsible sleeves to remove and simultaneously package glass tanks for disposal and developed a rigid sheath to safely remove sharp contaminated objects from within gloveboxes. Specialized scrapers, brushes, and core drill bits for surface decontamination were developed. These novel implements and associated methods vastly improved task safety, efficiency, and speed.

The PFP achieved its mission via chemical processes. Therefore, hazards associated with chemicals also had to be addressed. The explosion at PFP of a tank holding non-radioactive hydroxylammonium (or hydroxylamine) nitrate, HAN, on May 14, 1997 (Harlow et al. 1998), epitomized such chemical hazards. The HAN was used routinely to chemically reduce Pu(IV) to Pu(III) and thus shift Pu from the organic to the aqueous phase in solvent extraction. A dilute HAN solution in nitric acid was prepared in June 1993 and stored in a 1,500-L stainless-steel tank in anticipation of PRF restart. However, the PRF soon was placed on standby and plant restart canceled in December 1993. The HAN, essentially ignored, remained in the vented tank and evaporatively concentrated about 25-fold in the next four years. The increased concentration, warm storage conditions, and trace iron catalyst ultimately made the HAN unstable, culminating in a rapid and autocatalytic redox reaction. The reaction produced explosive internal pressure that ruptured the vessel itself, vastly damaged the chemical makeup room, breached the facility roof, and locally released a toxic nitrous oxide plume. Fortuitously, no one was injured.

The 1997 explosion helped instigate surveys of other chemical hazards, both radioactive and non-radioactive, at the PFP by creation of a database that listed process vessels, engineering drawings, and process histories, supplemented by engineering and design authority personnel interviews and physical inspections by subject matter experts (Hopkins et al. 2007a). With this information, chemical hazards were assessed to identify items of highest concern and were then updated through physical inspections and document reviews. Items and systems judged of greatest impact to the plant operations and safety included process vessels, ventilation and vacuum services, gloveboxes, Pu process and transfer lines, laboratories, and chemical storage facilities. Assessments included the risks of chemical and radiolytic hydrogen gas generation. This knowledge of hazards to be encountered in PFP deactivation thus was in place to prioritize and address the hazards most safely and expeditiously.

### 9.1.3 Decommissioning

A microcosm of the PFP decommissioning process is found in dealing with the 232-Z Contaminated Waste Recovery Process Facility, formerly used to recover Pu by incineration of process trash (gloves, rags, filters) and leaching of the resulting ash (Minette et al. 2007). The incinerator ash located in process equipment and attached to the walls was found to contain greater than 1,300 g of Pu. This ash had to be removed, packaged, and disposed of. The decontaminated process equipment then was removed and packaged for disposal with one large glovebox that was further size-reduced manually at Hanford's T Plant. Asbestos and other hazardous materials (e.g., batteries, mercury, hydraulic oils, and refrigerants) were removed and interior surfaces painted to fix loose contamination. Floor penetrations and underground ductwork to the 291-Z ventilation system were grouted and sealed before building demolition. The building debris was buried on-site at Hanford's Environmental Restoration Disposal Facility (ERDF), while TRU materials were packaged for disposal at WIPP. Planning, as elsewhere, followed CERCLA processes in coordination with the DOE, Ecology, and EPA.

A similar methodological approach was used for the 241-Z Liquid Waste Treatment Facility (Mattlin et al. 2007). This facility consisted of five below-grade ~19,000-L stainless-steel tanks, pumps, and piping within a concrete containment vault. The tanks staged dilute waste solutions for their in-ground disposal to the 216-Z-12 crib via the 241-Z-361 settling tank. An above-grade corrugated metal enclosure provided weather protection, ventilation, and other equipment. Removal of Pu-bearing solids from the five tanks was accomplished by first cutting the tank walls and then manned entry in anticipation of the later demolition of the entire facility.

Subgrade structures within the PFP complex were evaluated for potential remediation through a rational decision-making process based on effectiveness (protectiveness, long-term performance, compliance with regulations, ability to achieve the desired ends), feasibility (including availability of suitable technology), and cost (Hopkins et al. 2007b). Structures related to in-ground waste disposal (e.g., the 216-Z-9 trench), however, were outside of these considerations. Four alternatives were considered: (i) No Action; (ii) Surveillance and Maintenance, that is, regular inspection and maintenance of building slabs and contamination control covers to ensure integrity; (iii) Stabilize and Leave in Place, wherein selected subgrade contaminated structures are stabilized while others are not, but all remain in place; and (iv) Remove, Treat, and Dispose. Criteria and weightings were assigned to each option, and subject matter experts were consulted. In the end, Surveillance and Maintenance was selected as best meeting the criteria for Pu-contaminated subgrade structures.

Some of the most challenging work within the PFP building enclosure included manually size-reducing two large, high-holdup gloveboxes (requiring workers to wear impermeable pressurized suits and breathe supplied air); removal of process vessels from 242-Z, the facility left severely contaminated by the explosion of an Am ion exchange column on August 30, 1976 (see Chapter 4 breakout section); and, for PRF, sectioning and removal of long process pencil tanks and cleaning of the canyon floor, burdened with years of Pu-rich process spills and leaks (Swartz 2016).

### 9.1.4 Demolition

Open-air demolition of the PFP began in November 2016 with the PRF estimated to constitute 98% of the radiological burden within the PFP (CHPRC 2018a, 2018b; Casper 2020). High workplace hazards from falling debris, extraordinary Pu contamination, and complexities in understanding Pu contamination in an unconfined environment challenged demolition progress. Measures to seal radioactive contamination to both the facility surfaces being demolished and the rubble created in its demolition were implemented using water cannons (foggers) and chemical fixatives. Airflows and particle dispersion were modeled to anticipate potential contamination spread with respect to demolition actions, and application of fixatives was sequenced in planning to minimize the risk to the environment. Despite these measures, a spread of airborne contamination west of the PRF occurred in January 2017, caused by a pause in dust suppression during debris relocation and accentuated by wind eddies (wake effect) from the larger, main 234–5 building. To address the wake effect, the southeast corner of 234–5 (i.e., that closest to the PRF) was demolished before resumption of PRF demolition. The radiological boundaries also were expanded. Nevertheless, a subsequent but larger contamination spread with airborne radioactivity was detected on June 8, 2017, again shaped by the 234–5 building wake. This incident precipitated further work practice changes (in use of exhausters, fixative application, and controls for foggers), altered radiological boundaries and demolition sequence, revision of the air model, and added radiological monitoring locations.

PRF demolition resumed on November 3, 2017, with gallery glovebox removal (the alleged most hazardous step). Airborne radioactivity and removable contamination observed in the control areas stayed within expectations. Glovebox removal allowed access to, and removal of, the metal frameworks that supported the previously removed pencil tanks. Again, airborne radioactivity and removable contamination stayed within limits. Based on these successes, the last step, demolition of the PRF canyon structure to slab-on-grade foundation, was believed to be attainable to meet the scheduled December 31, 2017, completion target. Meeting this end-of-year target would also decrease the time that the demolition would be vulnerable to environmental contamination spread by impending winter weather, including high winds. Thus, plans were made to demolish the PRF canyon structure faster than originally outlined in the air model. To attain the goal safely, plans were made to apply fixatives, increase air monitoring, and cover the demolition debris with soil.

However, routine radiological surveys on December 14–19, 2017, found contamination well outside the PFP demolition zone control areas – near PFP offices, on seven personal and 29 government vehicles, and even outside the Hanford Site (CHPRC 2018a) – and included personnel Pu uptakes. Work was stopped to determine the cause and impact of the spread with the intent of remediating deficiencies to resume demolition. Significantly, prior to the discovered surface contamination, no airborne contamination had been detected, and thus, no attempt to improve contamination control by increased fixative or water fogging had occurred. By direction of the contamination plume, the source of the contamination was identified as the PRF (Casper 2020). A windstorm on the night of December 15, 2017, had occurred with gusts up to 31 m/s (~112 km/h) that seems to have spread contamination. The severity of exposure further increased following a windstorm early on December 18, 2017. Contaminated particles examined by electron microscopy were found to be from concrete, apparently wind-driven. Because larger building and debris piles were left exposed during accelerated demolition, the extent of contamination increased. The fixative treatment was found to be inadequate, because it had been applied only from one direction, meaning the leeward sides of the debris piles had no or lower coverage. Additionally, the fixative was weaker, because it had been applied at a 1:1 dilution with water rather than undiluted or at low dilution as had been used previously. Furthermore, it is surmised that the high wind speed dispersed the fine contaminants, making them undetectable by the control area air monitors, while carrying the larger (concrete) particles along the ground by strong eddies and saltation (skipping at the surface) where they were detected by surface surveys.

An analysis team composed of independent subject matter experts, program representatives, and mentors, advised by the Hanford Atomic Metal Trades Council Safety Representative and under DOE-RL oversight, was convened to perform a causal analysis of the contamination events (CHPRC 2018c). The team examined events prior to and during the December 2017 contamination spread outside the PFP control area to identify causes and develop corrective actions. The team identified two root causes to the December 2017 contamination events – overreliance on selective empirical data to guide demolition activities and a complementary inadequate evaluation of risks and consequences from evolving conditions within an ambitious pursuit of schedule. Selective use of empirical data gave a falsely optimistic appreciation of the increasing risks manifest in the early contamination events. The project cycled from physical demolition to rubble size reduction to loading rubble. Although size reduction and load-out limited the rate of progress, demolition in December 2017 proceeded apace to outstrip the latter two steps and left increasing amounts of contaminated material exposed to the environment. The desire to meet the December 31, 2017, completion target caused a "summit fever" that clouded judgment (Casper 2020). The team identified two further contributing causes. First, the assumption that the prior efficacy of the fixative would extend to containerized debris resulted in insufficient application of fixative. Second, faulty radiological information did not compel expanded application of fixative or increased radiological surveys. Finally, the team found that PFP management did not communicate evolving goals nor adequately acknowledge employee insights, concerns, and suggestions. Instead, they remained overly focused on completing demolition.

The remediation plan for the decontamination of the PRF canyon (within Section 2.1.3.3 of Toebe 2011) established that "The activities chosen will strive to reduce contamination to ALARA compatible with planned demolition methods," where ALARA means "as low as reasonably achievable." In retrospect and in light of the environmental outcomes in 2017, further decontamination within and exterior to the PRF canyon could have been performed while ventilation and structural enclosure (ceiling and walls) still were in place. Such decontamination, achieved via controlled and contained demolition methods, could have included removal of the dunnage (strongbacks) not integral to the canyon structure. The dunnage was used to support the PRF canyon extraction columns and tanks and likely contained much of the Pu material at risk. In prior routine plant operations, the dunnage and associated concrete-filled shield plugs had been cut, removed, and loaded into waste boxes when failed tanks were removed. As in those past operations, the existing overhead crane and the large bagout port present in the ceiling could have been used to transport the size-reduced pieces outside for disposal. Likewise, the access gallery gloveboxes could have been removed from the walls and size-reduced in situ when the building structure and exhaust ventilation still existed. Further removal of the remaining extensive at-risk Pu on the canyon floor could have been done while enclosure and ventilation were present, rather than grouting it in place (an option entertained in Section 2.1.3.3 of Toebe 2011 and performed on January 2016, CHPRC 2018c). Non-destructive assay (NDA) methods were used to locate hotspots and estimate residual Pu for safeguard termination (Attachment 10 of CHPRC 2018c). Such surveys could have been used to target particular areas for demolition and removal while ventilation and structural enclosure still were present.

Demolition of the remaining PRF structure resumed in August 2018 after the additional controls were implemented and concurrence was obtained from regulatory agencies and the U.S. DOE (Casper 2020). Supplemental measures enacted to resume demolition included work stoppage under windy conditions, enhanced radiological surveys and monitoring, more rapid debris removal, better use of fixatives, and more detailed and closely monitored work packages. Demolition work was suspended from March 2020 until April 2021 because of the COVID-19 pandemic. Demolition was completed in November 2021 (WA-DOE 2022; Bloom 2021). Aerial images of the PFP site before and after demolition, Figure 9.2, show the scale of the ~100×150 m 234–5 building, as well as the 63 buildings on 6 hectares comprising the entire site (Charboneau et al. 2006a).

### 9.1.5 Cultural Artifacts

The National Historic Preservation Act (NHPA) requires federal agencies to evaluate the historic significance of properties under their jurisdiction before altering or demolishing. The DOE Richland Operations Office, Washington State Historic Preservation Office, and the Advisory Council on Historic Preservation concurred that the Hanford Site Historic District is eligible for listing on the

**FIGURE 9.2** Aerial images of PFP before and after demolition – PRF is circled in white. (Courtesy of the U.S. DOE.)

National Register of Historic Places. Three major commitments were outlined in a programmatic agreement (DOE-RL 1996; Hopkins et al. 2006):

1. DOE will prepare a treatment plan to identify important buildings.
2. A comprehensive history of the plutonium production facilities will be prepared.
3. Historic items from the buildings with educational or interpretive value will be collected and maintained.

The first two steps have been documented broadly for the Hanford Site (Marceau 1998; Marceau et al. 2002). Because Pu production was the major mission of the Hanford Site and PFP conducted the final step of Pu metal production and finishing, the PFP complex was evaluated for buildings and items of historic and cultural significance. Eleven buildings were identified as having historic significance with four making the list for inclusion in the National Register. However, all buildings were so contaminated that they could not be preserved.

Thus, an extensive written record was prepared to describe the activities and artifacts within the PFP complex (Gerber 1997) and specific artifacts identified for possible preservation (Hopkins et al. 2006). Items included Radio Flyer wagons modified at PFP and similar shop-built wagons to transport Pu-laden articles in a critically safe manner (i.e., the Pu-bearing item locations in the wagon are sufficiently distant from each other that a criticality cannot occur; Figure 9.3). Some designs had cages built into the wagon beds to restrict the payload to five spaced locations accepting ~6-inch-diameter Pu scrap and button cans. Use of wagons allowed more items to be conveyed by one worker and kept the worker at some distance from the dose. Allegedly, in early PFP operations, workers carried canned Pu buttons directly by hand, in buckets, and even in their work coverall pockets, at significant radiological exposure and criticality risk.

**FIGURE 9.3** Radio flyer (or equivalent) wagons used to transport Pu-bearing items at PFP. Left top – five-position wagon used to transport Pu button and scrap cans. Left bottom – two-position wagon used to transport large dilute Pu-bearing scrap cans (lard cans). Right – shop-built five-position wagon used to transport Pu button and scrap cans; see drawing number. (Left images courtesy of the Hanford Site photo archive; right image courtesy of Donald L. Sorenson, Central Plateau Cleanup Company.)

Other items identified for possible preservation included a balance used to weigh Pu metal buttons, a sample ceramic crucible used in creating the Pu buttons, and control room panels (Hopkins et al. 2006). In the end, only a few of the ceramic crucibles and shop-made transport wagons have been retained for curation.

## 9.2 POTENTIAL REMEDIATION MEASURES AND SUBSURFACE GROUTING OF PFP WASTE DISPOSAL STRUCTURES

The current regulatory requirement as determined by the Record of Decision (ROD) is to return groundwater to its beneficial uses (as a potential source of domestic groundwater) wherever practicable and within a reasonable time frame given site-specific conditions, although this is not the anticipated future land use as it is expected to continue to be an industrial site into the future (EPA, Ecology, and DOE 2011). A remedy has been selected for PFP's Pu-contaminated process waste solution disposal sites (containing approximately 200 kg of Pu; Charboneau et al. 2006a) in accordance with CERCLA to address source contamination that poses threats to human health and the environment (EPA, Ecology, and DOE 2011). The remedy consists of the following: (i) removing a portion of contaminated soil, structures, and debris; (ii) treating these removed materials as required to meet disposal requirements at Hanford's ERDF or waste acceptance criteria for off-site disposal at WIPP; and (iii) disposal at ERDF or WIPP, currently the only deep geologic, long-lived radioactive waste repository in the United States. Specifically, for the high-salt waste in the 216-Z-1A tile field, 216-Z-9 trench, and 216-Z-18 crib, the Removal, Treatment (as needed), and Disposal (RTD) approach will be used to excavate contaminated soils and debris to a minimum of 2 ft below the bottom of the disposal structure (6.1–7.0 m below ground surface), with disposal at ERDF or WIPP, as appropriate. After the excavations are filled, an evapotranspiration barrier will be constructed over the remaining waste in these waste sites. A soil vapor extraction system was previously used to address carbon tetrachloride ($CCl_4$) contamination. About 80 tons of $CCl_4$ was removed between 1992 and 2012 from the vadose zone in regions underlying these disposal sites. The vapor extraction ceased after reaching asymptotically low removal rates. The $CCl_4$ concentrations in the groundwater beneath 216-Z-9 in 2016 were ~500 ppm (Byrnes et al. 2016). The $CCl_4$ plume within the underlying groundwater extends >18 km$^2$ north, south, and east from the source areas and is being addressed by long-term pump-and-treat technologies (Byrnes et al. 2016). A similar remedy has been selected for other Pu-contaminated waste sites, but the depth to which soils and debris are removed depends on the extent of contamination and is different for the Z-ditches and the low-salt waste group.

In addition to protecting human health and the environment, the selected remedy balances removal of contaminated soil with evapotranspiration barriers, soil covers, worker safety, and cost and uses permanent solutions to the maximum extent practicable. Several key factors influenced the selection of the remedy including the following: (i) location of the waste sites within the Central Plateau of the Hanford Site where they are adjacent to long-term waste disposal facilities; (ii) depth to groundwater at the waste sites; (iii) a low average annual precipitation of 17 cm; and (iv) anticipated industrial land use for these waste sites.

Based on a formal public comment period, the public is generally accepting of this remedy that removes contaminated soil. However, the remedy does not include treatment because there is no feasible technology to practicably treat radionuclide contamination that will not result in larger volumes and make disposal more impracticable. Because PFP's Pu-contaminated waste is TRU, it must be disposed of at WIPP, which has limited capacity. Furthermore, not all Pu contamination would be removed by the proposed remedy, which leaves significant risk from $^{239}$Pu's long half-life, radiotoxicity including carcinogenic risks, and potential migration to the groundwater and the Columbia River. After evaluation, it was concluded by the TPA agencies that minimal risk reduction is gained by removing Pu-contaminated soils at greater depths meaning long-term monitoring and institutional land-use controls will be required for these waste sites where contamination is left in place.

**FIGURE 9.4** View of the north interior of the Z-9 trench during inspection. Reprinted with permission from Hopkins et al. (2008).

A subsequent remedial design/remedial action work plan describes the technical approach that was developed to achieve the selected remedy and protect worker safety and the environment (DOE 2016). The work was estimated to take approximately 20 years and cost approximately $1.05 billion, including $40 million for long-term stewardship. However, it should be noted that decisions can change if new information is collected about the distribution and behavior of the subsurface contamination or if additional remediation technologies are developed.

As described in Section 4.2.2 of Chapter 4, experience in excavating Pu-contaminated soils and debris was obtained during historical mining operations to retrieve the top ~30 cm of the Z-9 trench sediments. The mining equipment, cover slab support columns, the sloped crib sediment walls, and the cubicle in the upper right from which the mining operations were conducted are shown in Figure 9.4. The ~50 $m^3$ of Z-9 soil mined from 1976 to 1978, including 58 kg of Pu (Ludowise 1978), still resides on the Hanford Site awaiting WIPP disposal.

The supplemental mining operations raised technical issues for the removal action at Z-9 in terms of structural stability because the soil removal structures extend over the reinforced concrete cover that provides Z-9 containment. A detailed structural characterization of the Z-9 trench cover in 2008 using a robotic "crib crawler" (Figure 9.5; see the tile-strewn debris field) and high-resolution photography concluded that the concrete roof of the trench was sufficient to meet the needs for the removal activities, although tile failure spallation from the underside brought into question the long-term cover stability. At that time, trench re-inspection was recommended at least every five years (Hopkins 2008).

In May 2017, a collapse occurred at one of the Plutonium/Uranium Extraction Facility (PUREX) storage tunnels used to dispose of failed equipment, a waste disposal technique and structure type not used at the PFP Implementation Area. An expert panel was chartered to evaluate the tunnel collapse and provide recommendations to DOE. The panel concluded that filling with grout was the preferred stabilization method, providing maximum protection to workers, the public, and the environment, while allowing future disposition options including in situ disposal or removal of materials. After stabilization of both the failed tunnel and a similar adjacent PUREX tunnel in 2017, evaluations were conducted for other aging structures at risk of collapse on the Site. This evaluation identified three structures in the PFP Implementation Area at "high risk" of collapse: (i) 216-Z-9 trench; (ii) 241-Z-361 settling tank; and (iii) 216-Z-2 crib. In response to the evaluation, a time-critical action was approved to support the interim stabilization of these below-grade structures (Cathel 2020). The outcome of this action was to use engineered grout to provide interim stabilization and

**FIGURE 9.5** Clamshell excavator (left) with Crib Crawler (right) inside 216-Z-9 trench bottom. (Courtesy of Jake Tucker, PNNL.)

prevent potential subsidence events and contamination release that could affect human health and the environment before the final remedial actions took place.

A low-strength grout, with a 28-day compressive strength of 50–600 pounds per square inch (psi), was selected for Z-9 to simplify future excavation (Smith 2021; Hanford 2020). The 28-day compressive strength tests for Z-9 proved higher than targeted, at 610 psi to 1960 psi. The grout used in Z-361 had a 28-day compressive strength requirement of greater than 2,000 psi. Traditional structural concrete has a compressive strength of 2,500–5,000 psi. The grout slurry was introduced to Z-9 at two locations and mounded to the ceiling circling those locations. Thus, at the northeast and southwest corners of the crib, grout levels were 22 inches (56 cm) from the ceiling to give 3,058 $m^3$ of emplaced grout. A diffuser was used to emplace the first 2 feet of grout in the Z-361 tank. The diffuser was blocked at that point and was abandoned in place. The grout that had been introduced had self-leveled and set when the remaining grout was added via another riser at the top of Z-361. Although visual confirmation is not possible, added grout volumes and the backing of grout up the addition hose suggest that the tank was filled completely with about 107 $m^3$ of grout (Farabee and Savior 2021). The stabilization action began at the end of 2020 and was completed in 2021. The Z-2 crib had a thick overlying layer of soil that already had slumped and was determined to be sufficient to prevent the spread of contamination. Thus, it was not grouted (Smith 2021). These time-critical actions for Z-9 do not preclude the remedial action involving the removal of Pu-contaminated soils as specified in the EPA's Record of Decision (ROD) (EPA, Ecology, and DOE 2011).

## 9.3 CONCLUSIONS

The successful demolition of the 70-year-old PFP and its ancillary surface structures was unprecedented in radiological risk and in quantities of Pu requiring disposition. The early collaborations forged between the TPA parties – the DOE, the EPA, and the Ecology – and the concerned Hanford Site contractors were key to structuring the work to meet requirements from both the DOE and CERCLA while monitoring health and safety, managing data, operating within cost and schedule constraints, and managing and verifying completion milestones. The creative engagement of the facility personnel fostered the invention of clever tools and methods that sped work and improved efficiency and safety. At the same time, the choreographed maintenance of the PFP capabilities and services was essential even as the plant was being prepared for its demolition. Despite extensive planning, open-air demolition of the PRF, which contained an estimated 98% of the radiological burden of the PFP, resulted in widespread environmental contamination events in January, June, and especially December 2017. Subsequent analyses identified overreliance on selective empirical data to guide demolition and inadequate risk evaluation under evolving and degrading conditions in ambitious pursuit of schedule deadlines. With supplementary and upgraded controls, PRF demolition was completed in November 2021.

However, nearly 200 kg of Pu remains in various engineered ground disposal facilities with substantial contamination present in other subgrade structures following demolition. Significant lessons have been learned with regard to soil and groundwater remediation and the importance of maintaining or reinforcing aging subsurface structures, which led to the recent interim stabilization of the 216-Z-9 trench and 241-Z-361 settling tank. However, the future disposition of Pu, and the associated Am, in these sites must consider the mobilities of Pu and Am in their current conditions to the underlying groundwater, evaluate potential means to limit their mobilities by in situ treatments, and consider the risks of excavating, treating, and disposing the contaminated sediments and materials versus the anticipated gains in limiting their hazards.

## NOTES

1 Criticality is the state in which a mass of fissionable material sustains or is able to sustain a nuclear chain reaction.
2 Pu-bearing scrap arising from impure process intermediates and off-site sources was retained during PFP operations for later recovery. The consignment of Pu-bearing materials to recoverable scrap streams or irrecoverable waste underwent shifting guidelines during plant operations, dictated by the current value of the Pu compared with the costs of its recovery. Scrap items within a certain stream type (e.g., Pu oxides received from off-site sources) were selected for processing based on Pu assay, that is, high-graded to choose those of higher assay for immediate processing and leaving leaner residues for later processing. Much of the Pu-bearing material that had been retained as recoverable scrap (e.g., incinerator ash, slag, and crucible residues) became waste in the de-inventorying of PFP, while richer Pu-bearing materials were stabilized, as will be discussed, for retention as a national resource.
3 For example, although the International Commission on Radiological Protection was established in 1928 under its former name as the International X-ray and Radium Committee, the first recommendations document was not released until 1951 and recommendations continue to be updated to this day (https://www.icrp.org).

## REFERENCES


Bloom RW. 2021. *234-5Z Plutonium Finishing Plant Demolition Report - As Left Characterization*. CWR-PFP-00026, Revision 0, Central Plateau Cleanup Company, Richland, WA. https://pdw.hanford.gov/document/AR-18440.

Boak J, EA Conrad, CH Delegard, AM Murray, GD Roberson, and TJ Venetz. 2003. *Recommendations on Stabilization of Plutonium Material Shipped to Hanford from Rocky Flats*. LA-14070, Los Alamos National Laboratory, Los Alamos, NM.

Byrnes, M, V Rohay, M Truex, and C Johnson. 2016. *Endpoint evaluation for 200-PW-1 Operable Unit soil vapor extraction, Hanford Site, Washington - 16478*. Waste Management Conference Proceedings, March 6-10, 2016. Phoenix, AZ. https://archivedproceedings.econference.io/wmsym/2016/pdfs/16478.pdf

Casper J. 2020. "Lessons Learned from Demolition of Hanford's Plutonium Finishing Plant." *WM2020 Conference*, March 8–12, 2020, Phoenix, AZ. https://s3.amazonaws.com/amz.xcdsystem.com/A464D2CF-E476-F46B-841E415B85C431CC_abstract_File498/FinalPaper_20507_0123100525.pdf.

Cathel RL. 2020. *Action Memorandum for the Interim Stabilization of 216-Z-2 Crib, 216Z9 Trench, and 241Z361 Settling Tank*. DOE/RL-2020-11, Revision 0, CH2M HILL Plateau Remediation Company, Richland, WA. https://pdw.hanford.gov/document/AR-03741.

Charboneau S, B Klos, R Heineman, and A Hopkins. 2006a. "The Deactivation, Decontamination and Decommissioning of the Plutonium Finishing Plant, a Former Plutonium Processing Facility at DOE's Hanford Site." *WM'06 Conference*, February 26–March 2, 2006, Tucson, AZ. https://archivedproceedings.econference.io/wmsym/2006/pdfs/6385.pdf.

Charboneau S, B Klos, R Heineman, B Skeels, and A Hopkins. 2006b. "The Creative Application of Science, Technology, and Work Force Innovations to the Decontamination and Decommissioning of the Plutonium Finishing Plant at the Hanford Nuclear Reservation." *WM'06 Conference,* February 26–March 2, 2006, Tucson, AZ. https://archivedproceedings.econference.io/wmsym/2006/pdfs/6442.pdf.

CHPRC. 2018a. *PFP Contamination Spread Discussion.* CH2MHill Plateau Remediation Company, Richland, WA. https://www.hanford.gov/files.cfm/PFP_Recovery_Briefing_to_Employees_Feb_13_2018.pdf

CHPRC. 2018b. *Plutonium Finishing Plant Resumption of Work. CH2MHill Plateau Remediation Company,* Richland, WA. https://www.hanford.gov/files.cfm/July_2018_PFP_Briefing.pdf.

CHPRC. 2018c. *Discovery of Contamination Spread at the Plutonium Finishing Plant during Demolition Activities - Root Cause Evaluation.* EM-RL--CPRC-PFP-2017-0018 CR-2018-0022, CH2MHill Plateau Remediation Company, Richland, WA. https://www.osti.gov/servlets/purl/1899309/.

DOE. 2016. *Remedial Design/Remedial Action Work Plan for the 200-CW-5, 200-PW-1, 200-PW-3, and 200-PW-6 Operable Units.* DOE/RL-2015-23, U.S. Department of Energy, Richland Operations Office, Richland, WA.

DOE. 2018. "DOE Standard Stabilization, Packaging, and Storage of Plutonium-Bearing Materials." DOE-STD-3013-2018, US Department of Energy, Washington, DC. https://www.standards.doe.gov/standards-documents/3000/3013-astd-2018/@@images/file.

DOE-EM. 2022. "Hanford Completes Historic Cleanup at Iconic Plutonium Finishing Plant." Article collected 13 May 2022, US Department of Energy, Office of Environmental Management, Washington, DC. https://www.energy.gov/em/articles/hanford-completes-historic-cleanup-iconic-plutonium-finishing-plant.

DOE-RL. 1996. *Programmatic Agreement among the U.S. Department of Energy, Richland Operations Office, the Advisory Council on Historic Preservation, and the Washington State Historic Preservation Office for the Maintenance, Deactivation, Alteration, and Demolition of the Built Environment on the Hanford Site, Washington.* DOE/RL-96-77, US Department of Energy, Richland Operations Office, Richland, WA. https://www.osti.gov/biblio/341257.

EPA, Ecology, and DOE. 2011. *Record of Decision Hanford 200 Area Superfund Site 200-CW-5 and 200-PW-1, 200-PW-3, and 200-PW-6 Operable Units.* U.S. Environmental Protection Agency, Washington State Department of Ecology, and U.S. Department of Energy, Olympia, WA. https://pdw.hanford.gov/arpir/index.cfm/viewDoc?accession=0093644.

Evans WJ. 2009. *200 Area Interim Storage Area Receipt of Slightly Irradiated Fuel Management Self Assessment.* CHPRC-00213, Rev. 1, CH2MHill Plateau Remediation Company, Richland, WA.

Ewalt J, A Hopkins, M Minette, C Simiele, R Scheele, P Scott, D Trent, and S Charboneau. 2005. "The Use of a Treatability Study to Investigate the Potential for Self-Heating and Exothermic Reactions in Decontamination Materials at the Plutonium Finishing Plant." *WM'05 Conference*, February 27–March 3, 2005, Tucson, AZ. https://archivedproceedings.econference.io/wmsym/pdfs/5277.pdf.

Farabee A and D Savior. 2021. "Stabilization of Aging Belowground Structures on the Hanford Site." *WM2021 Conference*, March 8–12, 2021, Phoenix, AZ. https://www.xcdsystem.com/wmsym/2021/143.html#21334.

Gerber MS. 1997. *History and Stabilization of the Plutonium Finishing Plant (PFP) Complex, Hanford Site.* HNF-EP-0924, Fluor Daniel Hanford, Inc., Richland, WA. https://www.osti.gov/biblio/325360-history-stabilization-plutonium-finishing-plant-pfp-complex-hanford-site.

Hanford. 2020. *Stabilizing Disposal Structures at Risk of Failure - Planned Actions for Stabilization of 216-Z-2 Crib, 241-Z-361 Settling Tank and 216-Z-9 Crib.* Fact Sheet. US Department of Energy, Richland, WA. https://www.hanford.gov/files.cfm/Aging_Structures_Fact_Sheet_REVISION(April)1.pdf.

Harlow, DG, RE Felt, S Agnew, GS Barney, JM McKibben, R Garber, and M Lewis. 1998. *Technical Report on Hydroxylamine Nitrate.* DOE/EH-0555, US Department of Energy, Washington, DC. https://www.osti.gov/servlets/purl/1374990.

Hebdon J, J Yerxa, L Romine, AM Hopkins, R Piippo, L Cusack, R Bond, O Wang, and D Willis. 2003. "Collaborative Negotiations a Successful Approach for Negotiating Compliance Milestones for the Transition of the Plutonium Finishing Plant (PFP), Hanford Nuclear Reservation, and Hanford, Washington." *WM'03 Conference*, February 23–27, 2003, Tucson, AZ. https://archivedproceedings.econference.io/wmsym/2003/pdfs/520.pdf.

Hopkins A. 2008. "A Plutonium Finishing Plant Model for the CERCLA Removal Action and Decommissioning Construction Final Report." *WM2008 Conference*, February 24–28, 2008, Phoenix, AZ. https://archivedproceedings.econference.io/wmsym/2008/pdfs/8411.pdf.

Hopkins A, B Klos, M Minette, A Sherwood, J Teal, S Charboneau, and E Mattlin. 2007a. "Assessing Chemical Hazards at The Plutonium Finishing Plant for Planning Future Decontamination and Decommissioning." *WM'07 Conference*, February 25–March 1, 2007, Tucson, AZ. https://archivedproceedings.econference.io/wmsym/2007/pdfs/7379.pdf.

Hopkins AM, DB Klos, AR Sherwood, SL Charboneau, EM Mattlin, C Negin, and JA Teal. 2007b. "An Approach to Characterizing and Evaluating Alternatives for the Decommissioning of Sub-Grade Structures at the Plutonium Finishing Plant." *WM'07 Conference*, February 25–March 1, 2007, Tucson, AZ. https://archivedproceedings.econference.io/wmsym/2007/pdfs/7361.pdf.

Hopkins AM, M Minette, D Sorenson, R Heineman, and M Gerber. 2006. "The Challenges of Preserving Historic Resources During the Deactivation and Decommissioning of Highly Contaminated Historically Significant Plutonium Process Facilities." *WM'06 Conference*, February 26–March 2, 2006, Tucson, AZ. https://archivedproceedings.econference.io/wmsym/2006/pdfs/6411.pdf.

Hopkins AM, C Sutter, G Hulse, and J Teal. 2003. "Disposal of TRU Waste from the Plutonium Finishing Plant in Pipe Overpack Containers to WIPP Including New Security Requirements." *WM'03 Conference*, February 23–27, 2003, Tucson, AZ. https://archivedproceedings.econference.io/wmsym/2003/pdfs/562.pdf.

Hopkins A, C Sutter, DB Klos, JA Teal, L Oates, FW Bond, E Mattlin, and S Clarke. 2008. "Structural Characterization of the 216-Z-9 Crib Prior to Decontamination and Demolition Using Robot Crawler and High Resolution Photography." *WM2008 Conference*, February 24–28, 2008, Phoenix, AZ. https://archivedproceedings.econference.io/wmsym/2008/pdfs/8117.pdf.

Jones SA. 1999. "White Paper Supporting Decision to Stabilize Polycubes Using Direct Oxidation in Muffle Furnace." letter 15F00 99 168 to RK Leugemors, Fluor-Daniel Hanford, Incorporated, Richland, WA. https://ehss.energy.gov/deprep/2000/TB00F16B.PDF.

Ludowise JD. 1978. *Report On Plutonium Mining Activities At 216-Z-9 Enclosed Trench.* RHO-ST-21, Rockwell Hanford Operations, Richland, WA. https://pdw.hanford.gov/arpir/pdf.cfm?accession=0076183H.

Marceau TE. 1998. *Hanford Site Manhattan Project and Cold War Era Historic District Treatment Plan.* DOE/RL-97-56, Rev. 1, Bechtel Hanford, Incorporated, Richland, WA. https://pdw.hanford.gov/download/78a88cf3-f945-432e-a849-30882ac5796a.

Marceau TE, DW Harvey, DC Stapp, SD Cannon, CA Conway, DH Deford, BJ Freer, MS Gerber, JK Keating, CF Noonan, and G Weisskopf. 2002. *History of the Plutonium Production Facilities at the Hanford Site Historic District, 1943-1990.* DOE/RL-97-1047, US Department of Energy, Richland, WA. https://www.osti.gov/servlets/purl/807939.

Mattlin E, S Charboneau, G Johnston, A Hopkins, R Bloom, B Skeels, and DB Klos. 2007. "The Integration of the 241-Z Building Decontamination and Decommissioning Under CERCLA with RCRA Closure at the Plutonium Finishing Plant." *WM'07 Conference*, February 25–March 1, 2007, Tucson, AZ. https://archivedproceedings.econference.io/wmsym/2007/pdfs/7447.pdf.

Minette M, A Hopkins, B Klos, S Charboneau, and E Mattlin. 2007. "Contaminated Process Equipment Removal for the Deactivation and Decontamination of the 232-Z Contaminated Waste Recovery Process Facility at the Plutonium Finishing Plant." *WM'07 Conference*, February 25–March 1, 2007, Tucson, AZ. https://archivedproceedings.econference.io/wmsym/2007/pdfs/7378.pdf.

Scheele RD, MJ Minette, BK McNamara, and JM Schwantes. 2016. "Thermochemical Reactivity Hazards of TRU Waste Constituents." *WM2016 Conference*, March 6–10, 2016, Phoenix, AZ. https://archivedproceedings.econference.io/wmsym/2016/pdfs/16075.pdf.

Smith DT. 2021. *Completion Report for Structural Stabilization of Crib 216-Z-9 and Tank 241-Z-361.* CPCC-00049 Revision 0, Central Plateau Cleanup Company, Richland,WA.

Swartz M. 2016. "Progression of Safe and Compliant Demolition of Hanford's Plutonium Finishing Plant." *WM2016 Conference*, March 6–10, 2016, Phoenix, AZ. https://archivedproceedings.econference.io/wmsym/2016/pdfs/16566.pdf.

Toebe WE. 2011. *Removal Action Work Plan for the Deactivation, Decontamination, Decommissioning, and Demolition of the Plutonium Finishing Plant Complex.* DOE/RL-2011-03, Rev. 1. CH2M HILL Plateau Remediation Company, Richland, WA. https://pdw.hanford.gov/download/001b2feb-0063-4968-a625-c4d65f17ee61.

Venetz TJ. 2007a. *Plutonium Finishing Plant Project to Disposition Excess Nuclear Materials.* HNF-32629, Fluor Hanford, Richland, WA.

Venetz T. 2007b. *Disposition of Sources, Standards and Miscellaneous Nuclear Materials at the Plutonium Finishing Plant (PFP).* NMS-29658-VA, Rev 0, Fluor Hanford, Richland, WA.

Venetz T. 2009. *Submittal for Project of the Year: Plutonium Finishing Plant Project to De-Inventory Nuclear Materials.* CHPRC-00625-VA, CH2MHill Plateau Remediation Company, Richland, WA.

WA-DOE. 2022. *Cleanup Work Concludes at Plutonium Finishing Plant.* Article collected 13 May 2022, Washington State Department of Ecology, Olympia, WA. https://ecology.wa.gov/Waste-Toxics/Nuclear-waste/Hanford-cleanup/Plutonium-Finishing-Plant.

# Part IV

## Remedial Action

Christian D. Johnson, Katherine A. Muller, and Hilary P. Emerson

Once remedies are selected, communicated to the public, and formalized in a decision document (Part III), the next step is to undertake selected remedial actions such as removal or cleanup of environmental media, processing of legacy waste, and decommissioning and demolition of legacy facilities. Detailed design, contracting, construction, and operational implementation of the remedies are completed at this stage. (Refer to Section 1.2 of Chapter 1 for how this step fits into the broader remediation process.)

Remediation technologies, engineering controls, and institutional controls can all play a role in the overall remediation strategy, as described in Part III. Frequently, more than one of these elements must be used in a specific area to meet the objectives for protection of human health and the environment during and after active remediation. These three elements may be included in different combinations in the final remedy selection and implementation.

### IV.1 REMEDY SELECTION DECISION

A remedy selection decision may be informed by public comment and approved by applicable regulatory agencies. The decision is formalized through documents such as a statement of basis or record of decision, depending on the regulatory framework. The Environmental Protection Agency provides guidance on the preparation, processing, and requirements for these types of regulatory documents (https://www.epa.gov/superfund/record-decision-rod-guidance).

Typically, the remedy selection decision document will define the timeframe, overall cleanup objectives, and the remediation approaches. However, in some cases, interim decision documents

DOI: 10.1201/9781003329213-13

may be used for incremental remedial actions. This book contains several examples of decision documents generated at the Hanford Site, including the following:

- The Tri-Party Agreement detailed in Chapter 1 formulated the overall remediation framework selected for Hanford and identifies timelines and milestones for key decisions.
- Groundwater remediation at 100-HR-3 OU is described in Chapter 10. The pump-and-treat remedy was selected to decontaminate subsurface groundwater of hexavalent chromium as part of an interim action decision prior to final approval of the remedy selection.
- Chapter 8 describes records of decision documenting the selection of glass as the desired waste form for specific tank wastes at several complex waste sites in the U.S.

The remedy selection decision document should also include a summary of the nature and extent of contamination, associated risks and hazards, and the rationale for the selected remedy. While remedy selection decision documents in the past have been prescriptive, a more prudent practice is to include flexibility for remedy assessment and optimization by applying an adaptive site management approach.

## IV.2 REMEDY DESIGN, IMPLEMENTATION, AND OPERATIONS

A detailed design of the remedy system and an implementation work plan is prepared after the remedy selection decision is approved. The remedy system design includes all the elements needed for treating contaminants. For example, plans for waste disposition generated before or arising from remedy implementation, as well as institutional controls (e.g., access controls), must be included. Construction of the remedy system follows the approval of its design. Subsequent system operation and maintenance, monitoring, and reporting will be ongoing for a period depending on the remedial approach and site conditions.

Remedy system performance assessment is also an important part of ongoing remedy operations. Remediation system and site monitoring provides information for strategic performance assessments that help verify the expected remedy performance and update the CSM. The assessments are used to determine progress toward achieving the cleanup objectives delineated in the regulatory documents describing the remedial decisions. The updated CSM and system performance information may lead to a need for additional site characterization or remedy optimization. Guidance on performance assessment (e.g., Truex et al. 2013, 2015, 2017; Johnson et al. 2022) can provide a structured approach for gathering information, updating the CSM, evaluating performance and impacts, and determining appropriate next steps. The next steps may include continued operation, optimization, transition to another technology (either passive, like MNA, or active), or site closure and long-term stewardship (described in Part V of this book).

## IV.3 REMEDY PROCESS OPTIMIZATION AND REVISION

Remedy system operations information and its subsequent assessment may lead to the system's optimization to decrease the remedy timeframe and reduce costs.

Revisions to the remedy system may be recommended because of a remedy system performance assessment and optimization. Typically, such revisions need to be documented. For example, CERCLA specifies that significant differences from the remedy defined in the decision document need to be documented in an Explanation of Significant Differences (ESD), whereas a fundamental change to the remedy is documented in a ROD Amendment (EPA 1999). The adaptive site management approach builds in flexibility for a remedy to adapt and evolve as new information is obtained (e.g., from characterization, remedy operations, or performance assessments). One example is the optimization and revision process described in Chapter 9 for demolition of a contaminated building within a complex waste site. The building demolition operations were placed on hold following the identification of contamination outside of the demolition zone control area and only resumed following a causal analysis and improved procedures being established.

## IV.4 HANFORD SITE REMEDY ACTIONS

Part IV of this book includes case studies focused on remedy selection, design, construction, and implementation for different areas and processes across the Hanford Site, including Chapter 9 "PFP Demolition and Subsurface Grouting," Chapter 10 "Groundwater Remediation with the Pump-and-Treat Technology," Chapter 11 "Polyphosphate for Enhanced Attenuation of Uranium," Chapter 12 "Tank Waste Retrieval," Chapter 13 "Tank Waste Treatment – Integrated Flow Sheets," and Chapter 14 "Legacy Waste Disposal in the Hanford Mission."

The process of decommissioning and demolishing the former plutonium processing facility, as well as the ongoing remediation of subsurface plutonium contamination at the Hanford Site, is described in Chapter 9 of this book. The discussion includes the challenges presented by a building with over 40 years of plutonium processing operations followed by over 30 years in a non-operational state. Such challenges included overcoming the unknowns with little to no historical precedence to draw upon. In addition, the ongoing subsurface cleanup operations are described with recent interim actions to stabilize historical underground waste disposal structures. For additional context, Chapter 4 includes a discussion of the history of plutonium processing at the Hanford Site, as well as the building and subsurface characterization efforts.

The ongoing pump-and-treat operations in the 100 Area of the Hanford Site are detailed in Chapter 10 of this book. This chapter covers the decision-making process for selecting the pump-and-treat technology for cleanup of hexavalent chromium from groundwater and the design and implementation of the remedy. In addition, the tools developed for monitoring and performance assessment are presented as examples for managing the significant amounts of data that are often gathered as part of this process, including both modeling and data visualization approaches.

Chapter 11 continues the discussion of remediation of uranium contamination in the 300 Area of the Hanford Site initiated following characterization and assessment as described in Chapter 4, which covers the historical waste releases and subsurface characterization efforts. The initial regulatory decision for monitored natural attenuation and access restrictions was re-evaluated based on uranium concentrations measured in groundwater and river water over time. The re-evaluation led to implementation of an alternate remedy. To date, two campaigns of liquid polyphosphate injection have been conducted in the 300 Area for enhanced attenuation of uranium. The effectiveness of these campaigns was assessed utilizing multiple methods including characterization of paired sediment samples from before and after treatment, regular monitoring of aqueous contaminant concentrations in groundwater and nearby river water, and geophysical monitoring of the subsurface during liquid polyphosphate injection.

Retrieval of the legacy liquid and sludge waste from underground storage tanks is explored in Chapter 12. This waste was generated from production activities undertaken at the Hanford Site from 1943 through 1990 and subsequently stored in large underground tanks. Several factors have contributed to the complex physical and chemical nature of the waste including the variety of production processes implemented and transfers of waste between tanks leading to unintended physical and chemical consequences. The chemical and physical complexity has necessarily led to development and implementation of a variety of technologies to retrieve the waste for subsequent treatment and disposal.

Treatment of the retrieved tank waste is covered in Chapter 13. The complexity of the waste and the scale of the infrastructure and the timeframe needed to process it led to breaking up the cleanup mission into manageable scope elements performed by different specialist contractors. Management techniques have been developed to integrate and administer the treatment processes to facilitate their overall effectiveness as an integrated flowsheet. Integration and management extend across the functional areas of the integrated flowsheet, as well as its depth from an overall summary level to that of individual unit processes. Furthermore, the longevity of the waste treatment project means special importance is placed on documenting and justifying technical assumptions and bases to ensure generations of scientists and engineers have the knowledge to progress it.

Finally, Chapter 14 is concerned with disposal of the treated and immobilized waste. The longevity of the treatment project and the evolution of stakeholder expectations and environmental regulations led to commensurate development of disposal concepts. Treatment processes necessarily generate primary and secondary waste streams that must be dispositioned sometimes with different requirements yet in the same disposal facility. Often overlooked but nonetheless equally important is the managed discharge of large quantities of water to the arid environment at the Hanford Site. Decommissioning and demolition of legacy facilities also generates solid waste requiring managed disposal compliant with specific requirements at a specific disposal facility.

## REFERENCES

EPA. 1999. *A Guide to Preparing Superfund Proposed Plans, Records of Decision, and Other Remedy Selection Decision Documents*. Environmental Protection Agency, Washington, DC. https://www.epa.gov/superfund/record-decision-rod-guidance.

ITRC. 2017. *Remediation Management of Complex Sites*. RMCS-1. Interstate Technology and Regulatory Council, Washington, DC. https://rmcs-1.itrcweb.org or https://rmcs-1.itrcweb.org/RMCS-Full-PDF.pdf.

Johnson, C.D., K.A. Muller, M.J. Truex, G.D. Tartakovsky, D. Becker, C.M. Harms, and J. Popovic. 2022. "A Rapid Decision Support Tool for Estimating Impacts of a Vadose Zone Volatile Organic Compound Source on Groundwater and Soil Gas." *Groundwater Monitoring and Remediation*, 42(1):81–87. https://doi.org/10.1111/gwmr.12468.

Truex, M.J., D.J. Becker, M.A. Simon, M. Oostrom, A.K. Rice, and C.D. Johnson. 2013. *Soil Vapor Extraction System Optimization, Transition, and Closure Guidance*. PNNL-21843. Pacific Northwest National Laboratory, Richland, WA.

Truex, M.J., C.D. Johnson, D. Becker, M.H. Lee, and M.J. Nimmons. 2015. *Performance Assessment for Pump-and-Treat Closure or Transition*. PNNL-24696. Pacific Northwest National Laboratory, Richland, WA.

Truex, M.J., C.D. Johnson, T. Macbeth, D.J. Becker, K. Lynch, D. Giaudrone, A. Frantz, and H. Lee. 2017. "Performance Assessment of Pump-and-Treat Systems." *Groundwater Monitoring and Remediation*, 37(3):28–44. https://doi.org/10.1111/gwmr.12218.

# 10 Groundwater Remediation with Pump-and-Treat Technology

*Sarah Saslow and Christian D. Johnson*

In this chapter, we will explore how the Hanford Site uses pump-and-treat (P&T) technology to treat contaminated groundwater aquifers and contain the existing plumes. Six P&T systems currently operate at the Hanford Site; however, to illustrate application of this technology, we will draw upon specific data from the DX and HX P&T systems implemented for the 100-HR-3 Operable Unit. This chapter covers the decision-making process for selecting the P&T technology for cleanup of hexavalent chromium from groundwater and the design and implementation of the remedy. In addition, the tools developed for monitoring and performance assessment are presented as examples for managing the significant amounts of data that are often gathered as part of this process, including both modeling and data visualization approaches.

## 10.1 INTRODUCTION

At the Hanford Site, several decades of nuclear reactor operation and plutonium processing activities, with associated nuclear waste management, have led to the contamination of groundwater, including both perched water and the water table aquifer.[1] Contaminated groundwater in the Hanford subsurface poses a potential risk to the Columbia River, in terms of protection of aquatic species (e.g., salmon) and because the river serves as a water source for downstream communities and industries (including agriculture). Therefore, a key remediation goal for the site is to protect the Columbia River by remediating the groundwater contamination to levels that meet the remedy decision document remedial action objectives (referred to as regulatory maximum contaminant levels, MCLs) and subsequently minimize contaminant migration into the river via both treatment and containment goals.

Figure 10.1 illustrates the existing groundwater contaminant plumes at the Hanford Site as of the 2021 groundwater monitoring report (DOE/RL 2022c). Groundwater contaminant plumes exist in Hanford's River Corridor – including the 300 Area – and the Central Plateau (see Chapter 3, Figure 3.3 for depiction of the areas at Hanford). These plumes consist of a range of contaminants of (potential) concern (COCs), including radionuclides (e.g., iodine-129, strontium-90, technetium-99, tritium, and uranium), organics (e.g., trichloroethene and carbon tetrachloride), inorganics (e.g., nitrate and cyanide), and heavy metals (e.g., hexavalent chromium). The plume map in Figure 10.1 shows where several contaminant plumes colocate to create complex remediation challenges, for example, in the Central Plateau's 200 East and West areas. These contaminants often persist in the subsurface because many remediation technologies that effectively remove or immobilize contaminants (i) cannot be directly applied to the contamination source (e.g., too deep for effective in situ treatment), (ii) are cost prohibitive, and/or (iii) are designed to treat one, not several, contaminants. To address these challenges, the Hanford Site relies on several remediation approaches, including P&T.

P&T is a widely used remediation technology for containing and removing contaminants from aquifers. Implementation of P&T involves pumping groundwater from extraction wells in the

DOI: 10.1201/9781003329213-14

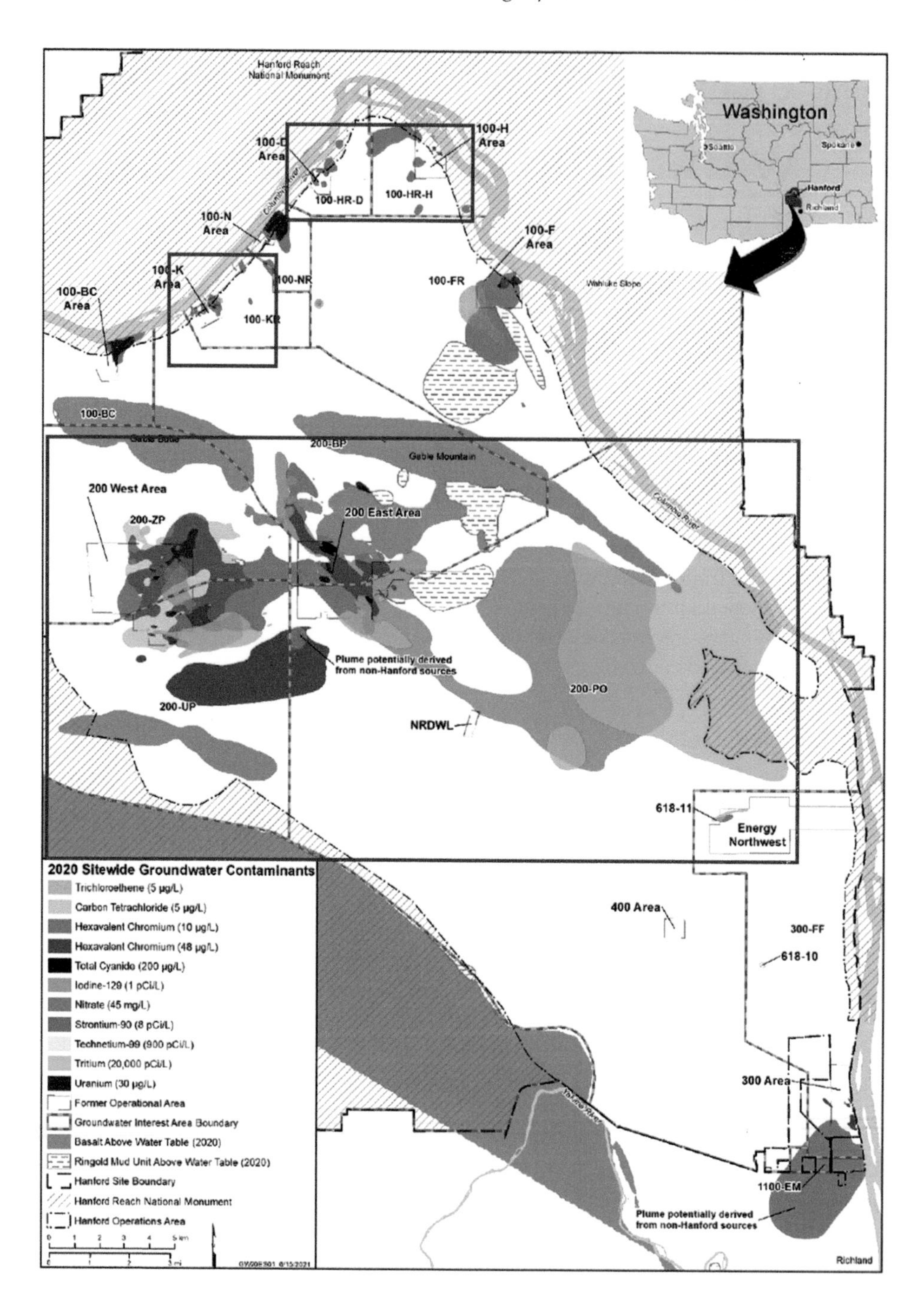


**FIGURE 10.1** Groundwater plumes at the Hanford Site as of 2021 monitoring. The numbers in parentheses next to legend indicate contours for groundwater plumes that are above their regulatory maximum contaminant levels (MCLs). Annotated red outlined regions indicate plume areas with pump-and-treat systems and are also shown in Figure 10.3. Note that there are different cleanup MCLs for Cr in different regions: Cr(VI) MCLs are different in the River Corridor versus the Central Plateau of the Hanford Site due to their potential impacts. Within the Central Plateau, MCLs are 48 μg/L in groundwater. However, in the River Corridor, MCLs are lower at 10 μg/L based on limits for surface water or groundwater that discharges to a surface water as contamination migrates offsite to the Columbia River. RUM refers to the Ringold Formation upper mud geologic unit. (Figure adapted from DOE/RL 2022c.)

subsurface, removing COCs from the groundwater in an aboveground treatment system, dispositioning treated water by injection back into the aquifer or by discharge to surface waters, and monitoring the groundwater. The general P&T process is illustrated in Figure 10.2.

The Hanford Site currently uses six P&T systems to help decrease contaminant concentrations in groundwater to the targeted cleanup levels and to hydraulically contain the groundwater plumes reducing plume migration. Each Hanford Site P&T system is designed to address the COCs found in specific site operable unit(s) (OU(s)). There are five P&T systems located in the 100 Areas of the Hanford Site that treat groundwater contaminated with hexavalent chromium (Cr(VI)). These include the DX and HX P&T systems located in the 100-HR-3 OU (Figure 10.3b) and the KR4, KW, and KX P&T systems located in the 100-KR-4 OU (Figure 10.3a) (DOE/RL 2022b). The sixth system is the 200 West (200W) P&T facility located in the Central Plateau (or 200 Areas, Figure 10.3c). These Hanford P&T facilities treat billions of gallons of contaminated groundwater each year. To put this volume into perspective, consider that an Olympic-size swimming pool holds approximately 660,000 gallons of water, which is enough water to fill more than 10,000 large bathtubs (60-gallon capacity) or to provide a single person a bath every day for 27 years. At Hanford, the P&T systems treat at least one, and in some cases more than five, Olympic-size swimming pools worth of contaminated groundwater every day.

P&T operations are complicated not only by overlapping COC plumes, but also by the site's geochemical heterogeneity, persistent contamination sources and uncertainty or lack of information about these sources, seasonal hydrogeology, and the scale of the operation. These challenges are taken into consideration when preparing the Record of Decision (ROD), which identifies the selected remedy. Implementation of the remedy is supported by a remedial design/remedial action (RDRA) work plan that details the P&T system (treatment facility and infrastructure) design. Operations governed by these documents are periodically reviewed for potential optimization efforts using monitoring data collected in the facility and the surrounding area. Effective monitoring and current and future performance assessment are critical to understanding progress toward successfully achieving remedial action objectives, leading to closure of the P&T remedy.

The following sections outline the process for recommending, installing, maintaining, and eventually closing a P&T remedy. The HX and DX P&T systems for Cr(VI) treatment in the 100-HR-3 OU are much closer to reaching closure than the OUs addressed by the 200W P&T system in the Central Plateau. Thus, these two P&T systems provide good examples to illustrate these phases, starting with site history to set the context.

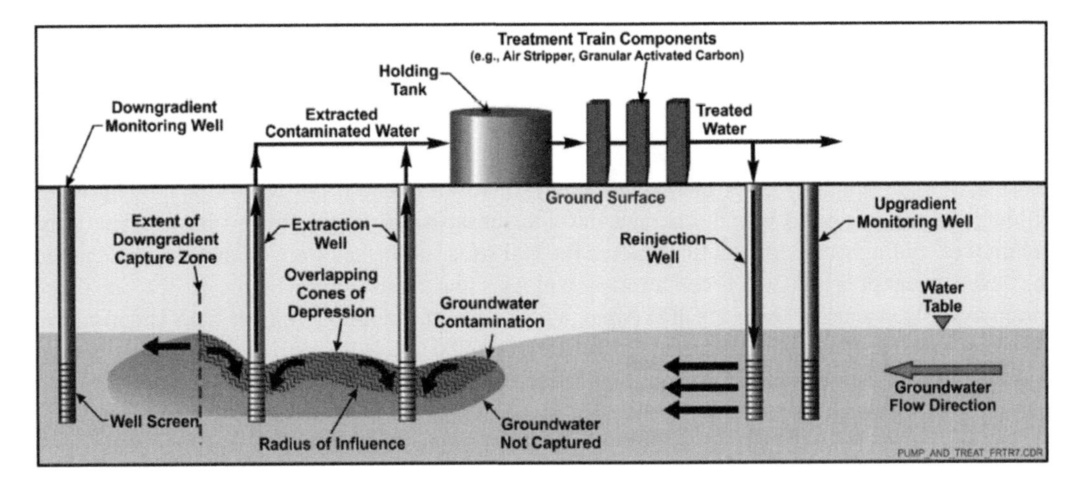


**FIGURE 10.2** General illustration of the pump-and-treat groundwater remediation process. Reprinted with permission FRTR (2020).

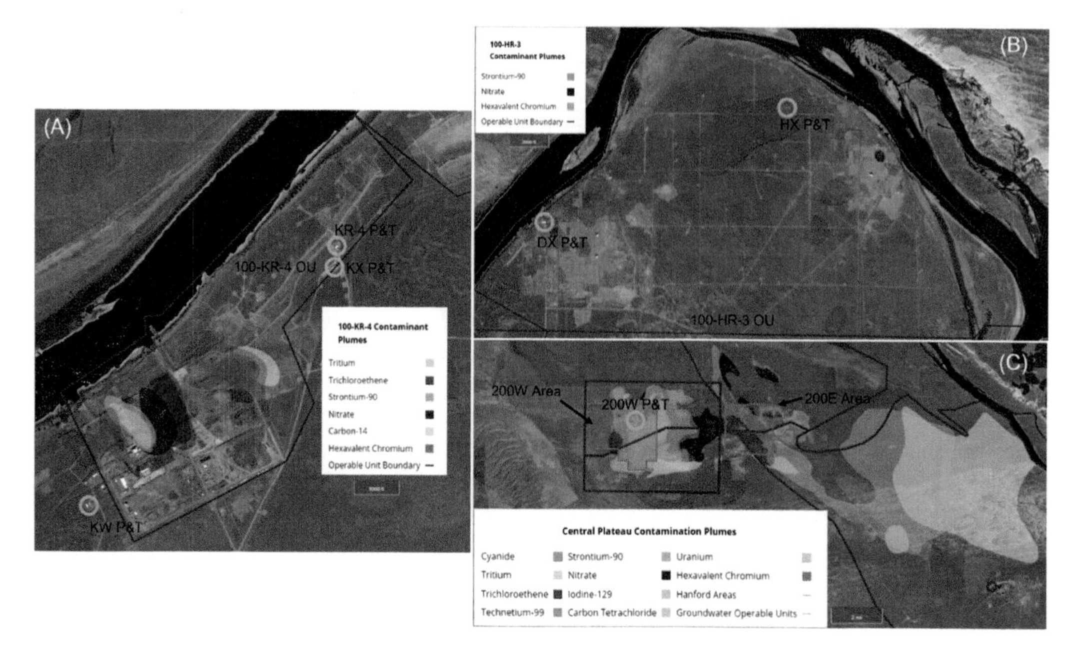


**FIGURE 10.3** Plume maps for the 100-KR-4 OU (a), 100-HR-3 OU (b), and a large section of the Central Plateau (c). Plume maps were generated from data as of January 31, 2023, using the PLATO module of the SOCRATES software application. Locations of the six Hanford P&T aboveground treatment facilities (yellow circles) are identified across the panels a–c. Courtesy of Sarah Saslow (Royer et al. 2021).

## 10.2 A BRIEF HISTORY OF HANFORD SITE OU 100-HR-3

The 100-HR-3 OU, located at the north-central portion of the Hanford Site along the Columbia River at what is casually referred to as the "Horn," is comprised of both the 100-D and 100-H Areas, and contains two groundwater interest areas (GWIA), 100-HR-D and 100-HR-H (Figures 10.1 and 10.3). There are two reactors located in the 100-D Area, and an additional reactor in the 100-H Area that operated within the time span of 1944–1967 (EPA/DOE/Ecology 2003, CHPRC 2014). During operation, sodium dichromate ($NaCr_2O_4$, comprised of sodium and Cr(VI)) was added to cooling water drawn from the Columbia River to prevent corrosion of the piping infrastructure. Once used in the single-pass reactor configuration, the cooling water effluent was stored in retention basins, allowed to cool, and then discharged to the bottom of the Columbia River using outfall pipes. However, the initial high temperatures of the cooling water (some near boiling) caused thermal expansion and cracking in the concrete that lined the retention basins (CHPRC 2014). These cracks provided pathways for large volumes of Cr(VI)-contaminated cooling water to enter the subsurface and eventually enter the groundwater. There were also times when this cooling water effluent (or portions of it) was discharged into the subsurface through unlined basins. The large volumes of cooling water effluent that entered the 100 Area subsurface were often dilute, but caused the development of broad Cr(VI) contamination plumes that persist today (Figure 10.3b). Localized groundwater concentrations in the 100 Areas with orders of magnitude higher Cr(VI) concentrations (e.g., $>400$ μg/L Cr(VI) (EPA et al. 2018)) are attributed to releases during the transfer, mixing, and handling of the sodium dichromate liquid stock solution and powder material that was mixed into the cooling water.

The use of P&T for Cr(VI) remediation in 100-HR-3 dates back to 1997 with the installation of the HR3 P&T system in the 100-H Area, followed by the DR5 system in the 100-D Area in 2004 (DOE/RL 2022b). In 2010 and 2011, these original systems were replaced with the larger and more efficient DX and HX P&T systems that are still used today. In accordance with the 100-HR-3

ROD (EPA et al. 2018), the present P&T systems are used to treat and hydraulically contain inland groundwater to Cr(VI) concentrations less than 48 μg/L and prevent impact on surface water (i.e., the Columbia River) at levels above 10 μg/L. Since 1997, at least 31,026 million L (8,196 million gallons) of groundwater has been treated, removing 2,709 kg of Cr(VI). In 2021 alone, P&T at the DX and HX P&T systems processed 2,402 million L (635 million gallons) of groundwater and removed 50.7 kg of Cr(VI). These P&T systems have decreased the plume sizes and overall Cr(VI) concentrations (Figure 10.4). A snapshot of the DX and HX performance metrics was published in the 2021 annual summary report for 100-HR-3 P&T operations and is provided in Table 10.1.

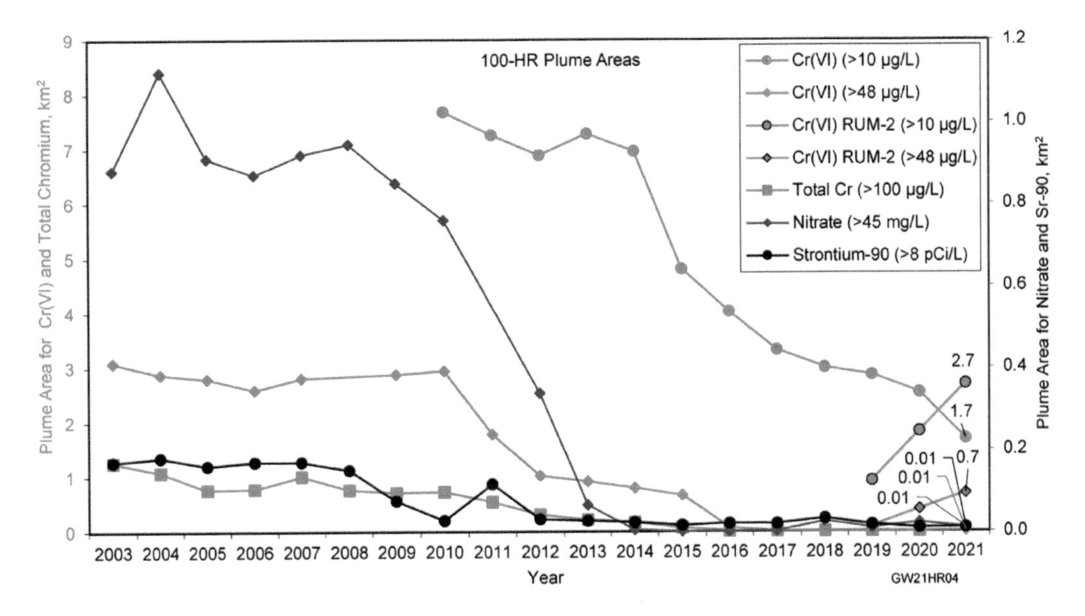


**FIGURE 10.4** Decrease in plume size as a function of time and plume concentration for 100-HR-3. Reprinted with permission from DOE/RL (2021).

**TABLE 10.1**
**P&T Performance Summary for DX and HX P&T Systems in the 100-HR-3 OU for Calendar Year 2021**

| **Groundwater Operable Unit** | **100-HR-3** | |
|---|---|---|
| P&T system | DX | HX |
| Design throughput (L/min [gal./min]) | 2,936 (775) | 3,407 (900) |
| Extraction wells | 45 | 43 |
| Injection wells | 14 | 20 |
| Volume treated (million L [million gal.]) | 1,334 (352) | 1,068 (282) |
| Average flow rate (L/min [gal./min]) | 2,538 (670) | 2,032 (538) |
| Cr(VI) mass removed (kg) | 17.05 | 33.65 |
| Average Cr(VI) influent concentration (μg/L) | 14.1 | 33.5 |
| Average Cr(VI) effluent concentration (μg/L)[a] | <2 | <2.3 |

*Source:* Reprinted from DOE/RL (2022b).

[a] The U.S. Environmental Protection Agency (USEPA) drinking water standard for total Cr is 100 μg/L (40 CFR 141 2023); however, the Hanford Site targets inland groundwater to Cr(VI) concentrations less than 48 μg/L (EPA et al. 2018).

## 10.3 PUMP-AND-TREAT SITE PLANNING

As outlined in Chapter 3, a significant amount of effort goes into characterizing a contaminated site. Feasible technologies are evaluated and tested. Regulatory documentation is issued to outline the objectives, approach, and success criteria before a remedial action can be implemented. This process (schematically illustrated in Chapter 3) is officially documented in a ROD (for CERCLA) and, as needed, in a ROD amendment or an ESD as remedial action continues. After the ROD has been approved, the P&T facility and system of extraction, injection, and monitoring wells can be constructed.

For the 100-HR-3 OU, the final ROD to document the approval of P&T as the remedy to treat Cr(VI) contamination was officially issued in 2018 (EPA et al. 2018). However, P&T remediation began much earlier because a ROD for an interim action was issued for the 100-HR-3 and 100-KR-4 OUs in 1996 (EPA/DOE/Ecology 1996). In this way, the site could begin treating Cr(VI) contamination while working through the process that ultimately led to the approval of the final ROD for the 100-HR-3 OU in 2018. The steps of this process were as follows:

1. **CERCLA Release:** Under CERCLA, the release and/or detection of a hazardous substance, like Cr(VI), is reported and initiates the remediation process (CFR 2022). The 100-HR-3 OU was listed on the National Priorities List (NPL) in 1989 for contaminated groundwater.
2. **Conceptual Site Model (CSM):** The CSM captures the nature and extent of contamination, as well as the (sub)surface characteristics, hydrogeology, and fate and transport mechanisms that may impact the remediation effort. An unconfined aquifer that is hydraulically connected to the Columbia River is the primary hydrogeological unit. Below this unconfined aquifer are several deeper aquifers in the Ringold Formation upper mud (RUM) that are also contaminated (CHPRC 2014, DOE/RL 2022c). During reactor operations, high-volume liquid discharges to the vadose zone elevated the water table in the unconfined aquifer and created a downward hydraulic gradient that drove contamination into these deeper aquifers (CHPRC 2014).
3. **Risk Assessment:** A risk assessment considers the current and potential future use of the land and resources, as well as potential pathways for exposure to the contamination. Due to the sodium dichromate used during reactor operations and the geochemical conditions of the Hanford Site, chromium introduced into the groundwater primarily persists as Cr(VI), which is toxic, carcinogenic, and significantly more mobile than other Cr oxidation states, for example, Cr metal (Cr(0)) or Cr(III) (Ertani et al. 2017). The maximum Cr(VI) level detected in the 100-HR-3 OU groundwater was 443 μg/L in 1996 when the interim action was issued (EPA/DOE/Ecology 1996), an order of magnitude higher than the cleanup level stated in the final ROD (EPA et al. 2018): 48 μg Cr(VI)/L for groundwater and 10 μg Cr(VI)/L for surface water or groundwater that discharges to surface water.
4. **Basis for Action:** COCs are identified that require a response action based on the current and future land and resource use. In the case of the 100-HR-3 OU, identified groundwater COCs include total chromium, Cr(VI), nitrate, and strontium-90, which are collocated in the 100-HR-3 OU.
5. **Remedial Action Objectives (RAOs):** RAOs are developed, identifying and addressing the media (e.g., groundwater) and COCs (e.g., Cr(VI)) that require remediation and defining cleanup goals. Three of the seven RAOs listed in the 100-HR-3 ROD (EPA et al. 2018) pertained to groundwater remediation that is addressed by P&T:
   - **RAO #1:** Prevent unacceptable risk to human health from ingestion of and incidental exposure to groundwater containing contaminant concentrations above federal and state standards and risk-based thresholds.

- **RAO #2:** Prevent unacceptable risk to human health and ecological receptors from groundwater discharges to surface water containing contaminant concentrations above federal and state standards and risk-based thresholds.
- **RAO #7:** Restore groundwater in the 100-HR-3 OU to cleanup levels, which include drinking water standards, within a timeframe that is reasonable given the particular circumstances of the site.

  The cleanup levels for the 100-HR-3 groundwater contamination are based on site-specific data, federal drinking water standards, state surface water quality standards (WAC 173-201A), and risk-based concentrations (WAC 173-340-720) (EPA et al. 2018).

6. **Remedial Alternative:** As a precursor to the final ROD for 100-HR-3, four remedial alternatives were developed for evaluation in the CERCLA process. A generalized summary of the four remedial alternatives is provided below:
   - **Alternative #1:** No Action. This is a required alternative that sets a baseline for comparison of other alternatives. Cr(VI) remediation for 100-HR-3 is not achieved in the 75-year period modeled.
   - **Alternative #2:** Incorporate an additional technology for Cr(VI) treatment. Specifically, this alternative would incorporate bioremediation treatment technology to convert Cr(VI) to less toxic and less mobile Cr(III), and would involve installing additional wells for improved access to contaminated groundwater. This approach would expect 100-HR-3 to reach Cr(VI) cleanup levels in ~25 years.
   - **Alternative #3:** Increase the P&T system capacity to treat more groundwater, with the intent of faster cleanup than with the interim remedy. This is achieved by increasing system throughput (gallons of influent pumped through the system per minute) and adding additional wells. This approach would expect 100-HR-3 to reach Cr(VI) cleanup levels in ~12 years.
   - **Alternative #4:** Expand overall P&T facility access to contaminated groundwater by adding new extraction and injection wells. This differs from Alternative #3 in that modifications to increase the P&T system throughput are not made. This approach would expect 100-HR-3 to reach Cr(VI) cleanup levels in ~39 years.
7. **Comparative Analysis:** The remedial alternatives were evaluated against the nine CERCLA criteria and against each other. These nine criteria include the following: (i) overall protection of human health and the environment; (ii) compliance with applicable or relevant and appropriate requirements; (iii) long-term effectiveness and performance; (iv) reduction of toxicity, mobility, or volume through treatment; (v) short-term effectiveness; (vi) implementability; (vii) cost; (viii) state acceptance; and (ix) community acceptance.
8. **Selected Remedy:** Alternative #3 was selected in the final ROD issued in 2018 (EPA et al. 2018) mainly due to higher ratings for short-term effectiveness, implementability, and community acceptance.
9. **Expected Outcome:** Under Alternative #3, upon implementation groundwater in 100-HR-3 is expected to reach cleanup levels within 12 years for Cr(VI) and will be returned to use as a drinking water source within approximately 44 years.

Remedial action progress is evaluated annually, and recommendations are made (if necessary) for changes to the P&T systems to improve performance and shorten the remediation timeframe. Changes recommended through a P&T performance assessment would need to be selected and enacted through a decision document such as a ROD amendment, explanation of significant differences (ESD), or equivalent. A remedial action completion report is needed for closure recommendations.

## 10.4 PUMP-AND-TREAT EXTRACTION OF GROUNDWATER

Design and implementation of an extraction and injection well network is fundamental to the P&T remedy. Extraction wells remove groundwater from the subsurface for treatment, and injection wells are used to return treated groundwater to the subsurface. When planning the location of these wells, key considerations include the contaminant plume distribution, the location of continuing sources, the location of receptors, and hydrologic conditions (including seasonal variations in aquifer hydraulic conditions). In the 100-HR-3 OU, there is an unconfined aquifer and several deeper semi- and fully confined aquifers in the RUM formation with Cr(VI) contamination. The unconfined aquifer is hydraulically connected to the Columbia River; thus, changes in river stage can impact water levels in the unconfined aquifer (DOE/RL 2022b). For instance, in the fall when the river stage is lower than inland groundwater levels, groundwater flow is in the direction toward the river. In the spring when the river stage increases due to snow melt, groundwater migration toward the Columbia River slows or horizontal hydraulic gradients are reversed and directed inland from the river (CHPRC 2014). In addition to these seasonal river stage changes, daily variation can occur from upstream dams that control the release of water down the Columbia River.

When water is extracted from a well, it creates a hydraulic head depression, whereas injection wells create groundwater mounds. Combined with an understanding of the seasonal hydraulic properties, the placement of injection and extraction wells can be designed to help control contaminant migration through hydraulic containment. For instance, in the 100-HR area, groundwater mounds created inland as a result of P&T reinjection can create outward flows that increase hydraulic gradients toward extraction wells. This can help increase access to contaminant mass for removal and treatment and hydraulically contain (or decrease migration of) a contaminant plume (DOE/RL 2021).

In addition to site hydrogeology, the contaminant plume location and migration will influence the design of the well network. Here, three different theoretical approaches to extraction well network design are summarized: (i) source control, (ii) plume containment, and (iii) plume reduction. See Truex et al. (2015) for additional details.

- *Source control* is used to minimize contaminant mass flux to downgradient areas, resulting in a detached dissolved-phase plume and facilitating remediation strategies such as monitored natural attenuation[2] or active remedies (Figure 10.5a). Extraction wells are positioned around and/or upgradient the contamination source to (i) prevent downgradient migration by extracting contaminated groundwater and (ii) alter the hydraulic gradient to reduce the driving force for downgradient migration.
- *Plume containment* may be used to minimize plume migration and protect receptors such as drinking water wells, the river, and aquatic habitats. Plume containment uses extraction wells positioned either within the plume (Figure 10.5b) or at the downgradient plume edge (Figure 10.5c). The scenario shown in Figure 10.5b is typically reserved for areas with known natural attenuation processes that can address the detached "bubble" of downgradient contaminant plume. In both cases, the placement of extraction wells is based on the objective of intercepting the dissolved-phase plume to prevent or minimize its continued expansion.
- *Plume reduction* is used to decrease the dissolved-phase plume volumetric extent and contaminant concentration (Figure 10.5d). Extraction wells are typically installed within the dissolved-phase plume to facilitate contaminant capture and mass removal.

The above strategies influenced the location of extraction and injection wells in the 100-H Area (Figure 10.6). Toward the riverfront, the number of extraction wells is much greater to prevent Cr(VI) contamination from reaching the Columbia River (plume containment). Additionally, areas where Cr(VI) concentrations are higher within the contaminant plume typically have additional extraction wells present for plume reduction.

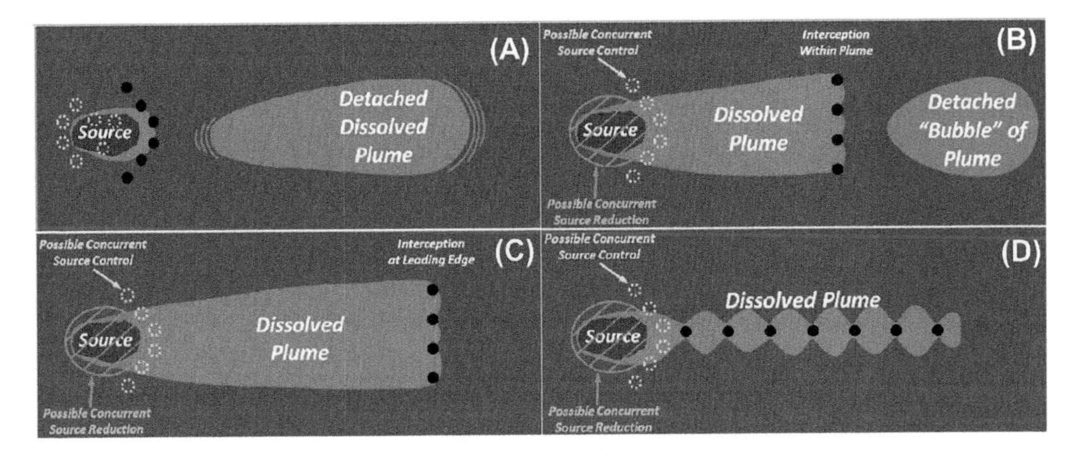


**FIGURE 10.5** Conceptual examples of P&T applied for contamination source control (a), plume containment (b and c), and plume reduction (d). In (b), extraction wells (black-filled circles) intercept within the plume area, thus creating a detached "bubble" of dissolved-phase plume that will diminish via natural attenuation or by alternative treatment technologies. Alternatively, (c) shows the placement of extraction wells at the downgradient edge of a plume to fully intercept the plume. The dashed circles indicate options for additional extraction wells. Figure from PNNL-24696. Reprinted with permission from Truex et al. (2015).

## 10.5 ABOVEGROUND TREATMENT

A variety of engineered treatment options exist for the aboveground treatment system, making that portion of the P&T system highly versatile. Selecting the most appropriate water treatment technology depends on the contaminant(s), effectiveness, cost, and other site-specific factors, for example, influent stream composition and variability. A summary of a range of engineered treatment options available for use in P&T facilities is provided in Table 10.2 (Marks et al. 1994, Casasso et al. 2020). In facilities that treat several COCs, more than one treatment technology may be used in a treatment train.

P&T aboveground water treatment technologies currently used in Hanford P&T systems, but not necessarily all, include the following:

- **Air Strippers:** For the removal of volatile organic compounds such as carbon tetrachloride and trichloroethene.
- **Granular Activated Carbon (GAC):** GAC adsorption is used to treat off-gases, including from air strippers.
- **Ion Exchange (IX):** For the removal of heavy metals, like Cr(VI), and radionuclides.

The HX and DX P&T systems both use IX to remove Cr(VI) from influent groundwater. IX involves passing contaminated water through vessels, referred to as columns, containing IX resin in a "bed." IX resins often consist of polymer beads designed with an open network structure to allow water and ions to pass through. As contaminated water flows through the IX column, the process involves a reversible chemical reaction in which contaminant ions in the groundwater are exchanged with ions on the IX resin (Abrams and Millar 1997). The ions that are selectively removed through IX can be controlled by the chemistry of the IX resin exchange sites themselves. In this way, IX resins can be designed to target specific contaminants based on ion properties such as size and charge (positive or negative, and magnitude).

Up until 2011, P&T in the 100-HR-3 OU used the DOWEX 21K IX resin (DuPont Water Solutions, Wilmington, DE), which exchanges chloride ($Cl^-$) anions for chromate ($Cr(VI)O_4^{2-}$) anions to treat Cr(VI) in groundwater. However, an evaluation of the SIR-700 weak base hybrid IX

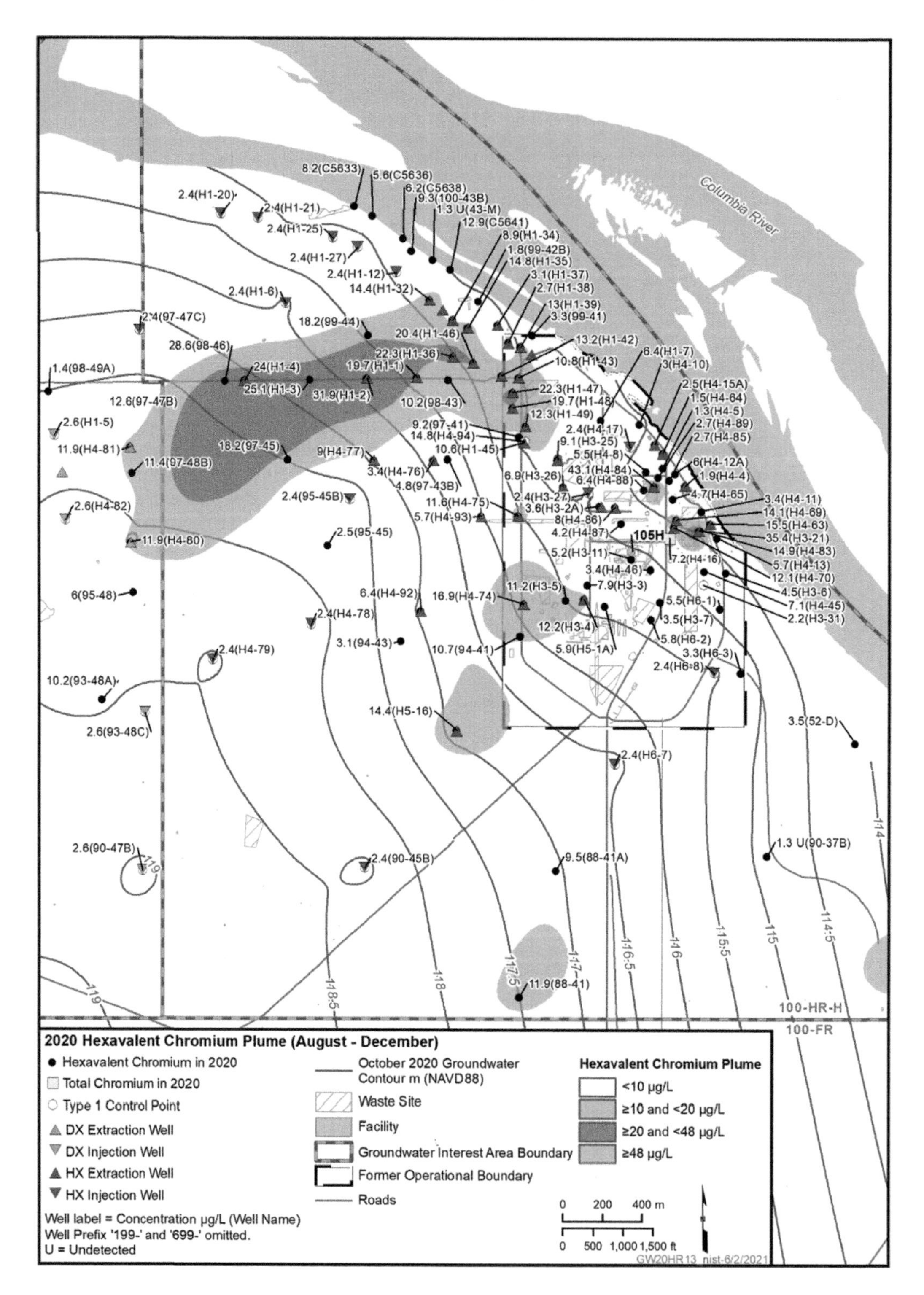


**FIGURE 10.6** 100-H area map of 2021 Cr(VI) plumes, along with October 2021 water-level contours (during low river stage). Extraction and injection wells are also shown. Reprinted with permission from DOE/RL (2022b).

**TABLE 10.2**
**Summary of Engineered Treatment Technologies Used to Treat Different Contaminants in P&T Facilities (Marks et al. 1994; Casasso et al. 2020)**

| | Heavy Metals | Hexavalent Chromium | Arsenic | Mercury | Cyanide | Corrosives | Volatile Organics | Ketones | Semi-volatile Organics | Pesticides | PCBs | Dioxins | Floating Products |
|---|---|---|---|---|---|---|---|---|---|---|---|---|---|
| Activated carbon | P | P | P | Y | N | N | Y | N | Y | Y | Y | Y | P |
| Air stripping | N | N | N | N | N | N | Y | Y | N | N | N | N | N |
| Biological | N | N | N | N | P | N | P | Y | Y | P | P | P | P |
| Chemical oxidation | N | N | P | N | Y | N | Y | Y | Y | Y | Y | P | N |
| Coprecipitation/coagulation | Y | N | Y | Y | N | N | N | N | P | P | Y | Y | Y |
| Distillation | N | N | N | N | N | P | Y | Y | Y | Y | Y | Y | Y |
| Electrochemical | Y | Y | N | N | P | N | N | N | N | N | N | N | N |
| Evaporation | Y | Y | N | N | Y | N | N | N | P | P | Y | Y | Y |
| Filtration | Y | N | Y | Y | N | N | N | N | N | Y | Y | Y | P |
| Flotation | N | N | N | N | N | N | N | N | P | P | Y | Y | Y |
| Gravity separation | Y | N | P | P | N | N | N | N | P | P | Y | Y | Y |
| Ion exchange | Y | Y | Y | Y | Y | N | P | N | Y | Y | Y | Y | Y |
| Membrane separation | Y | P | Y | P | Y | N | P | N | Y | Y | Y | Y | Y |
| Neutralization | N | N | N | N | N | Y | N | N | N | N | N | N | N |
| Precipitation | Y | Y | P | Y | N | Y | N | N | P | P | Y | Y | Y |
| Reduction | P | Y | N | Y | N | N | N | N | N | N | N | N | N |
| Steam stripping | N | N | N | N | N | N | Y | Y | Y | P | N | N | N |
| UV/ozone | N | N | P | N | Y | N | P | P | Y | Y | Y | Y | N |

Technology applicability for treating different contaminants is indicated with Y (Yes), P (Potentially), and N (No).

resin (ResinTech®, Inc.; Camden, New Jersey, USA) showed that a significantly higher capacity for Cr(VI) removal could be achieved. This resulted in the discontinued use of DOWEX 21K in favor of the SIR-700 IX resin in 2011 (CHPRC 2010). The SIR-700 resin can remove more Cr(VI) mass by combining the IX process with reduction of Cr(VI) to Cr(III). Specifically, Cr(VI) in the form of $HCr(VI)O_4^-$ exchanges for the resin anion ($Cl^-$ or $HSO_4^-$), and then undergoes a second step where Cr(VI) is transformed into the less toxic and less mobile Cr(III), thereby being immobilized within the resin (Meyers 2019, Saslow et al. 2023). In the case of the SIR-700 resin used in the 100 Area P&T facilities, the resin is purchased with chloride ($Cl^-$) as the exchange anion (communication from Central Plateau Cleanup Company[3]). The SIR-700 resin's unique epoxy backbone structure functionalized with mixed amine sites allows Cr(VI) removal to follow this hybrid, pH-dependent process that is optimized at pH 5. Because the mechanism for removal is pH-dependent, the influent groundwater, which typically has a pH above pH 7, is collected in a holding tank where it is pH-adjusted using sulfuric acid (Figure 10.7) prior to going through the IX column. The target final pH of the acidified influent was initially pH 5 when the Cr(VI) concentrations in the influent were very high. For example, the DX P&T system saw an average influent Cr(VI) concentration of ~800 µg/L in calendar year 2012 (PNNL 2021). Since 2015, the target pH has increased to ~pH 6.7 as Cr(VI) concentrations in the influent have decreased. The SIR-700 Cr(VI) removal mechanism is still achieved at this higher pH (also shown later in this chapter in Figure 10.10).

For the HX and DX IX treatment, four IX columns are grouped together (Figure 10.7) (Pryzbylski and Esparza 2010, Neshem 2011, DOE/RL 2018), with the system piping including options to route flow through two (lead-lag), three (lead-lag-polish), or four (lead-lag-lag-polish) columns. A split-flow configuration (two groups of two columns) is also used as shown in Figure 10.7. Sample ports are located after each treatment vessel to monitor the treatment performance with respect to

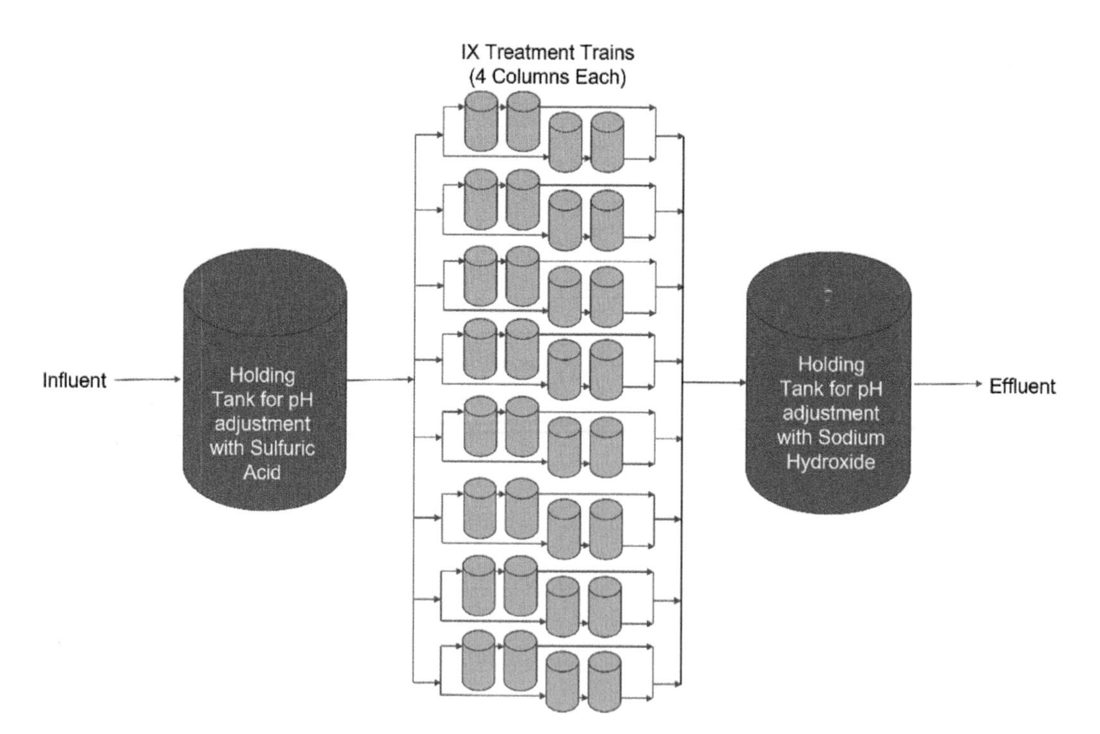


**FIGURE 10.7** Simplified schematic of the HX P&T facility that includes pre-IX pH adjustment, splitting the process stream across 16 lead-lag IX column pairs (in eight groups of four reconfigurable columns), followed by pH adjustment back to conditions suitable for injection into the aquifer. Each column contains ResinTech® SIR-700 IX resin for removal of Cr(VI). The setup is the same for the DX P&T facility, except there are six groups of four IX columns.

Cr(VI) concentration (breakthrough). Each IX process stream has a maximum flow rate of approximately 100 gallons per minute (gpm). In the HX P&T facility, with a total of 8 IX column groups, the maximum total capacity of the facility is 800 gpm. With 6 IX column groups for the DX P&T, the maximum total flow rate is 600 gpm. However, the P&T systems are not always operated at 100% capacity, but rather fall within the range of 50%–100% to account for variations in well pump operations. The multiple IX column groupings allow the P&T system to continue operations during any planned or unplanned maintenance of the system. Because the SIR-700 resin is a single-use resin and does not undergo regeneration, the resin is replaced at approximately four- to five-year intervals when Cr(VI) concentrations begin to increase in the effluent measured immediately after an IX vessel (CHPRC 2016a, DOE/RL 2021). Most IX resin replacements occur in the lead (or first) IX vessel of a treatment train and Cr(VI) that passes through the lead IX vessel before a resin change-out is captured by the next (lag) IX vessel. Spent SIR-700 resin is disposed of at the Hanford Environmental Restoration Disposal Facility (ERDF) (CHPRC 2010).

Influent groundwater is passed through the set of IX columns, with any groundwater not achieving the target cleanup levels being routed through a recirculation line for reprocessing. Once fully treated, the effluent is collected in a holding tank and adjusted to a neutral pH (6.5–8.5) using sodium hydroxide. The treated effluent is then delivered back to the subsurface through a network of injection wells.

As already mentioned, integral to the design of the P&T system is the ability to monitor system performance, which is achieved through sampling ports located before and after key steps in the groundwater treatment process and as outlined in the *100-HR-3 OU Pump-and-Treat System Operations and Maintenance Plan* (CHPRC 2016a). For evaluating system efficacy for Cr(VI) remediation, water samples are collected from extraction wells, influent tanks, after each IX resin vessel, and effluent tanks to assess system performance and to determine when the system is nearing a resin change-out event. These data also support groundwater monitoring and modeling efforts discussed in the next section. In addition to Cr(VI) concentrations, a network of sensors and transmitters collect data on key parameters like flow rates, water levels in the tanks or treatment vessels, well water levels, temperature, and pH, to understand if the system is operating effectively. These data feed into the Supervisory Control and Data Acquisition (SCADA) system, which is an automated control system for operations and equipment (CHPRC 2016a).

While remediation of Cr(VI) groundwater contamination is the main driver for P&T at the 100-HR-3 OU, the Atomic Energy Act of 1954 is the basis for sitewide radiation monitoring efforts at the Hanford Site (DOE/RL 2015). As part of the 2015 groundwater monitoring plan, effluent monitoring is required to prevent public and ecological exposure to radiation and for managing discharges that could create or increase the size of plumes requiring remediation. The effluent generated from the DX and HX P&T systems is sampled prior to injection back into the aquifer and is analyzed for radiological dose (target 100 mrem/yr or less to the public), primarily from the radionuclides tritium ($^3H$), technetium-99, strontium-90, and uranium. In 2021, DX and HX P&T effluents were sampled in February, April, July, and October. All effluents had an overall derived concentration standard (DCS)–based total effective dose (TED) that was <1 mrem/year. This value is significantly lower than both the target dose limit of 100 mrem/year and the yearly average dose for people living in the U.S., 620 mrem, according to the National Council of Radiation Protection and Measurements (NCRP 2009). The calculated beta/photon emitter drinking water dose was below the 4 mrem/yr maximum contaminant level (MCL) for both systems, and the total uranium mass concentration was below the 30 μg/L MCL; uranium was not detected in any of the DX samples and was found in only one of the HX sampling events. Therefore, no changes in the effluent monitoring sampling and analysis frequency or analytical approach were needed for 2022 operations.

Operation of the HX and DX P&T systems has resulted in year-over-year decreases in the Cr(VI) groundwater plume extent (Figure 10.4) due to effective treatment of high-concentration areas that results in less available Cr mass available for cleanup. Eventually, HX and DX P&T systems will reach a point of diminishing returns in terms of contaminant mass extracted from the groundwater aquifer. At this point, a performance assessment will be conducted to determine the next steps, which could include continued operation with optimization, transition to another treatment approach, or closure. This process is described after the next section.

## 10.6 GROUNDWATER MONITORING AND MODELING

Groundwater monitoring is "designed to track changing conditions, performance of the remedy, and effectiveness of ... remedial actions in meeting performance criteria required by the interim action ROD" (DOE/RL 2016). As with the P&T system itself, the *Sampling and Analysis Plan for 100-HR-3 Groundwater Operable Unit Monitoring* (DOE/RL 2016) serves as the governing plan for sampling and analysis activities that support groundwater monitoring objectives.

In addition to monitoring Cr(VI) concentrations in the P&T aboveground treatment facility, P&T remedy progress is also monitored by assessing contaminant concentrations in samples collected from extraction wells, monitoring wells, and aquifer tubes[4] located throughout the OU. Samples are periodically collected, and the results are discussed in groundwater monitoring reports (e.g., DOE/RL 2021), used to update plume maps, and used to qualitatively evaluate the effectiveness of remedial activities. For the 100-HR-3 OU, groundwater samplings occur several times throughout the year, but particular attention is paid to sampling during the fall when the river levels are low and the groundwater gradients toward the river are relatively higher (DOE/RL 2022c). Contaminant concentrations will vary seasonally; however, by testing in the fall, contamination is typically more concentrated downgradient of the contamination source relative to other seasons. Testing that takes place in the spring when water levels were higher could measure lower contaminant plume concentrations caused by mixing with river water. For the 100-HR-3 OU, in addition to total and hexavalent Cr, COC nitrate and strontium-90, along with detectable uranium (though not a COC for this OU), are also monitored in the aquifer (DOE/RL 2022c).

Monitoring also includes the hydraulic head (water level) in the remediation area to understand the effects of P&T on the groundwater flow directions and gradients (DOE/RL 2022b). Because an evaluation of hydraulic properties requires many wells to be sampled over a short period of time to compile a temporally consistent dataset, automated data-logging pressure transducers (automated water-level network [AWLN]) are used for some wells. The more wells that are sampled, the greater confidence in the monitoring system and technical data. In the 100-HR-3 OU, there are 81 AWLN stations, as well as two river gauges, that record water-level measurements hourly (DOE/RL 2022b). Water-level data are also collected at each extraction and injection well. These data collectively are used to prepare water elevation maps that track the effect of P&T on groundwater.

Groundwater monitoring data are used to evaluate how well the environment is protected from groundwater contamination. At Hanford, qualitative and quantitative river protection evaluations in the 100-HR-3 OU take into consideration the hydraulic effects of P&T operations, changes in the Columbia River discharge boundary conditions, and Cr(VI) distribution in groundwater to classify wells and shorelines as not protected, protected, or protected with action potentially required for long-term performance. A key difference between a qualitative and quantitative assessment for this application is that a quantitative assessment compares measurable contamination and hydraulic containment extents, whereas a qualitative assessment will also consider the transient effects of hydraulic capture, for example, duration/magnitude of hydraulic gradients, pumping well location(s), and concentration trends, in the evaluation. Table 10.3 presents the results of a qualitative assessment for shoreline protection in the 100-D and 100-H Areas in both 2020 and 2021 calendar years, which suggests shoreline protection is sustained due to P&T operations in this OU.

Data collected during groundwater monitoring activities are also used to support groundwater modeling efforts that predict future outcomes for remediation actions, as well as estimate time until remediation completion and P&T system shutdown. For example, earlier sections in this chapter discussed how Cr(VI) contamination can be controlled through hydraulic containment. The effects

**TABLE 10.3**
**Qualitative Assessment of Shoreline Protection in the 100-HR-3 OU in 2020 and 2021**

| 100-HR-3 | 100-D | | 100-H | |
|---|---|---|---|---|
| **Year** | **2020** | **2021** | **2020** | **2021** |
| Protected | 2,600 m (79% of shoreline) | 2,800 m (85% of shoreline) | 4,300 m (98% of shoreline) | 3,600 m (82% of shoreline) |
| Protected – action may be required | 600 m (18% of shoreline) | 400 m (12% of shoreline) | 100 m (2% of shoreline) | 800 m (18% of shoreline) |
| Not protected | 100 m (3% of shoreline) | 100 m (3% of shoreline) | 0 m (0% of shoreline) | 0 m (0% of shoreline) |

*Source:* From DOE/RL (2022b).

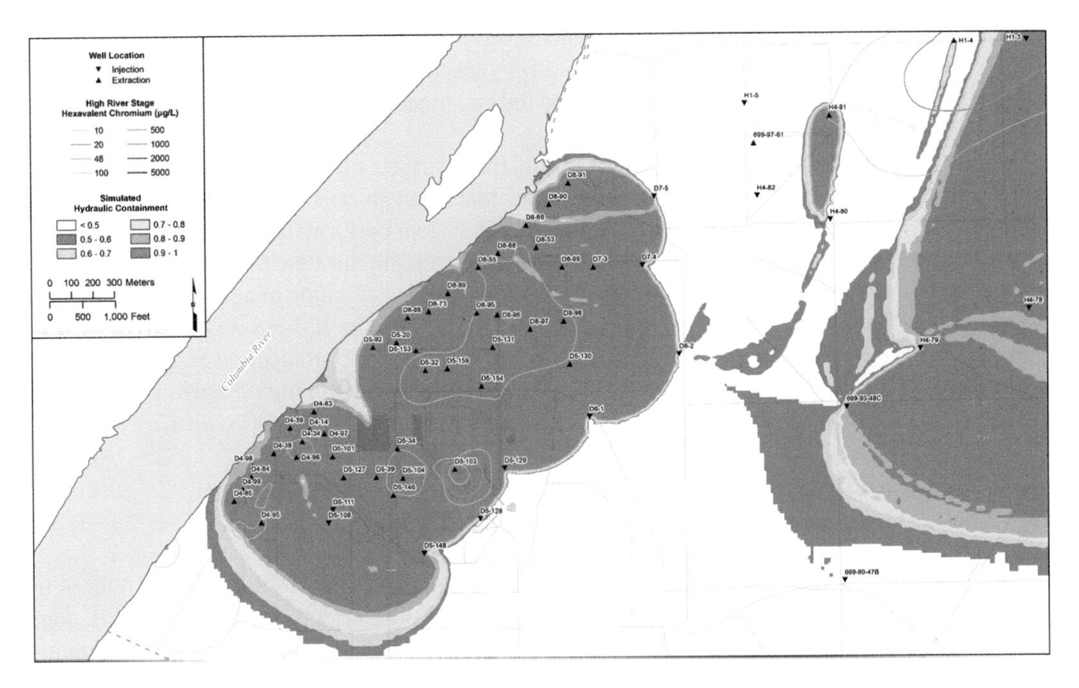


**FIGURE 10.8** Example of a simulated capture frequency map (SCFM) for the 100-D Area at high river stage that ranks hydraulic containment of Cr(VI) contamination on a scale of 0–1, where 0 indicates no containment and 1 indicates containment under all test conditions. Reprinted with permission from DOE/RL (2022b).

of this can be modeled using aquifer hydraulic head, extraction and injection flow rates, and water quality (Cr(VI) concentrations) data inputs to generate capture frequency maps (CFMs) (DOE/RL 2022b). A CFM visually captures how groundwater and contaminants move toward extraction wells. Over the course of one year, several of these maps are generated to represent different groundwater conditions caused by seasonal changes. For these CFMs, hydraulic containment is mapped on a scale of 0–1, where 0 represents no hydraulic containment under any condition encountered, and 1 represents hydraulic containment under all conditions simulated (groundwater was always moving toward an extraction well). Figure 10.8 provides an example of a CFM for the 100-D Area in the 100-HR-3 OU.

In some cases, there is more than one modeling approach that can be used to predict remediation performance. In such instances, multiple approaches may be compared to one another to increase confidence in the predicted outcome. Indeed, there are two methods of evaluation used to generate CFMs:

- Water-level mapping, which relies on monthly average groundwater elevations, pumping rates, and Columbia River data to produce what are referred to as interpolated capture frequency maps (ICFMs) using the methods described in SGW-42305, *Collection and Mapping of water Levels to Assist in the Evaluation of Groundwater Pump-and-Treat Remedy Performance* (CHPRC 2009).
- Groundwater modeling, which uses the 100-Area Groundwater Model (SGW-46279 *Conceptual Framework and Numerical Implementation of 100 Areas Groundwater Glow and Transport Model* (CHPRC 2016b)), monthly pumping rates, the Columbia River stage, and other time-varying boundary conditions to produce what are referred to as simulated capture frequency maps (SCFMs).

Specific details for development and use of these models are provided in report ECF-HANFORD-21–0030 (CPCCo 2021a). The two models thus consider both the P&T system and Columbia River impacts on the groundwater levels to identify areas where hydraulic containment is effective throughout the year and where other areas may need immediate or future attention. Confidence is gained in the modeled results when the two models produce similar maps. Where the maps differ, there is less confidence in the result.

Monitoring data are also used to support optimization efforts of the P&T systems themselves. An example of this is the pumping optimization model and interface, based on the 100-Area Groundwater Model, used by scientists to evaluate P&T system performance and identify potential alternative well configurations that may include changes to pumping flow rates, well realignment (extraction to injection well conversion, and vice versa), and/or installation of new wells to maintain or improve P&T performance. Recent optimization objectives for the 100-HR-3 OU are outlined in the report SGW-66695-VA, *FY22 Pump & Treat Remedial Process Optimization Scope, 100-HR-3 Groundwater Operable Unit*, which specifically aimed to use the pumping optimization model to evaluate expected extraction/injection well effects on plume capture in the OU (CPCCo 2021b).

## 10.7 PERFORMANCE ASSESSMENT FOR DECISIONS ABOUT P&T OPTIMIZATION, TRANSITION, OR CLOSURE

After a period of operation, a P&T remedy will typically reach a point of diminishing returns in terms of contaminant mass extracted from the groundwater aquifer. Diminishing returns in contaminant mass extraction result from trade-offs between bulk groundwater extraction and contaminant mass transfer limitations (e.g., due to desorption, dissolution, or diffusion, where the latter may be matrix diffusion or diffusion from contamination outside the swept area). As a P&T system approaches a point of diminishing returns, the remedy may be looking at many further years of operation while only removing small amounts of contaminant mass. Full-scale operation in this mode may not be desirable from the perspectives of cleanup timeframe or cost-effectiveness given, for example, at Hanford the total annual cost for operating and maintaining a single 100 Area P&T system from 2012 to 2019 ranged from $1.3M to $6.4M (DOE/RL 2022a).

Truex et al. (2015) describe a structured approach for P&T performance assessment. Initially, the conceptual site model (CSM) is updated to incorporate any new information from characterization, monitoring, and operations. Then, a set of decision elements are used to work through the decision logic (reprinted here in Figure 10.9) to determine the most appropriate outcome for the P&T system. The first step in the decision logic is to determine if RAOs have been met, for example, are Cr(VI) concentrations at or below cleanup levels? If yes, the P&T system may be considered for *closure*. However, if not, contaminant concentrations and trends, contaminant mass discharge, attenuation capacity of the aquifer, future plume behavior, time to reach RAOs, and P&T system cost are the

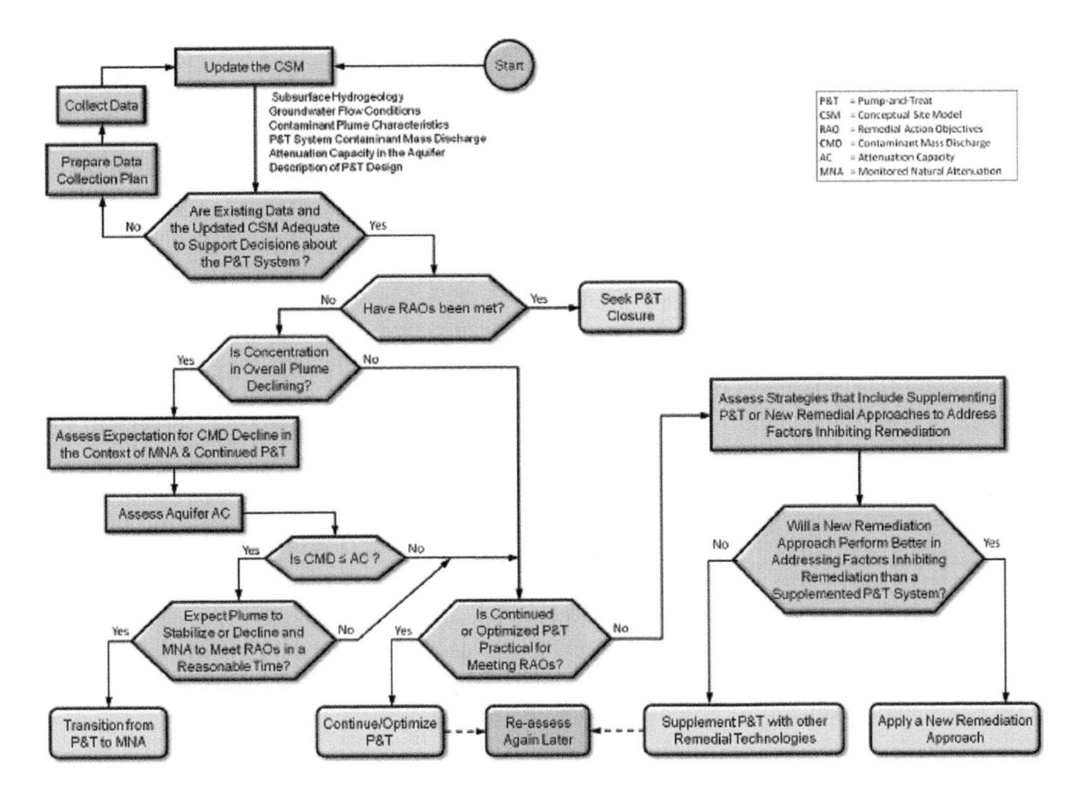


**FIGURE 10.9** Decision logic for determining P&T performance assessment outcomes of continued operation (perhaps with optimization), transition to another treatment approach, or closure. Reprinted with permission from Truex et al. (2015).

decision elements considered in establishing the best path forward. Contaminant concentration and trends would indicate if a contamination plume has declined during P&T. If it has, then the contaminant discharge (mass of contaminant per unit time) should be compared to the aquifer's natural attenuation capacity to determine which is greater. If mass discharge is less than attenuation capacity, then P&T could *transition* to monitored natural attenuation. If attenuation capacity is less than mass discharge, or if the contamination concentrations are overall not declining, an evaluation for P&T optimization is conducted. If RAOs can be met through continued operation of an optimized P&T system, then such optimization changes are planned an implemented. However, if continued operation or P&T optimization will not meet RAOs, the outcome would consider whether P&T can be supplemented by other remediation technologies or whether a different remedy from P&T is needed. Depending on the indicated outcome, the performance assessment may need to be revisited in the future. This performance assessment process could be incorporated into an adaptive site management approach (see sidebar).

## ADAPTIVE SITE MANAGEMENT (ASM)

As defined by the U.S. Environmental Protection Agency, ASM is a systematic technical approach to managing complex hazardous waste site projects or sites that is built upon rigorous site planning and understanding of the site conditions and uncertainties. ASM encourages routine re-evaluation and prioritization of site cleanup efforts that take into consideration new and changing site conditions to support effective and efficient site remediation progress (EPA 2020, 2022a, 2022b; Demirkanli 2021; ITRC 2017).

This decision logic and the groundwater monitoring data available support continued P&T operations in the 100-HR-3 OU because Cr(VI) contaminant plumes continue to decrease in size and concentration, although Cr(VI) concentrations remain >48 μg/L in some areas of the unconfined aquifer (DOE/RL 2022b). As P&T operations continue in the 100-HR-3 and other OUs across the Hanford Site, this decision logic approach or equivalent will be used to continue assessing the efficacy of P&T remedies.

The "Suite of Comprehensive Rapid Analysis Tools for Environmental Sites" (SOCRATES, https://socrates.pnnl.gov) application developed by Pacific Northwest National Laboratory (Truex and Johnson 2017) supports decision-makers by providing access, visualization, and analysis of Hanford environmental data. SOCRATES goes beyond simple data visualization by providing quantitative decision support through several modules built on computational algorithms to answer specific objectives relevant to the decision-making process and eventual remediation success. For instance, the Hydraulic Pump-And-Treat Information Analytics (HYPATIA) module (Figure 10.10 (PNNL 2021)) provides access to groundwater monitoring data from extraction wells, in-plant chemistry data, and plant sensor data that, together, can be used to assess P&T system performance as a whole or for a unit operation. This tool is unique because it combines these disparate data sources, with sensor data from the automated on-site SCADA system and chemistry data pulled from the separate Hanford Environmental Information System (HEIS) database (Rieger 2013). The aggregated data from these two sources allow for a more comprehensive analysis and visualization of the data that prove especially useful for developing and updating CSMs, for assessing treatment performance, and for system optimization. The intent of HYPATIA is to make data more accessible and digestible for decision-makers as they determine the future of P&T systems at the Hanford Site.

## 10.8 CLOSING REMARKS

The use of P&T systems as a groundwater contaminant plume remedy via both treatment and hydraulic containment is an established remediation approach that is used all over the world, including at the Hanford Site. All six of the P&T systems located across the Hanford River Corridor (100 Areas) and Central Plateau (200 Areas) rely on frequent evaluation and monitoring of facility and groundwater conditions to successfully operate the P&T systems and optimize groundwater treatment to meet remediation objectives. This involves a network of injection, extraction, and monitoring wells that connect to and/or inform what treatment technologies and operating conditions, for example, flow rate, are used at the aboveground P&T facility. Modeling tools that assess current P&T system performance and predict future P&T performance are used to inform site contractors and relevant stakeholders of remediation progress and guide decisions for continuing P&T operation, system optimization actions, transition to another remediation approach, or eventual closure.

The use of P&T for groundwater remediation at Hanford is not a fast treatment solution, primarily because subsurface and chemical properties in the aquifer control the rate of mass capture and removal. Under the selected remedy alternative for 100-HR-3 OU, it was predicted to take until about year 2030 to reach the target Cr(VI) cleanup levels. With decades-long operational plans, sustained monitoring, optimization, and adaptability of both the aboveground water treatment facility and supporting infrastructure, such as installing or decommissioning wells, are necessary to adapt to changing subsurface and contaminant conditions as treatment progresses. Adaptability across Hanford P&T systems has allowed the site to remove a significant amount of contamination from the site groundwater. Since the start of P&T remedial actions for the 100-HR-3 OU in 1997 through the end of calendar year 2020 (including both interim and final remedies), at least 28,000 million L (>7,500 million gallons) of groundwater has been treated and >2,600 kg of Cr(VI) removed (DOE/RL 2021). The three P&T systems in the 100-KR-4 OU (KR4, KW, and KX)

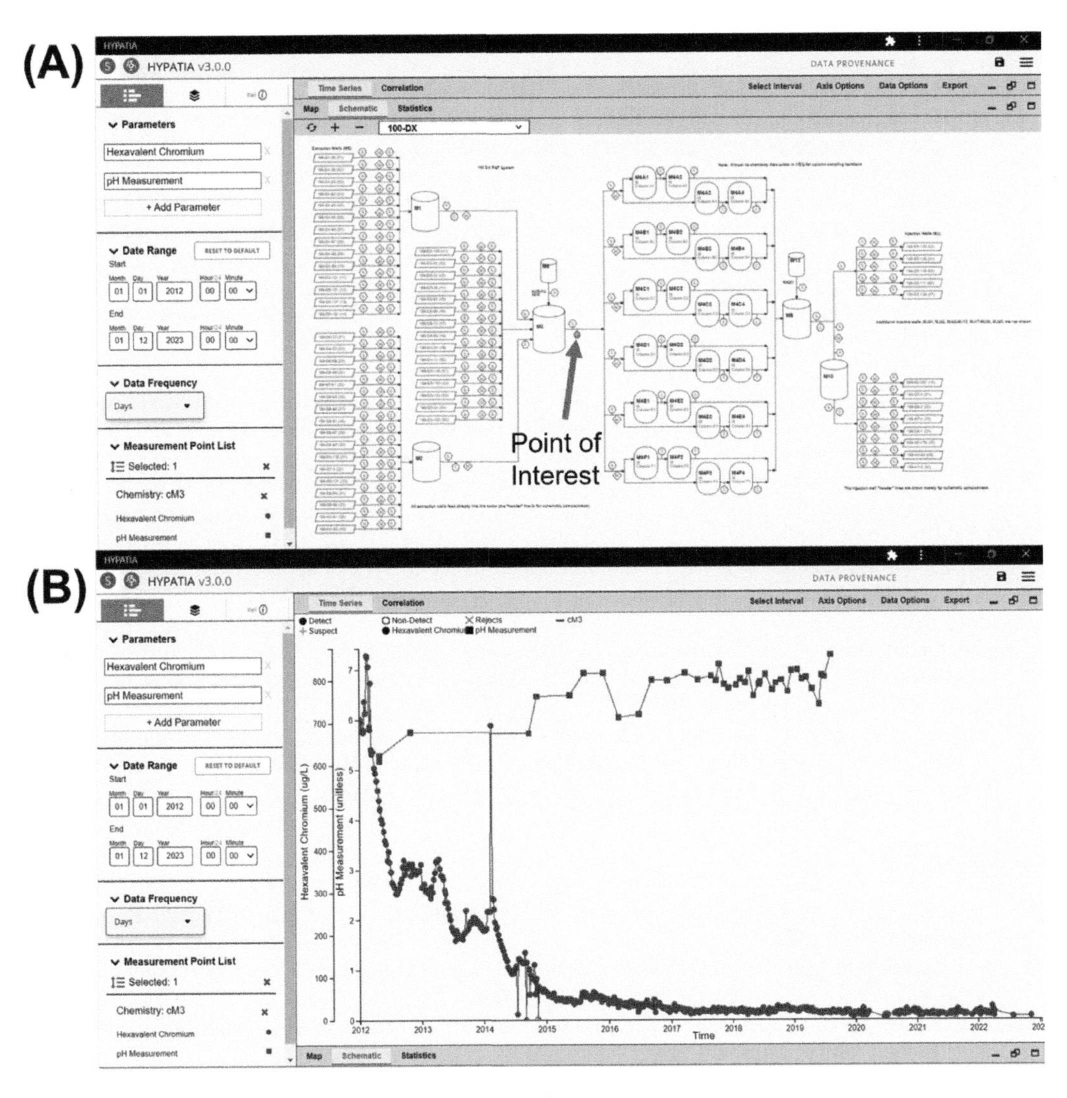


**FIGURE 10.10** Example of how HYPATIA may be used to interpret P&T facility performance over several years. (a) The DX P&T facility schematic with sensor and sample collection point locations that, when selected, allow the user to view relevant facility and contaminant concentration data. For this example, the plant chemistry icon after the acidification holding tank (identified with the large arrow annotation) is selected. (b) Data for the parameters of interest – hexavalent chromium and pH measurement – are plotted for the identified date range (January 2012 to January 2023). This presentation of the data shows that as the Cr(VI) concentration decreased, the system target pH for influent entering the SIR-700 treatment trains increased from ~pH 5 to ~pH 6.7. Courtesy of Sarah Saslow (PNNL 2021).

have removed an additional ~1,000 kg of Cr(VI) from Hanford-contaminated groundwater. The 200W P&T has treated nearly 30,000 million gallons of groundwater and perched water from the Central Plateau since 2012, and has removed approximately 19,000 kg of carbon tetrachloride, 90 kg of trichloroethene, 500 kg of Cr(VI), 900 g of technetium-99, and 700 kg of uranium (DOE/RL 2022a). These metrics highlight the magnitude of both the contamination challenges the Hanford Site faces and how the P&T remediation approach has been successfully used to move closer to cleanup objectives.

## NOTES

1 As defined by the U.S. Geological Service (https://www.usgs.gov/faqs/what-groundwater, https://or.water.usgs.gov/projs_dir/willgw/glossary.html), groundwater is water that exists underground in saturated zones beneath the land surface, with the uppermost surface of the saturated zone being called the water table. Saturated subsurface media that yields a usable amount of water is termed an aquifer. Aquifers may be referred to as unconfined (without an impermeable layer immediately above), confined (located underneath an impermeable layer and being under pressure), or a perched aquifer (a saturated zone situated on top of impermeable material in the vadose zone and not connected to the water table aquifer (Oostrom et al. 2013). The unconfined and lower confined aquifers extend across the Hanford Site. Perched water zones within the Hanford Site are areas that are not laterally extensive and are a result of natural infiltration and percolation of discharges from site operations.

2 Per EPA (1999c, 2012), monitored natural attenuation (MNA) "refers to the reliance on natural attenuation processes (within the context of a carefully controlled and monitored site cleanup approach) to achieve site-specific remediation objectives within a timeframe that is reasonable compared to that offered by other more active methods." Natural attenuation processes include a variety of physical (e.g., dispersion, dilution, volatilization, radioactive decay), chemical (e.g., sorption, stabilization, transformation), or biological (e.g., biodegradation, stabilization) processes that reduce the mass, toxicity, mobility, volume, or concentration of contaminants. MNA can apply to multiple contaminant types, including organics, inorganics, metals, and radionuclides (EPA 1999a, 1999b, 2007a, 2007b, 2010, 2015).

3 Email communication sent on February 22, 2022, from Central Plateau Cleanup Company to Rob D. Mackley at Pacific Northwest National Laboratory. This email was forwarded by Rob D. Mackley to Sarah A. Saslow on January 12, 2023.

4 Aquifer tubes are small-diameter, flexible tubes with a screen on one end that prevents sediment from clogging the tube. These tubes are installed along the Columbia River shoreline at different depths where the water table is shallow (DOE/RL 2022c). A pump is used to extract groundwater subsamples.

## REFERENCES


40 CFR 141 (2023). *National Primary Drinking Water Regulations.* U.S. Code of Federal Regulations, Washington DC, USA.

Abrams, I. M. and J. R. Millar (1997). "A history of the origin and development of macroporous ion-exchange resins." *Reactive and Functional Polymers* 35(1): 7–22.

Casasso, A., T. Tosco, C. Bianco, A. Bucci and R. Sethi (2020). "How can we make pump and treat systems more energetically sustainable?" *Water* 12(1): 67.

CFR, U (2022). *40 CFR Part 302.4.* Code of Federal Regulations, Washington, DC.

CHPRC (2009). *Collection and Mapping of Water Levels to Assist in the Evaluation of Groundwater Pump and Treat Remedy Performance.* SGW-42305. CH2M Hill Plateau Remediation Company, Richland, WA.

CHPRC (2010). *100 Area Groundwater Chromium Resin Management Strategy for ION Exchange Systems.* SGW-46621 Rev 0. CH2M Hill Plateau Remediation Company, Richland, WA.

CHPRC (2014). *Remedial Investigation/Feasibility Study for the 100-DR-1, 100-DR-2, 100-HR-1, 100-HR-2, and 100-HR-3 Operable Units.* DOE/RL-2010-95, REV. 0. CH2M Hill Plateau Remediation Company, Richland, WA.

CHPRC (2016a). *100-HR-3 Pump and Treat System Operations and Maintenance Plan.* DOE/RL-2013-49 Rev 0. CH2M Hill Plateau Remediation Company, Richland, WA.

CHPRC (2016b). *Conceptual Framework and Numerical Implementation of 100 Areas Groundwater Flow and Transport Model.* SGW-46279, Rev. 3. CH2M Hill Plateau Remediation Company, Richland, WA.

CPCCo (2021a). *Description of Groundwater Calculations and Assessments for the Calendar Year 2020 (CY2020) 100 Areas Pump-and-Treat Report.* ECF-HANFORD-21-0030, Rev 0. Central Plateau Cleanup Company, Richland, WA.

CPCCo (2021b). *FY22 Pump and Treat Remedial Process Optimization Scope, 100-HR-3 Groundwater Operable Unit SGW-66695-VA, Rev. 0.* Central Plateau Cleanup Company, Richland, WA.

Demirkanli, D. I. and V. L. Freedman (2021). *Adaptive Site Management Strategies for the Hanford Central Plateau Groundwater.* PNNL-32055 Rev 0. Pacific Northwest National Laboratory, Richland, WA.

DOE/RL (2015). *Hanford Atomic Energy Act Sitewide Groundwater Monitoring Plan.* DOE/RL-2015-56, Rev. 0. U.S. DOE Richland Operations Office, Richland, WA.

DOE/RL (2016). *Sampling and Analysis Plan for 100-HR-3 Groundwater Operable Unit Monitoring*. DOE/RL-2013-30. DOE Richland Operations Office, Richland, WA.

DOE/RL (2018). *100-HR-3 Pump and Treat System Operations and Maintenance Plan*. U.S. Department of Energy, Richland Operations Office, Richland, WA.

DOE/RL (2021). *Calendar Year 2020 Annual Summary Report for the 100-HR-3 and 100-KR-4 Pump and Treat Operations, and 100-NR-2 Groundwater Remediation*. DOE/RL-2020-61, Rev 0. U.S. DOE Richland Operations Office, Richland, WA.

DOE/RL (2022a). *Calendar Year 2020 Annual Summary Report for Pump and Treat Operations in the Hanford Central Plateau Operable Units*. DOE/RL-2020-62, Rev. 0. U.S. Department of Energy Richland Operations Office, Richland, WA.

DOE/RL (2022b). *Calendar Year 2021 Annual Summary Report for the 100-HR-3 and 100-KR-4 Pump and Treat Operations, and 100-NR-2 Groundwater Remediation*. DOE/RL-2021-52, Rev 0. U.S. DOE Richland Operations Office, Richland, WA.

DOE/RL (2022c). *Hanford Site Groundwater Monitoring Report for* 2021. DOE/RL-2021-51. U.S. DOE Richland Operations Office, Richland, WA.

EPA (1999a). *Monitored Natural Attenuation of Chlorinated Solvents*. EPA/600/F-98/022. U.S. Environmental Protection Agency Office of Research and Development, Ada, OK.

EPA (1999b). *Monitored Natural Attenuation of Petroleum Hydrocarbons*. EPA/600/F-98/021. U.S. Environmental Protection Agency Office of Research and Development, Ada, OK.

EPA (1999c). *Use of Monitored Natural Attenuation at Superfund, RCRA Corrective Action, and Underground Storage Tank Sites*. Directive 9200.4-17P. U.S. Environmental Protection Agency Office of Solid Waste and Emergency Response, Washington, DC.

EPA (2007a). *Monitored Natural Attenuation of Inorganic Contaminants in Ground Water, Volume 1: Technical Basis for Assessment*. EPA 600/R-07/139. U.S. Environmental Protection Agency Office of Research and Development, Washington, DC.

EPA (2007b). *Monitored Natural Attenuation of Inorganic Contaminants in Ground Water, Volume 2: Assessment for Non-Radionuclides Including Arsenic, Cadmium, Chromium, Copper, Lead, Nickel, Nitrate, Perchlorate, and Selenium*. EPA 600/R-07/140. U.S. Environmental Protection Agency Office of Research and Development, Washington, DC.

EPA (2010). *Monitored Natural Attenuation of Inorganic Contaminants in Ground Water, Volume 3: Assessment for Radionuclides Including Americium, Cesium, Iodine, Plutonium, Radium, Radon, Strontium, Technetium, Thorium, Tritium, and Uranium*. EPA 600/R-10/093. U.S. Environmental Protection Agency Office of Research and Development, Washington, DC.

EPA (2012). *A Citizen's Guide to Monitored Natural Attenuation*. U.S. Environmental Protection Agency Office of Solid Waste and Emergency Response, Washington, DC.

EPA (2015). *Use of Monitored Natural Attenuation for Inorganic Contaminants in Groundwater at Superfund Sites*. Directive 9283.1-36. U.S. Environmental Protection Agency Office of Solid Waste and Emergency Response, Washington, DC.

EPA. (2020) Adaptive Management Site Management Plan for the Bonita Peak Mining District, San Juan County, Colorado. 20-034(E)/112320. U.S. Environmental Protection Agency: Washington, DC.

EPA. (2022a) Considerations for Adaptive Management at Superfund Sites. OLEM Directive 9200.3-123. U.S. Environmental Protection Agency Office of Superfund Remediation and Technology Innovation: Washington, DC.

EPA. (2022b) Adaptive Site Management – A Framework for Implementing Adaptive Management at Contaminated Sediment Superfund Sites. OLEM Directive Number 200.1-166. U.S. Environmental Protection Agency Office of Superfund Remediation and Technology Innovation and Office of Research and Development: Washington, DC.

EPA, Ecology and USDOE (2018). *Record of Decision Hanford 100 Area Superfund Site 100-DR-1, 100-DR-2, 100-HR-1, 100-HR-2, 100-HR-3 Operable Units* July 2018. U.S. Environmental Protection Agency, Washington State Department of Ecology, and U.S. Department of Energy, Olympia, WA.

EPA/DOE/Ecology (1996). *EPA Superfund Record of Decision: Hanford 100 Area (USDOE)*. O.U. 100-HR-3 and 100-KR-4. Hanford Site, Benton County, WA; U.S. Environmental Protection Agency, Olympia, WA.

EPA/DOE/Ecology (2003). *Explanation of Significant Difference for the 100-HR-3 Operable Unit Record of Decision April 2003*. U.S. Environmental Protection Agency (EPA), U.S. Department of Energy (DOE), Washington State Department of Ecology (Ecology), Richland, WA.

Ertani, A., A. Mietto, M. Borin and S. Nardi (2017). "Chromium in agricultural soils and crops: A review." *Water, Air, & Soil Pollution* 228(5): 190.

FRTR (2020). "Groundwater Pump and Treat." Retrieved September 29, 2022, from https://frtr.gov/matrix/Groundwater-Pump-and-Treat/.

ITRC (2017). *Remediation Management of Complex Sites.* RMCS-1. Interstate Technology and Regulatory Council, Washington, DC. Retrieved from https://rmcs-1.itrcweb.org or https://rmcs-1.itrcweb.org/RMCS-Full-PDF.pdf.

Marks, P. J., W. J. Wujcik and A. F. Loncar (1994). *Remediation Technologies Screening Matrix and Reference Guide.* Weston (Roy F) Inc, West Chester, PA.

Meyers, P. (2019). "Chromate: In's and out's of available technology (with a focus on ion exchange)." *2019 National Groundwater Association Meeting*, Las Vegas, NV.

NCRP (2009). *NCRP Report No. 160: Ionizing Radiation Exposure of the Population of the United States.* National Council on Radiation Protection and Measurements, Bethesda, MD.

Neshem, D. (2011). *Functional Design Criteria for the 100-HX Pump and Treat System.* SGW-43616, Rev 4. CH2M HILL Plateau Remediation Company, Richland, WA.

Oostrom, M., M. J. Truex, K. C. Carroll and G. B. Chronister (2013). "Perched-water analysis related to deep vadose zone contaminant transport and impact to groundwater." *Journal of Hydrology* 505: 228–239.

PNNL (2021). Hydraulic Pump-and-Treat Information Analytics *(HYPATIA).* Retrieved from https://socrates.pnnl.gov/hypatia/.

Pryzbylski, M. A. and A. L. Esparza (2010). *Functional Design Criteria for the 100-DX Pump and Treat System.* SGW-40243, Rev. 4. CH2M HILL Plateau Remediation Company, Richland, WA.

Reiger, J.T. (2013) *HEIS Sample, Result, and Sampling Site Data Dictionary.* HNF-38155, Rev. 1. CH2M Hill Plateau Remediation Company: Richland, WA.

Royer, P. D., T. P. Franklin, J. J. Garza, C. D. Johnson, N. J. Huerta, J. Q. Wassing and E. A. C. B. Woodford (2021). *SOCRATES Software Release 2.0.* PNNL-SA-167052. Pacific Northwest National Laboratory, Richland, WA. Retrieved from https://www.pnnl.gov/projects/socrates.

Saslow, S. A., E. A. Cordova, N. M. Escobedo, O. Qafoku, M. E. Bowden, C. T. Resch, N. Lahiri, E. T. Nienhuis, D. Boglaienko, T. G. Levitskaia, P. Meyers, J. R. Hager, H. P. Emerson, C. I. Pearce and V. L. Freedman (2023). "Accumulation mechanisms for contaminants on weak-base hybrid ion exchange resins." *Journal of Hazardous Materials* 459: 132165.

Truex, M. J. and C. D. Johnson (2017). *Incorporating Pump-and-Treat Performance Assessment Into Hanford Remedy Documents.* PNNL-26930, Rev. 0. Pacific Northwest National Laboratory, Richland, WA.

Truex, M. J., C. D. Johnson, D. J. Becker, M. H. Lee and M. J. Nimmons (2015). *Performance Assessment for Pump-and-Treat Closure or Transition.* PNNL-24696. Pacific Northwest National Laboratory, Richland, WA.

# 11 Enhanced Attenuation of Uranium in the Subsurface

*Amanda R. Lawter and Michelle M.V. Snyder*

This chapter continues the discussion of remediation of uranium contamination in the 300 Area of the Hanford Site initiated following characterization and assessment as described in Chapter 4, which covers the historical waste releases and subsurface characterization efforts. Over time, the focus for the Hanford Site has moved from production to decommissioning and site remediation. Several hundred metric tons of near-surface vadose zone sediment has been excavated from the 300 Area to remove the source of the uranium contamination (DOE/RL-2010-99, 2011). Despite these efforts, uranium continued to be problematic, with concentrations greater than the maximum contaminant level for drinking water (MCL; 30 μg/L) across much of the plume. Next, the regulatory decision for monitored natural attenuation and access restrictions was implemented and then re-evaluated based on uranium concentrations measured in groundwater and river water over time. The re-evaluation led to implementation of an alternate remedy. To date, two campaigns of liquid polyphosphate injection have been conducted in the 300 Area for enhanced attenuation of uranium. The effectiveness of these campaigns was assessed utilizing multiple methods including characterization of paired sediment samples from before and after treatment, regular monitoring of aqueous contaminant concentrations in groundwater and nearby river water, and geophysical monitoring of the subsurface during liquid polyphosphate injection.

## 11.1 RECORD OF DECISION FOR 300 AREA

Remediation of the 300 Area is covered by CERCLA following the guidance set by the 300 Area Record of Decision (ROD) (SGW-63113, 2020). When determining the best remediation strategy for the 300 Area, the 1996 ROD kept three specific goals in mind:

1. "Protect human and ecological receptors from exposure to contaminants in soils and debris by exposure, inhalation or ingestion of radionuclides, metals, and organics."
2. "Protect human and ecological receptors from exposure to contaminants in the groundwater and control the sources of groundwater contamination in the 300-FF-1 OU to minimize future impacts to groundwater resources."
3. "Protect the Columbia River such that contaminants in the groundwater or remaining in the soil after remediation do not result in an impact to the Columbia River that could exceed the Washington State Surface Water Quality Standards."[1]

As described in Chapter 2, the number of stakeholders, and the varied backgrounds they come from are abundant and diverse. Comments and concerns about the 300-FF-1 and 300-FF-5 ROD in 1996 were included as part of the 1996 ROD. Comments were received from groups including the Hanford Advisory Board, the Nez Perce Tribe, Heart of America Northwest, the Washington State Department of Health, and the Washington State Department of Fish and Wildlife (EPA, 1996). Several comments requested stricter cleanup standards or more active cleanup alternatives. However, because future site plans are for industrial uses only (Hanford Future Site Uses Working

DOI: 10.1201/9781003329213-15

Group, 1992) and the treatment was purposely chosen to be non-active (refer to the rationale below), these suggestions were not pursued. Several specific comments submitted by stakeholders were included in the ROD, with detailed responses. To encourage stakeholder engagement, notices of public comment periods and a summary of the proposed plans were published in several local newspapers and sent to an "interested in Hanford" mailing list consisting of several thousand people and anyone who requested information during the public comment period. These notices and informational flyers highlighted that a public meeting would be held upon request, but no meeting was requested.

Based on stakeholder engagement with Tribal Nations, archeological sites have been identified as an important consideration during site cleanup. As described in Chapter 2, the location of the Hanford Site includes places sacred to local groups, including Native Americans and those displaced from homestead sites within the former Hanford township. During archeological surveys, a few isolated archeological sites and one human remains site were located within the 300 Area (EPA, 1996). Isolated archeological sites containing three or less artifacts within 10 m of each other allow work to continue; larger sites or additional items of significance would require more protection with additional site surveys and a stop work (EPA, 1996). There were six larger sites found within 0.5 miles of the 300 Area. Prior to beginning remediation, additional site surveys were required and, if any artifacts or remains were unearthed during activities that included ground disturbances, work would halt and the contractor would follow the requirements included in the Native American Graves Protection and Repatriation Act (House Resolution 5237, 1990).

With the three ROD goals, stakeholder comments, and a three- to ten-year time frame in mind, the selected remedies in the 1996 ROD included (i) removal of contaminated soil/debris and (ii) groundwater cleanup. The first action was excavation and disposal of contaminated soil and debris from specified waste sites with backfill and revegetation of the excavated sites. For the groundwater, institutional controls were selected, meaning that groundwater would be monitored and exposure would be limited until cleanup criteria were reached. The data available at that time indicated that natural attenuation would reduce the uranium plume in groundwater from 150 µg/L to below the drinking water standard (DWS) of 30 µg/L over a period of five to ten years, so monitored natural attenuation (MNA) was selected as the remedy in the 300 Area (EPA, 1996). Cleanup targets are given in Table 11.1.

## 11.2 RE-EVALUATING THE REMEDIATION DECISIONS

The 1996 ROD identified remedial action for the uranium contamination in the groundwater (the 300-FF-5 OU), including the following: (i) groundwater monitoring to continue to determine whether active remedial measures were necessary, and (ii) controls in place that would restrict groundwater use in order to prevent accidental exposures. The intent of the interim action ROD was to restore the groundwater to the DWS. Groundwater monitoring indicated that the uranium contamination did not reach the DWS within five to ten years, as was predicted by modeling, due to sources in the deep vadose zone and the periodically rewetted zones (the part of the vadose zone that is saturated when the water table rises due to an increase in the elevation of the Columbia River; Figure 11.1) (DOE/RL-2010-99, 2011).

When contamination remains at concentrations above what is acceptable after remediation, the CERCLA process and the National Contingency Plan (NCP) require a review at least every five years to assess whether the environment and human health are at risk. The first five-year review occurred in 2001 and determined that remediation identified in the 1996 ROD for the 300-FF-1 and 300-FF-5 OUs was still appropriate, but more monitoring was added along the river shoreline (DOE/RL-2010-99, 2011). The second five-year review occurred in 2006 and determined that the 300-FF-1 OU waste sites met the objectives identified in the 1996 ROD, and therefore, cleanup of the 300-FF-1 OU was complete. The review also determined that the remediation identified in the

**TABLE 11.1**
**Various Cleanup Targets Applicable to the 300 Area**

| Cleanup Target | Total Uranium Concentration | Description |
|---|---|---|
| Cleanup levels (adjacent to 300 Area) in shallow zone (≤15 ft bgs) | 56.1 pCi/g (167 μg/g) | Cleanup levels are targets established in the ROD with agreement between DOE and EPA. These targets are based on levels that protect or restore the groundwater and Columbia River and also consider future use of each site |
| Cleanup levels within the 300 Area (≤15 ft bgs) | 350 pCi/g (1,040 μg/g) | |
| Groundwater cleanup levels | 30 μg/L (10 pCi/L) | |
| Drinking water standard – maximum contaminant level | 30 μg/L (10 pCi/L) | "The National Primary Drinking Water Regulations (NPDWR) are legally enforceable primary standards and treatment techniques that apply to public water systems. Primary standards and treatment techniques protect public health by limiting the levels of contaminants in drinking water." https://www.epa.gov/ground-water-and-drinking-water/national-primary-drinking-water-regulations |
| Drinking water standard – maximum contaminant level goal | 0 μg/L | |
| Surface water quality standards | In water, ranges from 8E$^{-8}$ to 9E$^{-4}$ μCi/mL (0.24–2,675 μg/L/mL) depending on the isotope | "We developed water quality criteria to provide protection for designated uses. Criteria may be numeric – for example, not to exceed some concentration – or narrative, described by words<br>Water quality criteria are set to meet the designated uses associated with every water body in the state. Our water quality standards contain numeric and narrative criteria for both marine and fresh waters"<br>Designated uses within Washington State include aquatic life, recreational, water supply, and miscellaneous (e.g., habitat, boating, or aesthetic values) uses<br>Washington Department of Ecology<br>https://ecology.wa.gov/Water-Shorelines/Water-quality/Water-quality-standards |

interim action EPA (2001) for the 300-FF-2 OU was still appropriate and could continue, but the 300-FF-5 OU uranium plume did not meet the goals set within the ten-year time frame required in the 1996 interim action ROD and found that:

> Predicted attenuation of uranium contaminant concentrations in the groundwater under the 300 Area has not occurred. DOE is currently performing additional characterization and treatability testing in the evaluation of more aggressive remedial alternatives.
>
> *(DOE/RL-2010-99, 2011).*

In order to address the continued uranium contamination, the following steps were identified: (i) conduct more characterization of the 300-FF-5 OU uranium contamination, (ii) create an updated conceptual site model, (iii) verify ecological consequences, and (iv) evaluate alternative treatments.

## 11.3 REMEDIAL INVESTIGATION AND FEASIBILITY STUDY

After finding that the uranium concentrations were not decreasing as quickly as expected (based on the 1996 ROD), the Tri-Parties set new milestones, including an update of the conceptual site model for the 300 Area uranium plume (Williams et al., 2007). In order to more accurately complete a feasibility study and identify how to remediate the uranium in the groundwater, the DOE

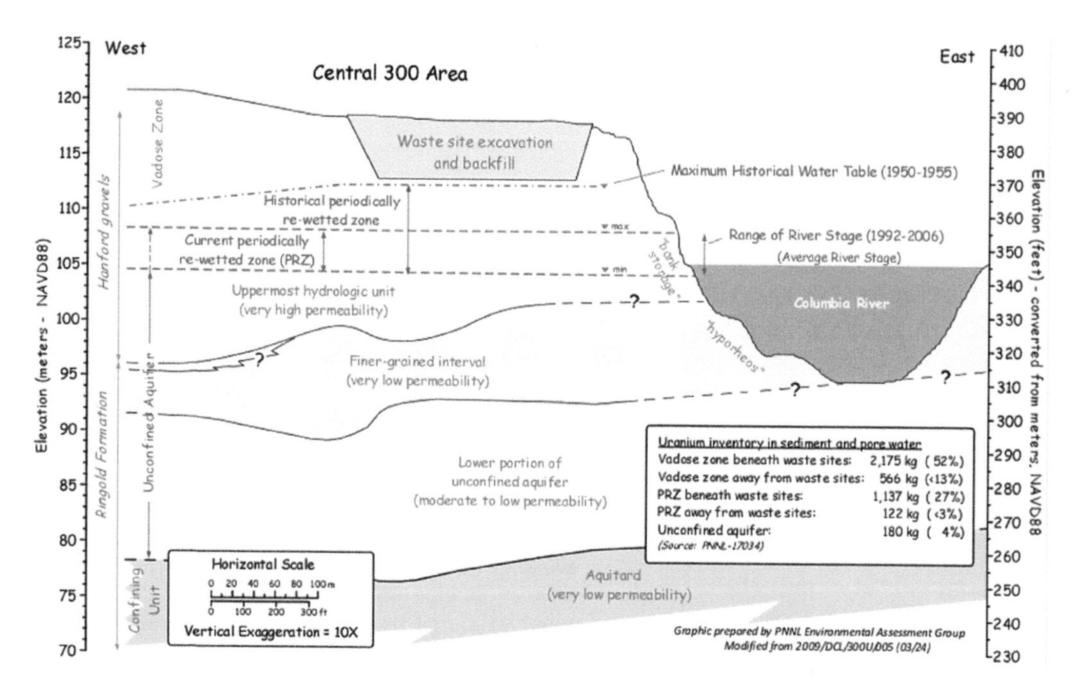


**FIGURE 11.1** 300 Area view showing the periodically rewetted zone (PRZ) that has been identified as a continuous source of uranium in the groundwater. Reprinted with permission from EPA and DOE (2013).

authorized a remedial investigation/feasibility study (RI/FS). RI/FSs are designed to screen technologies to determine effectiveness, implementability, and cost (see Chapter 1b). These studies are designed to be conducted concurrently as part of the CERCLA process, so the data collected during site characterization (part of the RI) can inform the treatability studies and field tests conducted under the FS (EPA, 1988). For the 300 Area, this study included laboratory, field, and modeling components, as well as additional sediment and groundwater characterization to determine a suitable remediation strategy to reduce the uranium concentration in the groundwater plume within the expected time frame and cost (Williams et al., 2007; Wellman et al., 2007; DOE/RL-2018-17, 2018b).

Laboratory experiments conducted at PNNL, including batch and column studies (Wellman et al., 2007; Vermuel et al., 2009; Szecsody et al., 2012; Pan et al., 2016), helped to identify the ideal mixtures of phosphate solutions, the expected reduction in mobile uranium, and the likelihood of remobilization post-treatment.

The conclusion of the feasibility study included a limited field investigation (LFI) (Williams et al., 2007) in 2006 to yield a better understanding of the processes and distribution of uranium in the 300 Area. The LFI included characterization of sediments from four new boreholes that were installed between May and July of 2006 (Figure 11.2). Characterization of these boreholes included hydraulic conductivity measurements, multiple extractions to determine the mobility of the uranium, geologic interpretation, and gamma energy analysis. Altogether, this characterization was used to update the uranium conceptual site model of the 300 Area by providing additional site-specific information (Zachara et al., 2012). One of the four wells installed in 2006 was selected as an injection well for a field study of phosphate injection. Other existing wells or new wells located around and downgradient from the selected injection well were used to track the movement of the injected phosphate and to monitor changes in uranium concentration (Wellman et al., 2007).

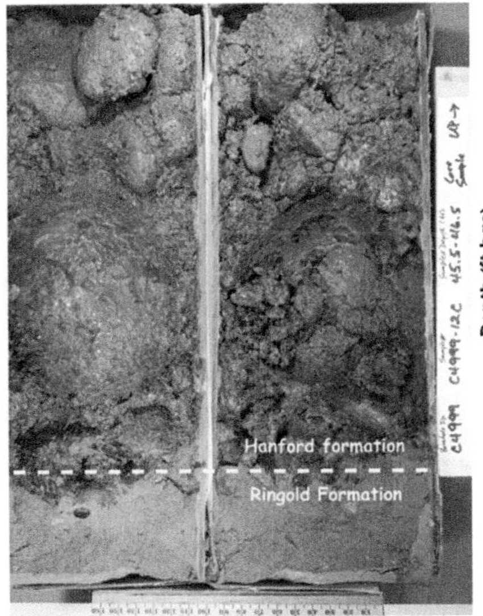


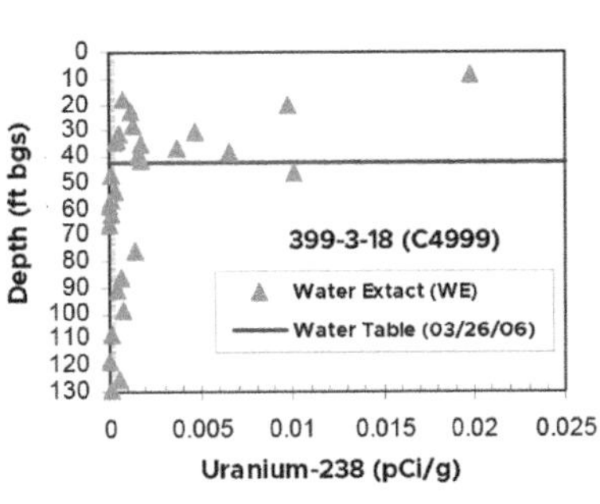


**FIGURE 11.2** Unopened core obtained from a drilled well (left), an open core (cut vertically to display geologic features; center), and water extraction results all from the same borehole (right). Reprinted with permission from Williams et al. (2007).

The 300-FF-5 OU Sampling and Analysis Plan (SAP) was updated in 2006 to include installation and results from the new monitoring wells and aquifer tubes, and to provide data for the next phase of the FS. At that time, the monitoring wells included 56 wells and 8 aquifer tubes. Another revision in 2009 to the 300-FF-5 OU SAP was made to include wells that were drilled for the LFI. This addition brought the numbers up to 60 wells and 12 aquifer tubes (DOE/RL-2018-17, 2018b).

At the conclusion of the 300-FF-5 feasibility study technology screening process, polyphosphate treatment was determined to be the most promising technology and was selected for field testing. The polyphosphate solutions injected into the subsurface immobilize uranium in multiple ways including formation of low-solubility uranyl phosphate minerals (e.g., autunite) and apatite, which can adsorb dissolved uranium and coat uranium-containing minerals (Vermuel et al., 2009). Apatite is a mineral that makes up our bones and is known to adsorb uranium. While adsorption of uranium to mineral surfaces is a reversible process, over time, apatite may coat adsorbed phases. In addition, the calcium, phosphate, and uranium may also form insoluble mineral autunite (Vermuel et al., 2009). The immobilized uranium will be released more slowly than natural conditions, even during the changing river stage, significantly reducing the uranium leach rate (Szecsody et al., 2012) (Figure 11.3). The RI/FSs provided information on the optimal phosphate formulation (using a mixture of orthophosphate and polyphosphate to form the mineral autunite), as well as surface infiltration data to show where the phosphate could be infiltrated from the surface and where it would need to be injected directly into the subsurface.

## 11.4 RESULTS

Polyphosphate injection was implemented in two stages, named Stage A and Stage B, targeting two different areas with some updates to the injection strategy for Stage B as it occurred after Stage A. The treatment focused on the areas known to be the highest contributors to the uranium contamination; these sites included the top of the groundwater aquifer, as well as the deep vadose zone and the periodically rewetted zone (EPA and DOE, 2013).

**FIGURE 11.3** Schematic showing the difference in uranium mobility in untreated and treated systems; after polyphosphate treatment, the uranium incorporated into autunite and/or apatite is very slowly released into the groundwater.

## HOW DOES POLYPHOSPHATE INJECTION WORK?

Formation of apatite can reduce contaminant mobility through several mechanisms: sorption onto calcium phosphate solids, formation of other contaminant-containing minerals, incorporation of contaminants into the calcium phosphate solids, or coating existing contaminant-containing minerals with a protective calcium phosphate coating (SGW-63113, 2020).

To promote formation of the calcium phosphate, two chemicals are mixed together in the subsurface. While the exact ratios, concentrations, and formulations may change due to site-specific needs, generally multiple phosphate sources are combined to reach an optimal pH for formation of apatite, which is then injected into the subsurface. Some calcium is available in the subsurface, but it must also be included in the injection solution to ensure enough is available for mineral formation. Calcium phosphate precipitates very quickly, so citrate can be added to slow down the reaction. The citrate quickly complexes with calcium, and then must first be degraded by microbes in the subsurface for the calcium to be available for the reaction with phosphate (Moore et al., 2007). While the uranium can be immobilized by phosphate addition through sorption or incorporation into the mineral structure during mineral formation, the predominant mechanism for uranium immobilization is expected to be formation of an amorphous monocalcium phosphate mineral that will form a coating over the uranium-containing sediments. Crystallization of this amorphous coating to the stable mineral hydroxyapatite, which has low solubility, will further immobilize the uranium (SGW-63113, 2020).

The method of formation of apatite from liquid phosphate solutions that was used at the Hanford Site used concentrated solutions of monosodium or potassium phosphate and pyrophosphate. These phosphate chemicals were mixed onsite with water from the Columbia River, then injected into wells or near-surface drip lines (SGW-60778, 2019; SGW-59455, 2016; SGW-63113, 2020) (Figure 11.4).

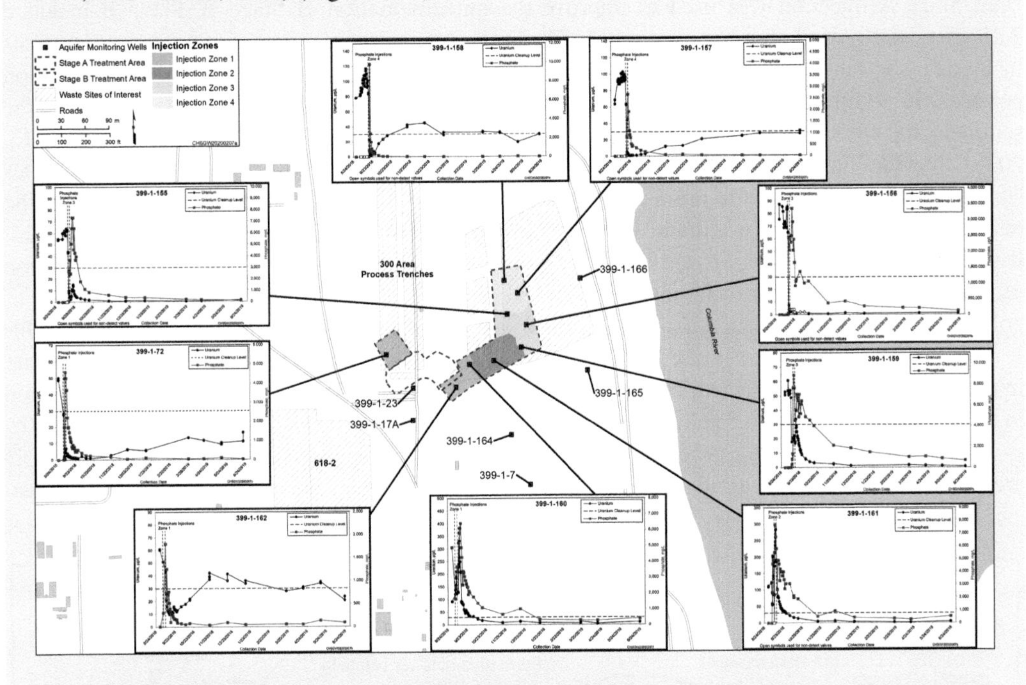


**FIGURE 11.4** Injected polyphosphate blend reacts with uranium-bearing minerals to form less soluble compounds that are less mobile in the subsurface.

## Polyphosphate to Immobilize Contaminants at Other Sites

The use of polyphosphate for contaminant immobilization is not unique to Richland. The use of calcium apatite–based permeable barriers for sequestration of radioactive elements was initiated by Robert Moore of Sandia National Labs (personal communication). Since then, the technology has been used for several contaminants, including strontium-90 at the Hanford Site (Szecsody et al., 2010); zinc, cadmium, and lead in mine tailings at the Success mine in Idaho (Conca and Wright, 2006); and uranium, molybdenum, and vanadium at the Rifle mining site in Rifle, CO (Rappe, 2021). The technology was also being considered for use at the Fukushima Dai-ichi nuclear plant in Japan (Holinka, 2014).

### 11.4.1 Stage A

Stage A was implemented in November 2015 and covered 0.75 acres (SGW-63113, 2020).

Following Stage A of phosphate injection, sediment characterization of pre- and post-injection core samples showed that for most boreholes, >50% of the uranium was still mobile (Last et al., 2016). The Stage A delivery strategy included near-surface infiltration of solutions, as well as injection of solutions into the vadose zone, periodically rewetted zone (PRZ), and directly into the aquifer. This was partly attributed to the delivery of the phosphate solution; precipitation occurred too quickly for some of the solutions and did not reach the targeted uranium in the subsurface for other areas. Using this knowledge, improvements were planned prior to starting Stage B.

### 11.4.2 Stage B

Data from analysis of the soil and groundwater, as well as electrical resistivity tomography (ERT), from Stage A injection was used to improve the implementation of Stage B. Stage B included 2.25 acres, and several changes were made after Stage A to improve injection. Some improvements included using injection for all solutions instead of surface infiltration and injecting more solution per area. In addition, to increase the reaction times between the polyphosphate solution and the sediment, the 48 injection wells injected into the PRZ first, then the lower vadose zone (SGW-63113, 2020). Stage B injection occurred in September 2018. Post-injection analysis of sediment samples showed a decrease in mobile uranium of approximately 60% (SGW-63113, 2020). Although the results show an increase in immobilization of uranium from Stage A injections, additional challenges remain. For example, based on laboratory-scale characterization of pre- and post-injection sediments, three of the 19 locations received very little phosphate suggesting that the delivery of polyphosphate solutions evenly across the subsurface was still a challenge (Szecsody et al., 2020).

Uranium concentrations in the monitoring wells decreased immediately following phosphate injection. The uranium concentrations remained low, and the phosphate levels remained elevated in these monitoring wells for at least 8–12 months post-injection (Figure 11.5) (SGW-63113, 2020).

Monitoring of 300 Area groundwater will continue until cleanup levels defined in the ROD are reached for each contaminant; cleanup levels include both groundwater concentrations and

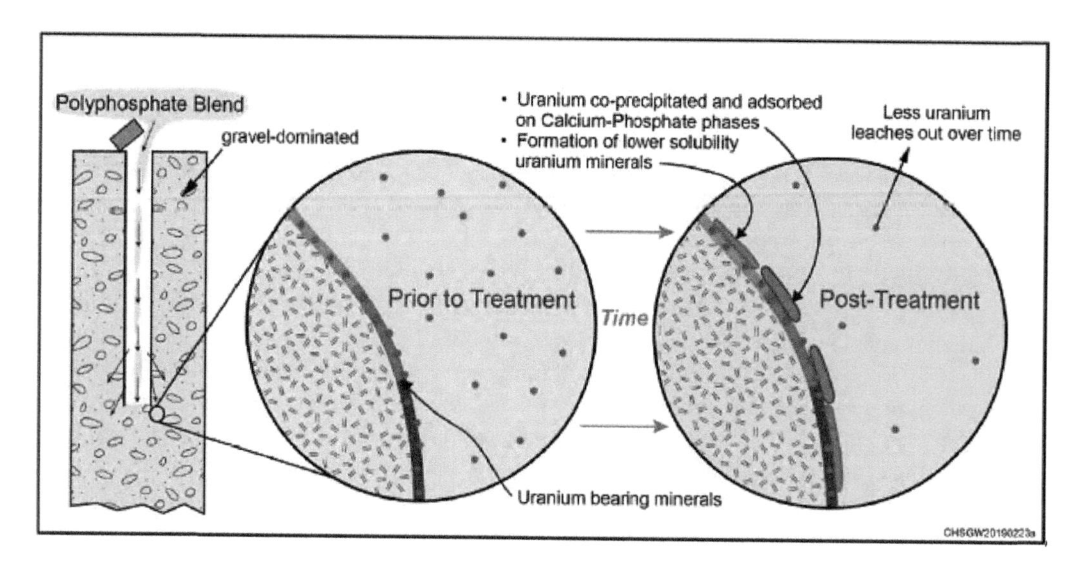


**FIGURE 11.5** Uranium and phosphate concentrations in wells around the Stage A and Stage B injection sites. Uranium is represented by the small dots with potential precipitated phosphate phases as cylindrical blobs on the grain surface. Reprinted with permission from SGW-63113 (2020).

sediment concentrations (EPA and DOE, 2013). Sampling and monitoring of river water is also executed annually according to the Hanford Site Environmental Monitoring Plan (DOE/RL-91-50, Rev. 8, 2018a). Predictions in the 2013 ROD show that the soil and groundwater uranium concentrations are expected to reach the cleanup levels in 22–28 years (EPA and DOE, 2013).

The 300 Area uranium plume is an example of why the CERCLA process includes five-year reviews and allows for improvements on the cleanup plan as new information becomes available. The CERCLA reviews led to further characterization and understanding of the processes controlling uranium within the 300 Area. Site-specific features, like the seasonally infiltrating river water, can have significant effects on the success of remediation at contaminated sites. The CERCLA reviews in 2001 and 2006 led to a better conceptual model, testing and execution of a remediation technology, and, overall, a safer and cleaner 300 Area.

During both Stage A and Stage B injections of polyphosphate in the Hanford Site 300 Area, ERT was used to monitor the delivery of solutions within the subsurface in near real time. Based on results from Stage A with ERT and soil sampling and analysis, the delivery conditions were optimized to improve the delivery process for solutions. Direct aquifer injections were not conducted in Stage B, as it was determined that delivery to the aquifer was more effective when initially injected at a higher elevation in the PRZ due to the slower flow rates in this area, which allowed for more time to infiltrate and react (Figure 11.6, DOE/RL, 2016). In addition, near-surface infiltration was also eliminated based on Stage A results.

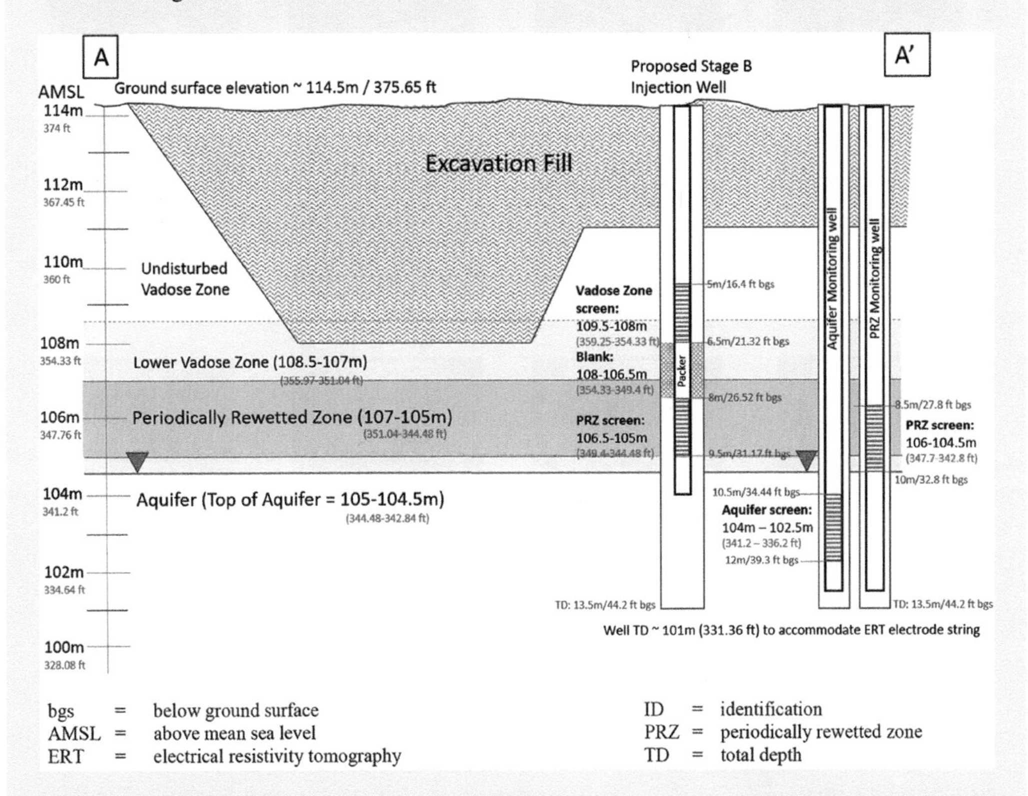


**FIGURE 11.6** Schematic of updated polyphosphate solution delivery from Stage A to Stage B. Reprinted with permission from DOE/RL (2014).

ERT measures the bulk resistivity via an array of electrodes. Resistivity is a measurement of how much a material or solution resists the flow of electricity. The inverse of resistivity is conductivity, which is also a parameter used to quantify the amount of ions in a solution (or ionic strength). When monitored over time, including before, during, and after injection of polyphosphate solutions, ERT measurements are able to show the changes in bulk conductivity due to injection of the polyphosphate, which has a high ionic strength (or high conductivity) as shown in Figure 11.7.

The solutions were injected at average rates of 300 gal/min during Stage A and 48 gal/min during Stage B. The injections were monitored using ERT to visualize the distribution of the solutions (SGW-63113, 2020; Figure 11.6).

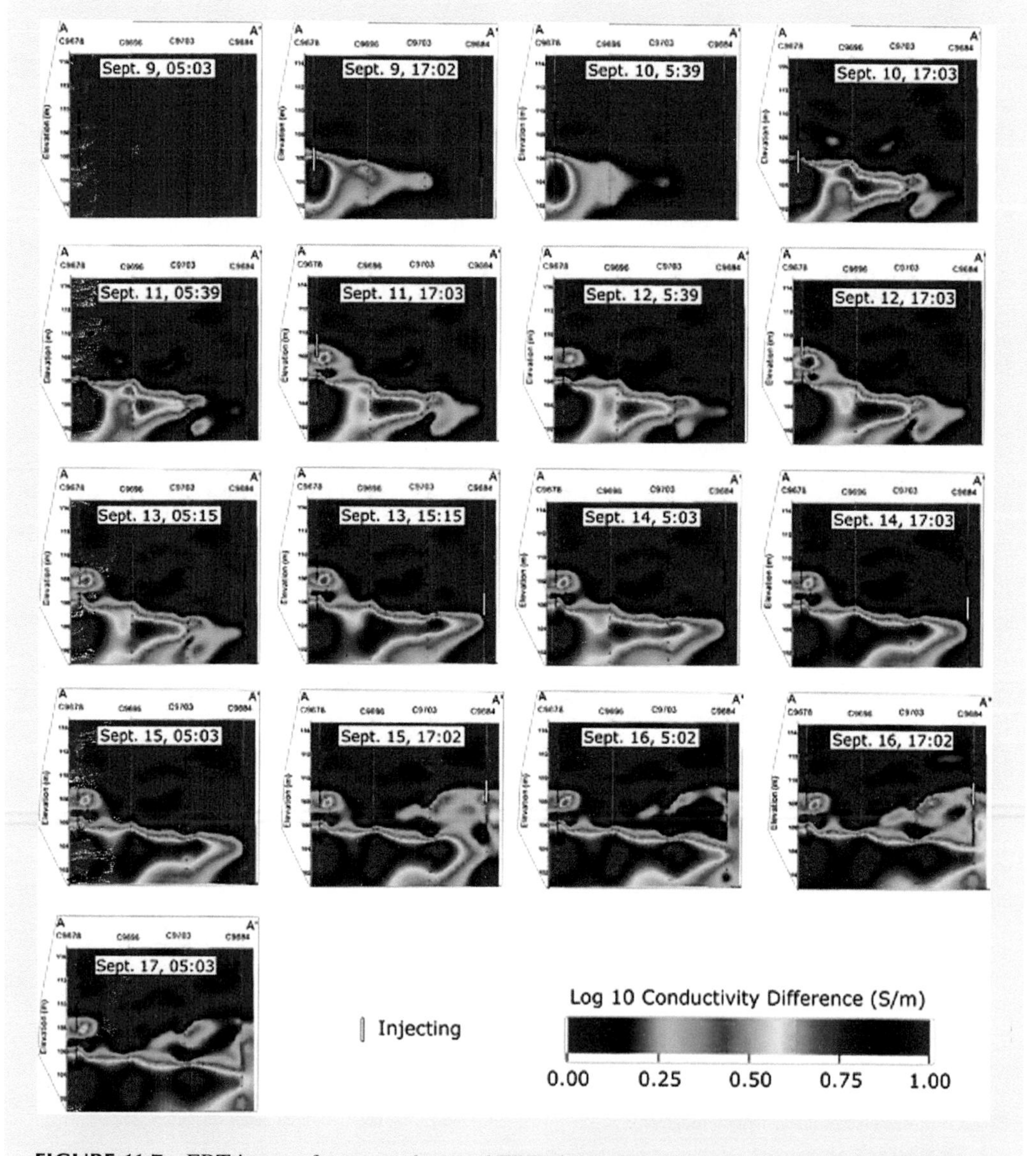


**FIGURE 11.7** ERT images from one cluster of ERT electrodes, using the change in conductivity to visualize the movement of phosphate solutions during Stage B injection. Reprinted with permission from Johnson et al. (2019).

## NOTE

1 Washington State Surface Water Quality Standards differ from the EPA MCLs and focus on several designated uses, while EPA MCLs focus on drinking water. Learn more about both standards at https://ecology.wa.gov/Water-Shorelines/Water-quality/Water-quality-standards and https://www.epa.gov/ground-water-and-drinking-water/national-primary-drinking-water-regulations

## REFERENCES


Conca, J.L. and J. Wright. 2006. An Apatite II permeable reactive barrier to remediate groundwater containing Zn, Pb and Cd. *Applied Geochemistry*, 21(8), 1288–1300.

DOE/RL-2014-42-ADD1. 2016. *300-FF-5 Operable Unit Remedy Implementation Sampling and Analysis Plan Addendum for Stage B Uranium Sequestration*. Department of Energy, Richland, Operations Office, Richland, WA. Available at: https://pdw.hanford.gov/document/0071642H.

DOE/RL-2010-99. 2011. *300 Area Remedial Investigation/Feasibility Study Report for the 300-FF-1, 300-FF-2, and 300-FF-5 Operable Units*. Prepared by CH2M-Hill Plateau Remediation Company for the U.S. Department of Energy, Richland, WA.

DOE/RL-2018-17. 2018b. *Remedial Action Report for the 300-FF-5 Operable Unit Interim Remedial Action*. Prepared by CH2M HILL Plateau Remediation Company for the U.S. Department of Energy, Richland, WA.

DOE/RL-91-50, Rev. 8. 2018a. *Hanford Site Environmental Monitoring Plan*. Prepared for the U.S. Department of Energy, Richland Operations Office, Richland, WA. Available at: https://www.hanford.gov/files.cfm/2018_EMP_estars.pdf.

EPA. 1988. *Guidance for Conducting Remedial Investigations and Feasibility Studies under CERCLA: US Environmental Protection Agency*. Office of Emergency and Remedial Response, Washington, DC.

EPA. 1996. *Record of Decision for USDOE Hanford 300-FF-1 and 300-FF-5 Operable Units Remedial Actions*. Agreement between U.S. Department of Energy and U.S. Environmental Protection Agency, with Concurrence by the Washington State Department of Ecology, July 17, 1996.

EPA. 2001. *Interim Action Record of Decision for the 300-FF-2 Operable Unit, Hanford Site*. U.S. Environmental Protection Agency, Region 10, Seattle, WA.

EPA and DOE. 2013. *Hanford Site 300 Area Record of Decision for 300-FF-2 and 300-FF-5, and Record of Decision Amendment for 300-FF-1*. U.S. Environmental Protection Agency and U.S. Department of Energy, Richland, WA.

Hanford Future Site Uses Working Group. 1992. *The Future for Hanford: Uses and Cleanup: The Final Report*. Hanford Future Site Uses Working Group. Westinghouse Hanford Corporation (WHC)

Holinka, S. 2014. An 'apatite' for radionuclides: Permeable reactive barriers may be deployed at Fukushima. *Sandia Labnews*, July 25, 2014.

House Resolution 5237. 1990. *Native American Graves Protection and Repatriation Act*. US Congress, Washington, DC.

Johnson, T.C., J.N. Thomle, J.L. Robinson, R.D. Mackley, and M.J. Truex. 2019. *Stage B Uranium Sequestration Amendment Delivery Monitoring Using Time-Lapse Electrical Resistivity Tomography*. No. PNNL-28619. Pacific Northwest National Laboratory (PNNL), Richland, WA.

Last, G.V., Z. Wang, J.R. Stephenson, B.D. Williams, M.M.V. Snyder, O. Qafoku, R.E. Clayton. 2016. *Analytical Data Report for Sediment Samples Collected from 300-FF-5: Boreholes C9580, C9581, and C9582*. PNNL-25420. Pacific Northwest National Laboratory, Richland, WA.

Moore, R.C., J. Szecsody, M.J. Truex, K. Helean, R. Bontchev, and C. Ainsworth. 2007. Formation of nanosize apatite crystals in sediments for containment and stabilization of contaminants. In *Environmental Applications of Nanomaterials, Synthesis, Sorbents, and Sensors*, pp. 89–109. Imperial College Press. https://www.amazon.com/ENVIRONMENTAL-APPLICATIONS-NANOMATERIALS-SYNTHESIS-SORBENTS/dp/1848168039

Pan, Z., D.E. Giammar, V. Mehta, L.D. Troyer, J.G. Catalano, and Z. Wang. 2016. Phosphate-induced immobilization of uranium in Hanford sediments. *Environmental Science & Technology*, 50(24), 13486–13494.

Rappe, M. 2021. *Using a Mineral "Sponge" to Catch Uranium*. Sandia National Lab Press Release . https://www.sandia.gov/labnews/2021/06/04/using-a-mineral-sponge-to-catch-uranium-2/#:~:text=The%20apatite%20sponge%20captures%20contaminants,onto%20captured%20contaminants%20for%20millennia.

SGW-59455. 2016. 300-*FF-5 Operable Unit Stage A Uranium Sequestration System Installation Report*. Rev. 0. CH2M HILL Plateau Remediation Company, Richland, WA. Available at: https://pdw.hanford.gov/arpir/pdf.cfm?accession=0077730H.

SGW-60778. 2019. 300-*FF-5 Operable Unit Stage B Uranium Sequestration System Installation Report*. Rev. 0. CH2M HILL Plateau Remediation Company, Richland, WA. Available at: https://pdw.hanford.gov/document/0063871H.

SGW-63113. 2020. *300-FF-5 Operable Unit Enhanced Attenuation Uranium Sequestration Completion Report*. Rev. 0. CH2M HILL Plateau Remediation Company, Richland, WA. Available at: https://pdw.hanford.gov/document/AR-04134.

Szecsody, J.E., H.P. Emerson, R.D. Mackley, C.T. Resch, B.N. Gartman, C.I. Pearce, S.A. Saslow, O. Qafoku, K.A. Rod, M.K. Nims, S.R. Baum, and I.I. Leavy. 2020. *Evaluation of the Change in Uranium Mobility in Sediments from the Hanford 300-FF-5 Stage B Polyphosphate Field Injection*. PNNL-29650. Pacific Northwest National Laboratory, Richland, WA.

Szecsody, J.E., V.R. Vermeul, J.S. Fruchter, M.D. Williams, M.L. Rockhold, N. Qafoku, and J.L. Phillips. 2010. *Hanford 100-N Area In Situ Apatite and Phosphate Emplacement by Groundwater and Jet Injection: Geochemical and Physical Core Analysis*. No. PNNL-19524. Pacific Northwest National Laboratory (PNNL), Environmental Molecular Sciences Lab. (EMSL), Richland, WA.

Szecsody, J.E., Zhong, L., Oostrom, M., Vermeul, V.R., Fruchter, J.S., and Williams, M.D. (2012). *Use of Polyphosphate to Decrease Uranium Leaching in Hanford 300 Area Smear Zone Sediments*. No. PNNL-21733; RPT-DVZ-AFRI-003. Pacific Northwest National Laboratory (PNNL), Environmental Molecular Sciences Lab. (EMSL), Richland, WA.

Vermeul, V.R., B.N. Bjornstad, B.G. Fritz, J.S. Fruchter, R.D. Mackley, D.R. Newcomer, and M.D. Williams. 2009. *300 Area Uranium Stabilization through Polyphosphate Injection*. No. PNNL-18529. Pacific Northwest National Laboratory (PNNL), Richland, WA.

Wellman, D.M., E.M. Pierce, E.L. Richards, B.C. Butler, K.E. Parker, J.N. Glovack, and Rodriguez, E.A. 2007. *Uranium Stabilization through Polyphosphate Injection: 300 Area Uranium Plume Treatability Demonstration Project*. Interim Rep. PNNL-16683. Pacific Northwest National Laboratory, Richland, WA.

Williams, B.A., C.F. Brown, W. Um, M.J. Nimmons, R.E. Peterson, B.N. Bjornstad, and D.C. Lanigan. 2007. *Limited Field Investigation Report for Uranium Contamination in the 300 Area, 300-FF-5 Operable Unit, Hanford Site, Washington*. No. PNNL-16435. Pacific Northwest National Laboratory (PNNL), Richland, WA.

Zachara, J.M., M.D. Freshley, G.V. Last, R.E. Peterson, and B.N. Bjornstad. 2012. *Updated Conceptual Model for the 300 Area Uranium Groundwater Plume*. No. PNNL-22048. Pacific Northwest National Laboratory (PNNL), Richland, WA.

# 12 Retrieval of Tank Waste from Storage

*Matthew Fountain and Beric Wells*

## 12.1 OVERVIEW

The Hanford Site has a unique history that started in the 1940s and evolved over decades with the development, implementation, multiple modifications, and completely new nuclear fuel reprocessing flowsheets to primarily recover uranium and plutonium. The operation of up to 9 plutonium-production nuclear reactors requiring waste storage of fuel reprocessing by-products, intra- and inter-tank farm transfers, and decades of secondary waste handling operations (e.g., evaporation, saltwell pumping, Cs-Sr recovery) has produced an estimated 54 million gallons of diverse radioactive waste inventory stored in 149 single-shell tanks (SSTs) and 28 double-shell tanks (DSTs). The waste in the tanks is a mixture of sludge (i.e., predominantly insoluble solids), saltcake (i.e., predominantly soluble solids), liquids, and gas. The insoluble sludge fraction of the waste consists of metal oxides and hydroxides and contains the bulk of many radionuclides. The saltcake, generated by extensive evaporation of aqueous solutions, consists primarily of dried sodium salts. The liquids consist of concentrated (5–15 M) aqueous solutions of sodium and potassium salts and contain the bulk inventory of technetium-99 and cesium-137. These wastes exhibit complex chemistries, physical properties, and behaviors. For example, as presented in detail in Wells et al. (2011), the solid size can vary several orders of magnitude, the liquid viscosity and solid density can vary by up to one order of magnitude, and the liquid density can increase by 50% from water to salt-saturated liquid. It is precisely the waste complexity and variability, worker and environmental safety concerns, nuclear safety, and regulatory restrictions that create so many operational challenges for Hanford and other similar DOE-complex sites.

The discussion to follow is focused on Hanford wastes, the existing infrastructure to store and transfer these wastes, a short history of operations in Hanford Tank Farms, the SST retrieval needs, and subsequent DST waste handling operations for retrieval, treatment, and closure. Waste retrieval operational examples are summarized.

## 12.2 SST RETRIEVAL NEEDS AND CHALLENGES

Hanford's 149 SSTs are not considered Resource Conservation and Recovery Act (RCRA)-compliant storage tanks[1] due to their single containment design and were placed in an interim-stabilized condition waiting for the waste to be retrieved into the RCRA-compliant DSTs for subsequent storage, blending, and eventual treatment in the Hanford Waste Treatment and Immobilization Plant (WTP) before disposition in a long-term repository. Efficient and effective waste retrieval is desired, but the waste properties, existing SST infrastructure, and available retrieval technologies hinder realization of these objectives. Furthermore, DST space to store the resulting supernatant liquids is nearly exhausted and continued retrieval operations will cease until treatment and immobilization operations begin to open up DST space.

The existing SST infrastructure at Hanford is described to illustrate retrieval application variability. SSTs at the Hanford Site were fabricated between 1945 and 1965 using carbon steel liners supported by reinforced concrete walls, 15-inch concrete domes, and multiple dimensions with

DOI: 10.1201/9781003329213-16

varying capacities. A total of 12 separate SST farms were constructed resulting in 149 individual SSTs with expected design lives of 20–25 years based on corrosion predictions (De Lorenzo et al., 1994). These tank farms were designed to provide waste storage for co-located nuclear fuel reprocessing canyons (i.e., U-Plant, T-Plant, B-Plant, PUREX, and REDOX) and are geographically separated by as little as thousands of feet and up to miles apart.

As described also in Chapter 6 but provided here for the reader's convenience, there were four different types of SSTs constructed. These types are depicted in Figure 12.1, with associated volumes of 55 kgal (Type I), 530 kgal (Type II), 758 kgal (Type III), and 1 Mgal (Type IV). Type IV SSTs also have three different configurations, but all with 1 Mgal capacities. The Type I tanks have a 20 ft diameter and 38 ft height, while Type II, III, and IV tanks all have 75-ft-diameter cylindrical sections with varying heights of 32, 39, and ~47 ft, respectively. The Type I, II, and III tanks have dished bottoms, while Type IV has flat with slight depression in the middle for Type IVA and flat for Types IVB and IVC. Finally, Type IV tanks were configured with airlift circulators: four in Types IVA and IVB, while Type IVC contained 22 (Boomer et al. 2012).

Retrieval access to SSTs is provided through tank dome risers that also vary across the tank types and tank farms. A majority of SSTs have central 42-inch diameter risers available for use that extend down from the soil surface, through the tank-soil overburden, through the tank dome, and into the tank. In A-Farm (six SSTs), the 42-inch center riser extends to within 20 ft of the bottom of the tank, which can limit clearances for inserted equipment and the ability to extend out to retrieve

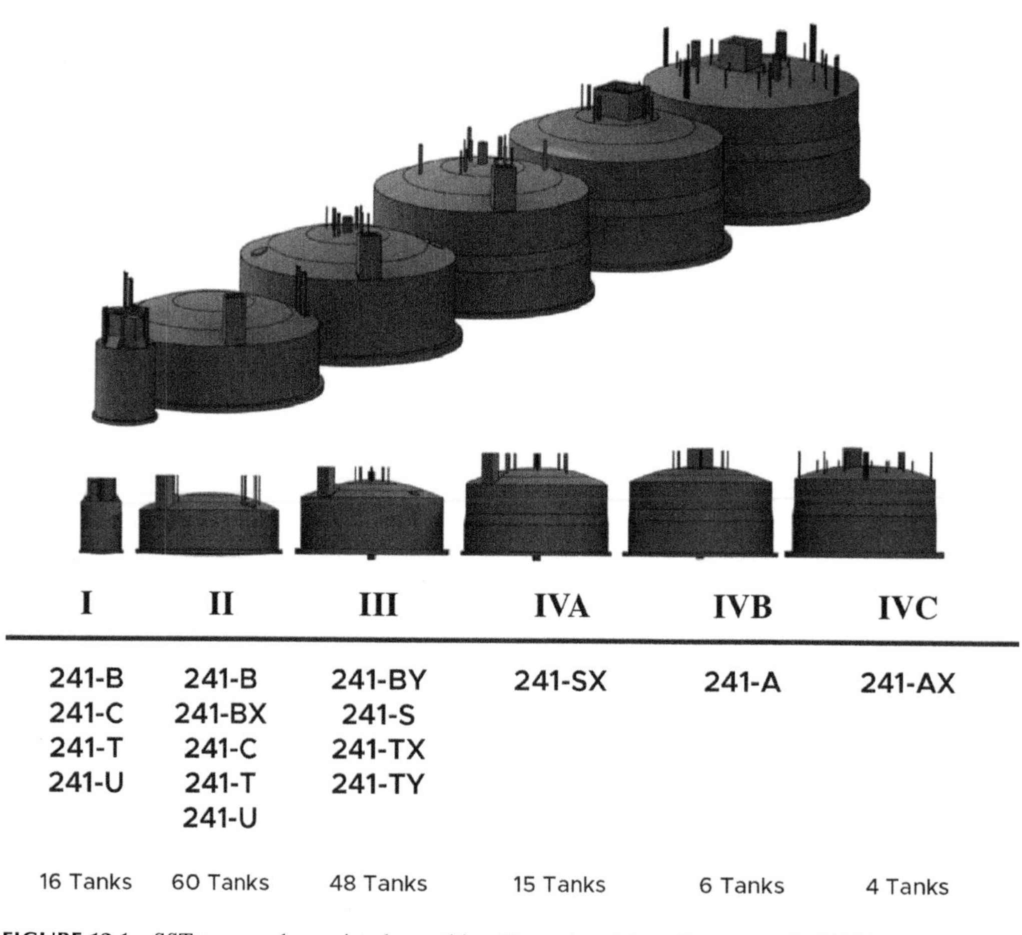


**FIGURE 12.1** SST types and associated quantities. (Reproduced from Boomer et al., 2012.)

waste in the 75-ft-diameter tank. Similar physical limitations exist in other SST farms, and this accessibility to the waste directly impacts the selection of retrieval equipment.

New waste additions to SSTs stopped in 1980 primarily because all processing canyons except PUREX were shut down by 1967 (Gephart 2003), multiple SSTs were known to leak and were already well beyond their 20-yr design life, and DSTs were available to accept and store new waste generated by PUREX. In 2004, the last remaining pumpable liquids were removed to declare the SST inventory as "interim-stabilized" where all drainable liquids have been removed and sealing of tank piping and openings were completed to avoid potential waste leaking to the surrounding soil and deter potential liquid intrusion, respectively (Smith 1991). In this condition, no transfer line infrastructure exists and pumping low-liquid slurries is difficult and conventional technologies are not readily applied to retrieval of this form of waste. As will be discussed, the baseline approach is to add liquids to the saltcake and sludge wastes to solubilize and mobilize the solids while minimizing the threat of liquids escaping containment.

In-tank obstructions from legacy equipment (e.g., airlift circulators, thermocouple trees) and limited tank risers all contribute to the challenges of retrieving the waste. Solidified waste forms "bathtub-type" rings on the tank walls, internal equipment, and hard crust layers making bulk waste layer access with retrieval technologies difficult as well. Finally, residual waste removal to regulatory compliance levels (e.g., 1% residual volume) can be prohibitively expensive, time-consuming, require the addition of reagents (e.g., sodium hydroxide, oxalic acid) for residual solids dissolution, and challenge the physical limits of the retrieval technology.

Multiple retrieval equipment options exist at Hanford (as described later) and were born out of the complex waste conditions and variability. The effectiveness of these retrieval technologies depends on waste chemistry and conditions, tank configuration, and the mobilizing/sluicing fluid selected for retrieval. The retrieval fluid for SSTs is primarily filtered river water and varying concentrations of dilute supernatant liquid feedstocks. During modified sluicing SST retrievals, soluble solid dissolution can occur at varying rates while simultaneously mobilizing insoluble solids and supernatant liquid. A transfer pump transports the slurry to a DST via a temporary, at-surface grade, hose-in-hose pipeline, bypassing the original capped/abandoned SST transfer line infrastructure. Solids received into the DST settle, and the supernatant liquid fraction is simultaneously effectively decanted off and recirculated back to the SST retrieval equipment for reuse. Periodic sampling, tank-level monitoring, and predictions of waste composition all factor into operational decisions made in SST retrieval planning and execution. A detailed discussion of SST retrievals at Hanford is provided by Brown et al. (2021).

The basis for planned retrieval operations and selection of tank waste retrieval technologies to implement at Hanford primarily relies on the Best Basis Inventory (BBI) waste volumes and composition, tank solid sampling for physical properties, liquid observation well (LOW) gamma and neutron scan data, and saltwell jet pumping information. However, in addition to waste variability, components of this information are often missing making it challenging to derive a sound retrieval basis. Core-drill sampling has been the method of choice to improve waste properties bases for SSTs. However, core-drill sampling is limited by riser locations, leading to non-representative samples of the bulk tank volume, as well as heterogeneous distribution of the waste radially and axially. These physical sampling activities and subsequent analytical measures are also costly to conduct.

The rate of SST retrievals is impacted not only by retrieval equipment capabilities and the aforementioned challenges but also by several additional factors including maintenance and operations stoppages to conduct safety analyses. For Hanford, a metric called retrieval duration factor (RDF), defined as the ratio of the actual retrieval operating time divided by the total time between retrieval startup and stopping with 100% operating efficiency, is used. Since July 2003 through July 2020, a gross RDF of 17% was reported by Brown et al. (2021) for SSTs undergoing retrieval operations. Tracking the RDF provides one quantitative means of measuring retrieval equipment performance.

### 12.2.1 Saltwell Pumping (Liquids)

Saltwell pumping at Hanford refers to the process of removing remaining drainable liquids (i.e., as supernatant liquid and interstitial liquid) from the SST waste inventory leaving predominately saltcake and sludge. These operations started in 1972 and continued until 1978 with the initial objective of maximizing SST storage space. However, active use of SSTs stopped in November 1980 (Swaney 2005) and the objective of any continued saltwell pumping turned to maximize removal of remaining SST drainable liquids to achieve an "interim-stabilized" tank status. Stabilization of the tanks had the intent of minimizing environmental risk from loss of tank integrity to the greatest extent technically and economically practicable. The program to stabilize SSTs defined this tank condition to contain less than 50,000 gallons of drainable interstitial liquid and less than 5,000 of supernatant liquid.

The SSTs were deactivated and stabilized by first pumping most of the supernatant liquid using the installed high-volume turbine pumps. Saltwell jet pumping (commonly referred to as saltwell pumping) occurred in 69 of the 149 SSTs at Hanford (Brown et al. 2021), while the remaining SSTs were declared "interim-stabilized" by either administrative review or supernatant liquid pumping only. The last pumpable liquids were removed in 2004 and all SSTs declared "interim-stabilized."

### 12.2.2 SST Retrievals for Tank Closure

As of April 30, 2022, 19 SSTs have been retrieved of sufficient waste to declare these SSTs "retrieved" (Hay 2022). Retrieval of SST tanks is conducted with the objective of achieving target residual waste volume limits less than or equal to 360 $ft^3$ for 75-ft-diameter tanks and 30 $ft^3$ for 20-ft-diameter tanks as required by the Tri-Party Agreement. The agreement requires that up to three retrieval technologies be deployed with the objective of meeting residual volume limits. However, requests to defer application of a third retrieval technology have been approved when cost and risk significantly outweigh low retrieved-volume benefits from further operations. It has been proposed at Hanford to modify this agreement to use tank radiological inventory as an environmental risk-based closure approach to guide the extent of SST retrieval instead of a volume-based metric.

Significant time and expense are expended on SST retrievals due to numerous operational challenges in the quest to achieve the volume limit targets. Multiple retrieval technologies (i.e., up to 3) are deployed to retrieve SST volumes in an attempt to achieve residual waste volume limits. In addition, cold chemical additions, such as 19 M NaOH to dissolve aluminum-bearing solids and/or oxalic acid ($C_2H_2O_4$) to dissolve predominately iron-bearing solids into solution, are often employed to reduce the residual waste volume by chemical dissolution. Aside from the worker safety considerations due to radiological and chemical hazards, criticality and flammable gas risks must also be mitigated. Equipment capability limits exist based on riser diameter, riser location, and physical properties of the waste. On the latter, sludge and saltcake agglomerates can be very large, plugging pump inlet screens and resisting mobilization to retrieval pump inlets. To overcome these challenges and limitations on retrieval rate and the extent of retrieval (i.e., residual waste volume), SST retrieval technology development is necessary to improve safety and reduce cost and schedule impacts. A suite of retrieval tools and experiences exist at Hanford that are instructive in understanding these SST retrieval operations.

### 12.2.3 SST Waste Retrieval Methods

Olander et al. (2016) describe SST waste retrieval processes that have either been deployed in Hanford tanks or been sufficiently tested. The identified SST waste retrieval processes are as follows:

- Insoluble Solids – Modified Sluicing-Sludge Removal (MS-SR)
  - With an in-tank vehicle (ITV) for hard-to-remove heel (HTRH) retrieval, or
  - With chemical dissolution (CD) for HTRH retrieval, or
  - In 200 series SSTs with no HTRH retrieval required (MS-200)

- Soluble Solids – Modified Sluicing-Saltcake Dissolution (MS-SD)
  - With high-pressure mixers or equivalent equipment for high phosphate salt, and
  - With an ITV for HTRH retrieval, or
  - With a continuation of the same process for HTRH retrieval
- Extended Reach Sluicing System-High-Pressure Water (ERSS-HPW)
  - Sluicing with separate supernate and high-pressure water nozzles attached to an arm with an extendable boom,
  - With CD for HTRH retrieval, or
  - With an ITV for HTRH retrieval
- Mobile Arm Retrieval Sluicing System (MARS-S) for
  - Sludge removal
  - Saltcake dissolution
- Mobile Arm Retrieval Vacuum System (MARS-V) for
  - Sludge removal
  - Saltcake dissolution
- Mobile Retrieval System (MRS) for
  - Sludge removal
  - Saltcake dissolution
- Vacuum Retrieval in 200-series tanks (VR-200)
  - 200-series tanks contain only sludge

Hanford wastes are designated as soluble (saltcake) or sludge (insoluble) based on a ratio R (Rasmussen 2017), which is the mass ratio of key soluble analytes to key insoluble analytes and represented by Equation (12.1).

$$R = \frac{K + Na}{Al + Bi + Fe + Ca + Mn + La + Ni + Si + U + Zr} \quad (12.1)$$

Each parameter is the mass concentration of the metal represented by its elemental symbol in the equation. Sludge (insoluble) designations are made for R values less than 2.5 and saltcake (soluble) for ratios above 2.5. For R values equal to 2.5, designation is made based on nearest core segments or grab samples. In addition to the solubility of the waste, the waste retrieval process selected for an SST is based primarily upon the tank status of either "sound" or assumed "to have leaked in the past." For an assumed leaking tank, Olander et al. (2016) specify the following:

- If the tank is an assumed leaker and it has a central 42-in. riser, MARS-V is used.
- If the tank is an assumed leaker and it does not have a central 42-in. riser, MRS is used when the waste volume is small enough to be removed within a nominal year or less. The maximum waste volume that can be retrieved with an MRS in an SST within a 12-month period is estimated to be approximately 64 kgal. No tanks with more than 62 kgal are assumed to use MRS.
- If the tank is an assumed leaker and does not have a central 42-in. riser, and the volume is large enough that it may take more than a year to complete using MRS, a new large central riser is installed and MARS-V is used.

Example SST waste retrieval processes that have been deployed are summarized for insoluble and soluble wastes.

#### 12.2.3.1 Insoluble Solids

The retrieval of SST C-106 sludge (i.e., primarily insoluble) waste into DST AY-102 was conducted to eliminate the high-heat safety issue for C-106, as well as demonstrate a retrieval technology (Carothers et al. 1999). The Waste Retrieval Sluicing System (WRSS) deployed for the C-106 used

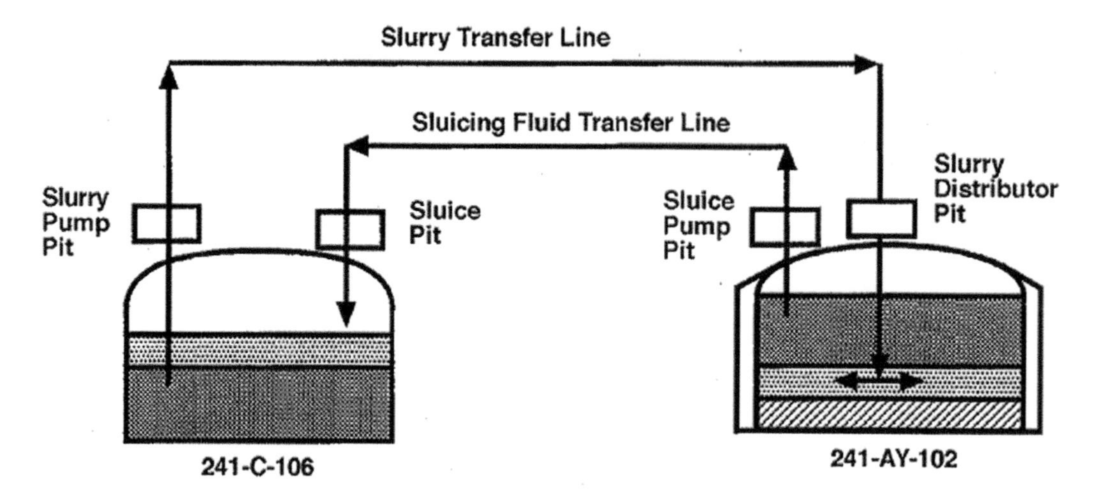


**FIGURE 12.2** C-106 to AY-102 retrieval WRSS schematic. Reprinted with permission from Carothers et al. (1999).

a modified past-practice sluicing process and was the precursor to the MS-SR. In the WRSS process, sluicing nozzles were used to suspend the solids in C-106 for subsequent transfer to AY-102. Supernatant from tank AY-102 was used as the sluicing fluid during the retrieval. The WRSS, schematically depicted in Figure 12.2, includes two nearly identical pump systems (one for delivery of the supernatant liquid sluice stream from AY-102 to tank C-106 and a second for delivery of the C-106 slurry to AY-102), a sluicer system in C-106, a slurry distributor in AY-102, and a dedicated closed-loop inter-farm piping system.

In the WRSS process, supernatant was withdrawn via the sluice pump from near the top of the waste in AY-102 and transported to C-106 to form a fluid stream at the sluicer nozzle. As a result of the sluicing operation in C-106, a slurry was formed and flowed to the suction of the slurry pump. The slurry pumping system transferred the slurry to AY-102, where it was discharged through the submerged slurry distributor into the supernatant layer and solids were settled under the force of gravity. Supernatant near the top of the waste in tank AY-102 continued to be pumped to the sluice nozzle to sustain the process.

Three retrieval campaigns were conducted for the retrieval of C-106 waste into AY-102. Cuta et al. (2000) report a best estimate total of C-106 sludge transfer volume to be 186,000 gallons, approximately 97% of the original waste inventory, with 16 vol% solid dissolution during the retrieval process.

#### 12.2.3.2 Soluble Solids

The retrieval of the predominantly saltcake (i.e., soluble solids) SST S-112 waste into DST SY-102 was conducted to demonstrate a saltcake dissolution retrieval technology (Voogd 2005). The technology used sluicing jets and the Fury™ tank washer to add dilution water to the S-112 waste as depicted in Figure 12.3. In addition to the water additions, the system included a centrally located transfer pump and transfer line to the receipt DST. The solubilized S-112 waste flowed to the transfer pump and was transferred to the receipt DST SY-102. 583,000 gallons of waste was retrieved from S-112, corresponding to 95% of the original waste inventory.

## 12.3 DST RETRIEVAL NEEDS AND CHALLENGES

Unlike SSTs, Hanford's 28 DSTs are considered RCRA-compliant storage tanks due to their double-containment design and serve as the only capability to store supernatant liquid. Saltcake and sludges are also commonly stored as well. Efficient and effective waste retrieval in DSTs is hindered by similar challenges to the SSTs; however, greater flexibility in risers and retrieval tools

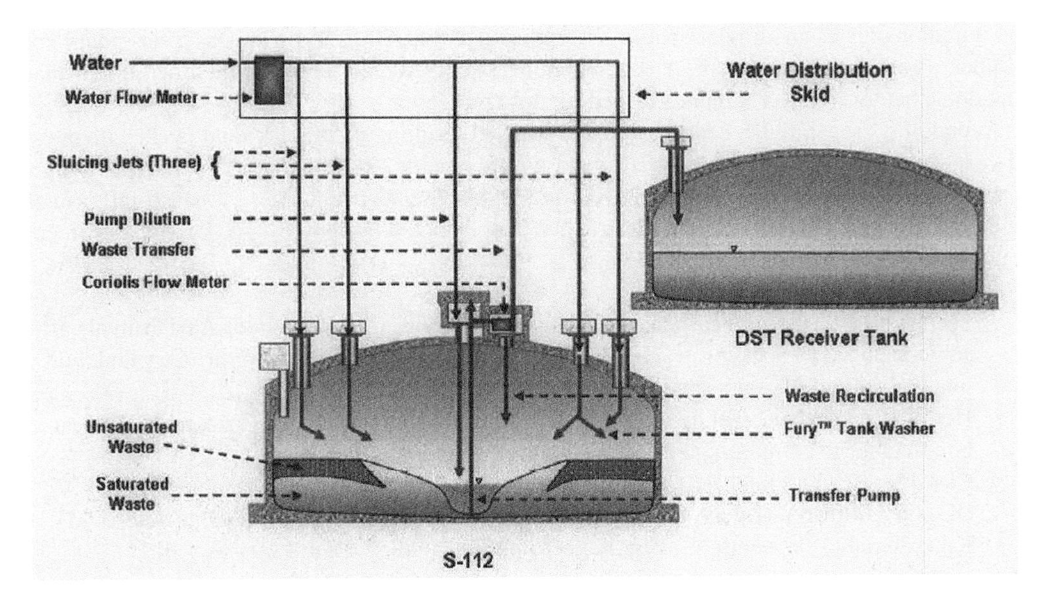


FIGURE 12.3 S-112 to SY-102 saltcake waste dissolution retrieval schematic. Reprinted with permission from Voogd (2005).

FIGURE 12.4 DST construction stages and attributes. Reprinted with permission from Venetz and Gunter (2014).

offer considerable advantages. As well, unlike SST infrastructure, much of DST infrastructure (e.g., transfer line piping, jumpers, tank monitoring equipment) has been maintained and remained in active service since 2004 when the last remaining pumpable liquids were removed from SSTs and, in subsequent years, during SST retrievals for purposes of tank closure.

The DSTs at Hanford are relatively consistent in configuration, and major elements are depicted in Figure 12.4. Built between 1968 and 1986, these tanks have secondary carbon steel liners built on concrete pads with castable refractory beds that each primary tank sits on. The tank-in-a-liner configuration creates an annular space for ventilation and cooling, and serves as a secondary containment if the tank should leak. A concrete dome is constructed on top of the liner and primary tank dome before soil overburden is placed for dose protection.

A total of six separate DST farms were constructed resulting in 28 individual DSTs with mostly 50-year service lives, except for AZ-Farm and AY-Farm with 20-, and 40-year service life designs, respectively (Campbell et al. 2020). All Hanford DSTs have 75-ft-diameter by ~46-ft-tall primary tanks with similar tank design and construction best depicted by Figure 12.5. Differences in construction did occur including the following:

- Secondary liner bottom carbon steel plate thickness began at ¼-inch for AY-Farm and all subsequent DST Farm construction used 3/8-inch thick plates. Same for primary tank bottom, but started 3/8-inch and then ½-inch thereafter.
- Use of A515 carbon steel for AY- and AZ-Farms, A516 type for SY-Farm, and A537 carbon steel for the remaining AW-, AN-, and AP-Farms.
- Castable refractory materials used varied almost with every farm; AY (Kaolite 2200LI), AZ (Kaolite 2000), SY (Lite Wate 50), AW and AN (Lite Wate 70), and AP (Litecrete-60M).
- Knuckle radius between the primary tank bottom and the wall did vary as well.

The entire tank structure was buried with overburden at a depth of 6–8 ft from the top of the tank dome.

Similar to SSTs, DST farms are geographically separated by thousands of feet and up to approximately 7 miles in the case of SY-Farm to AN-Farm. A system of interconnected waste transfer lines allows intra-farm, as well as inter-farm, transfers through a series of pump and valve pits.

Retrieval access to DSTs is enabled through tank dome risers that are somewhat consistent for all DSTs, but size and locations vary widely on the tank itself limiting access to the waste inside.

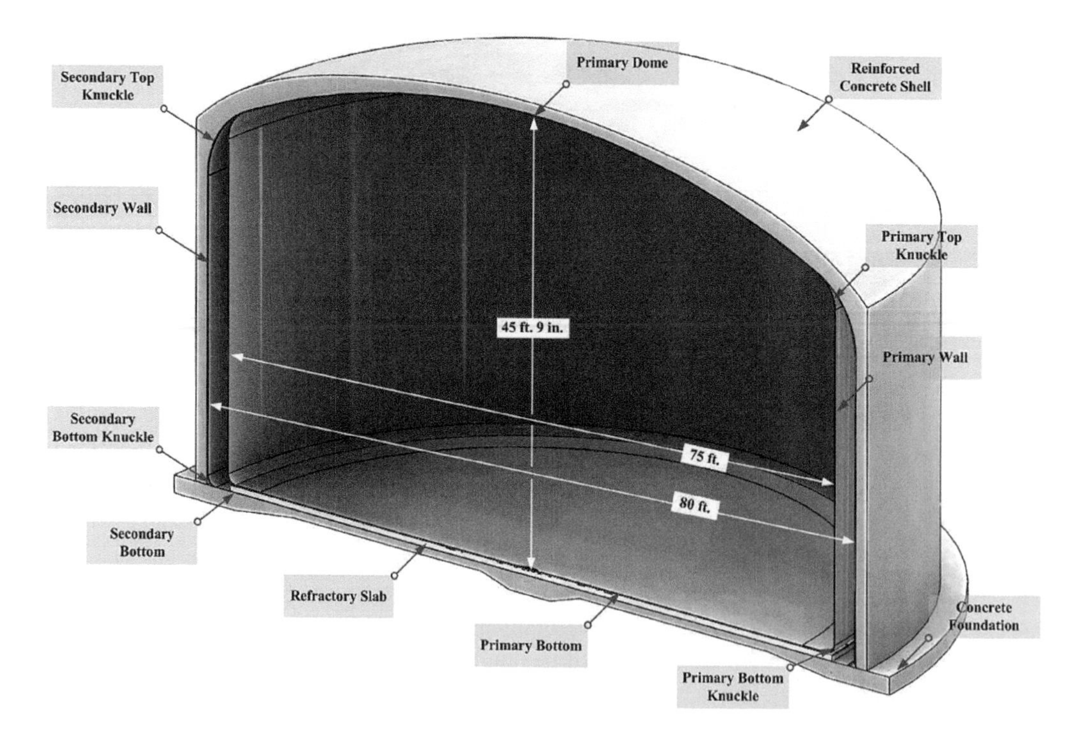


**FIGURE 12.5** General arrangement of the Hanford DST.

DSTs commonly have 42-inch-diameter risers available for use, in varied radii and circumferential locations, that extend down from the soil surface, through the tank-soil overburden, through the tank dome, and into the tank.

DST retrievals are planned using either sluicing or jet mixer pumps with a separate submerged transfer pump to convey the slurried waste mixture to a receipt DST and/or eventually to the WTP low-activity waste (LAW) or high-activity waste (HAW) facilities. The motive fluid for retrieval can be filtered river water or existing dilute supernatant liquid. Use of existing liquid inventory to retrieve settled solids preserves precious available DST space and can be more effective in solid mobilization due to greater viscosity and density compared to filtered water from the Columbia River.

Like SSTs, worker safety considerations affect DST operations, as well as flammable gas, nuclear safety, corrosion control, and other safety controls to prevent damage to the carbon steel materials, loss of containment, and/or radiological exposure. The rate of retrieval is similarly impacted directly by the waste rheological and physical properties, the existing tank infrastructure and configuration, and equipment selected to conduct retrievals. Any modifications to the pump pits, risers, and existing tank infrastructure require extended planning and execution before subsequent retrieval operations. Example DST waste retrieval processes that have been deployed are summarized for insoluble and soluble wastes.

### 12.3.1 Retrieval of Soluble Solids

Concentrated evaporator and complex concentrate waste in DST SY-101 generated, retained, and spontaneously released flammable gas. Three of the largest spontaneous gas release events resulted in the tank headspace exceeding the lower flammability limit for hydrogen (Kirch et al. 2000). To control the gas release behavior and reduce the flammability hazard, a mixer pump was installed in 1993 and operated through April 2000.

The initial nominally 5.33-m-thick sediment layer in SY-101 comprised predominantly of soluble salt solids was mobilized by the 150-horsepower mixer pump. The mixer pump fluid intake level was ~235 inches above the tank bottom, and horizontal discharge was from two opposed 2.6-inch-diameter nozzles at ~28 inches above the tank bottom. Initial operations of the mixer pump were carefully controlled to limit gas release, and during the majority of its operational period, the mixer pump was typically operated on a tri-weekly schedule of 30-minute operations with nozzle velocities of approximately 18 m/s, incrementally rotated by approximately 30° (Alleman et al. 1994, Johnson et al. 2000, Rassat et al. 2000). An unanticipated consequence of the mixer pump operations and mitigation of the spontaneous gas releases was excessive growth of the floating crust layer. The resulting waste surface level rise necessitated remediation in December 1999 through March 2000 via a sequence of waste transfers and back dilutions with water to dissolve the soluble solids. Following the transfers and back dilutions, operations of the mixer pump were conducted to mix settled solids and dilution water, to enhance solid dissolution, and to continue to release retained gas (Kirch et al. 2000). The bulk of the soluble solids remaining in the tank were dissolved (Mahoney et al. 2000), and the spontaneous flammable gas release hazard was remediated (Kirch et al. 2000).

### 12.3.2 Retrieval of Insoluble Solids

The retrieval of insoluble solids from DSTs has been evaluated with mixer pumps and conducted with sluicing operations.

#### 12.3.2.1 Mixer Pumps

Two 300-horsepower mixer pumps were installed in available risers in DST AZ-101 and used to mobilize the largely insoluble solid sediment layer in 2000 (Carlson et al. 2001). The mixer pumps take fluid in from ~12 inches above the tank bottom and discharge it horizontally from two opposed 6-inch-diameter nozzles at ~18 inches above the tank bottom.

Initial testing operations were conducted with a single mixer pump with the horizontal nozzle discharge at five fixed radial directions with typically incrementally increasing pump nozzle discharge speeds for approximately 3 hours at each speed. The same mixer pump was then operated in oscillation mode at each of the three nominal velocities totaling an operating period of almost 3.5 days. Both mixer pumps were subsequently operated at equivalent rates, again ramping up in speed over a total operating period of approximately 11 days.

The total flow for both mixer pumps during the test was nominally 285.5 million gallons, which is equivalent to 336 final AZ 101 waste volumes. The final operation of the mixer pumps, consisting of concurrent oscillatory operation (0.05 rpm) for about 2.5 days at nozzle velocities of approximately 18 m/s, resulted in the recirculation of 72 waste volumes through the mixer pumps. As reported in Carlson et al. (2001), 95%–100% of the approximately 18-inch-thick initial settled solid sediment was mobilized during the final mixer pump operation. The insoluble solids were segregated vertically within the tank during the mixer pump operation, with approximately 32% of the solid mass consisting of smaller size particles (median spherical size ~10–40 μm by volume depending on sample location) suspended above 38 inches in the tank and 68% of the solids mass consisting of larger particles (median spherical size ~ 550 μm by volume) remaining below that level (Carlson et al. 2001, Wells and Ressler 2009, Meacham et al. 2012).

A similar configuration of mixer pumps was tested at a reduced scale with waste simulants for solids transfer via a transfer pump located at the tank center and a prototypic tank bottom offset of 6 inches. A schematic of the dual mixer pump and transfer pump configuration and waste solid mobilization and suspension is shown in Figure 12.6. Kelly et al. (2013) reported that solid accumulation testing demonstrated that fast-settling solids are concentrated in mounds left behind when mixing is not capable of clearing all solids from the tank bottom. The test results indicate that there is little mixing between material previously deposited and new material added to the tank. For retrieval for waste treatment, the solid concentrations of the pre-transfer samples were generally within ±20% of the samples from the first transferred batch (a batch being nominally 1/5 of the DST waste volume) and the pre-transfer samples and the first batch had higher concentrations of fast-settling solids than later batches.

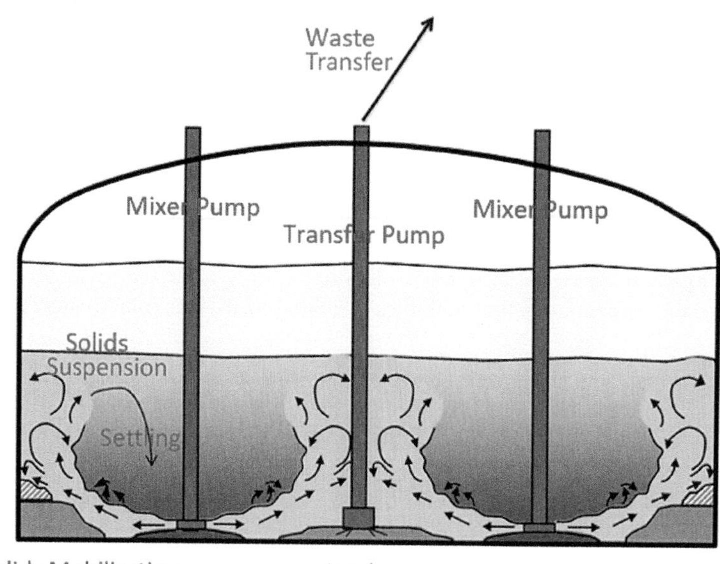


**FIGURE 12.6** DST waste feed delivery system example. Reprinted with permission from Wells et al. (2013).

#### 12.3.2.2 Sluicing

The sediment in DST AY-102 has been retrieved to AP-102 using a sluicing methodology similar to that described for insoluble solid retrieval from SSTs. A schematic of the AY-102 to AP-102 retrieval system is shown in Figure 12.7. Solids are mobilized in AY-102 using the sluicing nozzles. Supernatant from AP-102 was removed near the top of the waste via the supernatant transfer pump and transferred to AY-102 where it was discharged from the sluicing nozzles. Both standard and extended reach sluicers were employed. The mobilized waste in AY-102 was removed via the slurry transfer pump and returned to AP-102 where it was discharged from the slurry distributor lower in the tank. With approximately 490 hours of total sluicing time, approximately 135,000 gallons of sediment was retrieved.

## 12.4 PIPELINE TRANSFER

Safe, efficient, and compliant pipeline transport of radioactive wastes is essential to retrieval, staging, and treatment operations at any site. The integrity of the transfer system is the primary driver with multiple aspects garnering attention and consideration. The discussion here focuses on pipeline transfers conducted in support of SST and DST retrievals to delineate the major challenges and operational objectives. A more detailed assessment and discussion of all radioactive waste transfers at Hanford and industry approaches are provided by Nguyen et al. (2017).

Waste transfer pipelines and associated infrastructure (e.g., pumps, jumper pits, valves) have a functional objective to completely transfer the radioactive waste entering the pump inlet from a source tank and to a receipt tank without solid accumulation in the line, within pressure limits, and at reasonable flow velocities to avoid excessive pipeline erosion. Aside from a low probability failure of the pipeline due to excessive pressures or long-term erosion, the most likely process upset to experience is potential pipeline plugging due to solid accumulation. Limited capabilities exist that allow operators to recover from a plugged line due to pressure limits, poor pipeline access (buried in ground approximately 3 ft for shielding and freeze protection), radiological worker dose considerations, environmental release risks, and limited implementable mechanical tools (e.g., pipeline pig, steam lance).

Waste transfers at Hanford are managed through graded requirements of transfer velocity and follow-on line flushing based on waste-type definitions (i.e., supernatant, concentrated supernatant, or slurry) and experience-based rules of thumb documented in site-specific standards.

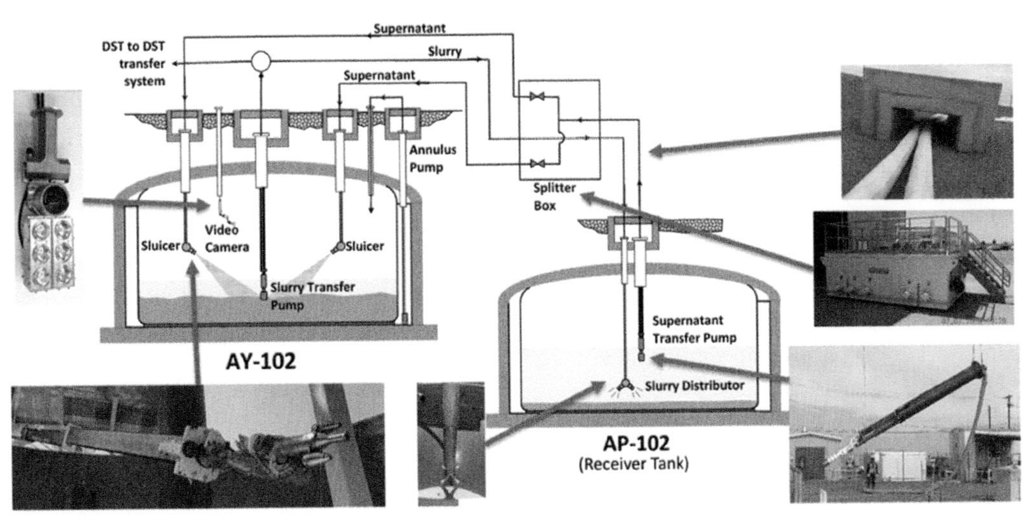


**FIGURE 12.7** DST sediment sluicing retrieval system, AY-102 to AP-102. Reprinted with permission from Hanson et al. (2017).

Established waste transfer lines are buried 2–3 ft below the ground for radiation and freeze protection and serve to link SST and DST pumps to valve pits, valve pit to valve pit, farm to farm, and tank to facilities (e.g., AW-102 to the 242-A Evaporator, 222-S Laboratory to SY-101). Hanford waste transfer pipelines are typically 2- or 3-inch diameter, schedule 40, constructed out of carbon or stainless steel. These transfer pipes are housed within an outer secondary pipe, 4 or 6 inch diameter, to serve as secondary containment. 2-inch-diameter lines are generally considered only for slurry transfers due to minimum transfer velocity requirements (i.e., driven by minimum critical deposition velocity), whereas supernatant (<5 wt% and <1.35 g/cc density) and concentrated supernatant (<5 wt% and >1.35 g/cc density) are transported in 3-inch-diameter lines not requiring the higher transfer velocities to avoid solid accumulation on the pipeline invert. Pipeline lengths vary from 100's of feet up to ~7 miles for the West Area Tank Farm to East Area Tank Farm. The pipelines have frequent bends to address line expansion, inline valves, required slopes for gravity draining to low points, and other process fittings interfacing pumps, valve pits, and facilities.

Routine use of hose-in-hose transfer lines (HIHTLs) for temporary (maximum service life of 3 years) at-surface transfers is commonly deployed to retrieve SST wastes into DSTs and other one-off type activities for non-permanent operations. HIHTLs are constructed of ethylene propylene diene monomer (EPDM) and similarly have inner hoses of 2 inch with a secondary encasement hose with 4 inch diameter. Requirements, technical bases, and operations details for HIHTLs are documented for reference in Erhart (2020), Goessmann (2016), and related internal standards, respectively.

Flushing of transfer pipelines is routine and required to maintain the long-term integrity of the pipeline and defend against solid deposition, which could lead to pipeline plugging. Flushing requirements and associated technical and historical bases are documented in internal company standards.[2] Flush velocities are conducted at a velocity at least equivalent to the transfer velocity just used and ideally some margin greater to resuspend any deposited solids and avoid solid accumulation. The flushing fluid can be raw filtered river water, deionized water, inhibited water, and/or dilute supernatant liquid (only if within 14 days of the transfer, where this limit is based on historic operations experience). The volume of flush fluid used is 1.5–3 times the pipeline volume, and flushing is required after every transfer unless the same supernatant transfer is conducted within 14 days of the prior. A flush of filtered river water, deionized water, or inhibited water is required if greater than 14 days will pass prior to the next waste transfer.

## 12.5 HANFORD TANK WASTE RETRIEVAL SUMMARY

The unique history of the Hanford Site has resulted in complex and varied multi-phased wastes stored in tank farms of varying configurations and RCRA compliance. The waste complexity and variability, worker and environmental safety concerns, nuclear safety, and regulatory restrictions create operational challenges for Hanford and other similar DOE-complex sites.

A short history of operations in Hanford Tank Farms Hanford wastes, the existing infrastructure to store and transfer these wastes, and SST and DST waste handling operations for retrieval, treatment, and closure together with waste retrieval operational examples has been provided. The use and adaptability of multiple technologies for the varied wastes and tank configurations have enabled successful progress to the waste retrieval and treatment mission, and thus provide a meaningful framework for application to similar sites.

## NOTES

1 RCRA is the public law that gives the Environmental Protection Agency the authority to control generation, transportation, treatment, storage, and disposal of hazardous wastes by establishing regulations, guidance, and policies for its proper management.

2 Most notable and applicable are current revisions of engineering standards: TFC-ENG-STD-26 *Waste Transfer, Dilution, and Flushing Requirements* and TFC-ENG-STD-21 *Hose-in-Hose Transfer Lines*.

## REFERENCES

Alleman RT, JD Hudson, ZI Antoniak, JJ Irwin, WD Chvala, NW Kirch, LE Efferding, TE Michener, JG Fadeff, FE Panisko, JR Friley, CW Stewart, WB Gregory, and BM Wise. 1994. *Mitigation of Tank 241-SY-101 by Pump Mixing: Results of Testing Phases A and B*. PNL-9423. Pacific Northwest Laboratory, Richland, WA.

Boomer KD, JB Johnson, TJ Venetz, and DJ Washenfelder. 2012. *Overview of Enhanced Hanford Single-Shell Tank (SST) Integrity Project -12128*. WRPS-51713-FP, Rev. 0. Washington River Protection Solutions Inc., Richland, WA.

Brown NL, AR Olander, and JD Scholkowfsky. 2021. *Single-Shell Tank Waste Retrieval Plan*. RPP-PLAN-40145, Rev. 7. Washington River Protection Solutions Inc., Richland, WA.

Campbell ST, CL Girardot, JR Gunter, JD Larson, JS Page, GE Soon, CJ Zeleny, and JS Garfield. 2020. *Double-Shell Tank Integrity Program Plan*. RPP-7574, Rev. 7. Washington River Protection Solutions Inc., Richland, WA.

Carlson AB, PJ Certa, TM Hohl, JR Bellomy III, TW Crawford, DC Hedengren, AM Templeton, HS Fisher, SJ Greenwood, DG Douglas, and WJ Ulbright Jr. 2001. *Test Report, 241-AZ-101 Mixer Pump Test*. RPP-6548, Rev. 1. Numatec Hanford Corporation, Richland, WA.

Carothers KG, SD Estey, NW Kirch, LA Stauffer, and JW Bailey. 1999. *Tank 241-C-106 Waste Retrieval Sluicing System Process Control Plan*. HNF-SD-WM-PCP-013, Rev. 2. Fluor Daniel Hanford Inc., Richland, WA.

Cuta JM, KG Carothers, DW Damschen, WL Kuhn, JA Lechelt, K Sathyanarayana, and LA Stauffer. 2000. *Review of Waste Retrieval Sluicing System Operations and Data for Tanks 241-C-106 and 241-AY-102*. PNNL-13319. Pacific Northwest National Laboratory, Richland, WA.

De Lorenzo DS, AT DiCenso, DB Hiller, KW Johnson, JH Rutherford, DJ Smith, and BC Simpson. 1994. *Tank Characterization Reference Guide*. WHC-DS-WM-TI-648 Rev. 0. Westinghouse Hanford Company, Richland, WA

Erhart MF. 2020. *Specification for Hose-in-Hose Transfer Lines and Hose Jumpers*. RPP-14859, Rev. 15. Washington River Protection Solutions Inc., Richland, WA.

Gephart RE. 2003. *A Short History of Hanford Waste Generation, Storage, and Release*. PNNL-13605, Rev. 4. Pacific Northwest National Laboratory, Richland, WA.

Goessmann GE. 2016. *Technical Basis Document for TFC-ENG-STD-21 Hose in Hose Transfer Lines*. RPP-RPT-28499, Rev. 2. Washington River Protection Solutions Inc., Richland, WA.

Hanson C, D Greenwell, S Guillot, and J Follett. 2017. Hanford's Double-Shell Tank AY-102 Recovery Project. *43rd Annual Waste Management Conference (WM2017),* Session 006 - Paper 17416. Washington River Protection Solutions Inc., Richland, WA.

Hay BD. 2022. *Waste Tank Summary Report for the Month Ending April 30, 2022*. HNF-EP-0182, Rev. 412. Washington River Protection Solutions Inc., Richland, WA.

Johnson GD, NW Kirch, RE Bauer, JM Conner, CW Stewart, BE Wells, and JM Grigsby. 2000. *Evaluation of Hanford High-Level Waste Tank 241-SY-101*. RPP-6517, Rev. 0. CH2MHILL Hanford Group, Inc., Richland, WA.

Kelly SE, RX Milleret, TA Wooley, and KP Lee. 2013. *One System Waste Feed Delivery Mixing Performance and Solids Accumulation Test Report*. RPP-RPT-53931, Rev. 0. Washington River Protection Solutions, LLC, Richland, WA.

Kirch NW, JM Conner, WB Barton, CW Stewart, BE Wells, and JM Grigsby. 2000. *Remediation of Crust Growth and Buoyant Displacement Gas Release Event Behavior in Tank 241-SY-101*. RPP-6754. CH2MHILL Hanford Group, Inc., Richland, WA.

Mahoney LA, ZI Antoniak, WB Barton, JM Conner, NW Kirch, CW Stewart, and BE Wells. 2000. *Results of Waste Transfer and Back-Dilution in Tanks 241-SY-101 and 241-SY-102*. PNNL-13267. Pacific Northwest National Laboratory, Richland, WA.

Meacham JE, SJ Harrington, JS Rodriquez, VC Nguyen, JG Reynolds, BE Wells, GF Piepel, SK Cooley, CW Enderlin, DR Rector, J Chun, A Heredia-Langner, and RF Gimpel. 2012. *One System Evaluation of Waste Transferred to the Waste Treatment Plant*. RPP-RPT-51652, Rev. 0, PNNL–21410. Washington River Protection Solutions LLC, Richland, WA.

Nguyen VC, MS Fountain, RC Daniel, CW Enderlin, and BE Wells. 2017. *One System River Protection Project Integrated Flowsheet - Critical Velocity for Slurry Waste Transfer Study*. RPP-RPT-59585, Rev. 0. Washington River Protection Solutions Inc., Richland, WA.

Olander AR, NL Brown, and JD Scholkowfsky. 2016. *Single-Shell Tank Waste Retrieval Plan*. RPP-PLAN-40145, Rev. 6. Washington River Protection Solutions Inc., Richland, WA.

Rasmussen JH. 2017. *Guidelines for Updating Best -Basis Inventory*. RPP-7625, Rev. 13. Washington River Protection Solutions Inc., Richland, WA.

Rassat SD, CW Stewart, BE Wells, WL Kuhn, ZI Antoniak, JM. Cuta, KP Recknagle, G Terrones, VV Viswanathan, JH Sukamto, and DP Mendoza. 2000. *Dynamics of Crust Dissolution and Gas Release in Tank 241-SY-101*. PNNL-13112. Pacific Northwest National Laboratory, Richland, WA.

Smith EH. 1991. *Overview of the Closure Approach for the Hanford Site Single-Shell Tank Farm*. WHC-SA-1306-FP. Westinghouse Hanford Company, Richland, WA.

Swaney SL. 2005. *Single-Shell Tank Interim Stabilization Record*. HNF-SD-RE-TI-178, Rev. 9. Babcock Services Inc., Richland, WA.

Venetz TJ and JR Gunter. 2014 *Double Shell Tank Construction: Extent of Condition*. TOC-PRES-14-1370. Washington River Protection Solutions, Richland WA.

Voogd JA. 2005. *Demonstration Retrieval Data Report for Single-Shell Tank 241-S-112*. RPP-RPT-27406, Rev. 1. CH2M HILL Hanford Group, Inc., Richland, WA.

Wells BE, JA Fort, PA Gauglitz, DR Rector, and PP Schonewill. 2013. *Preliminary Scaling Estimate for Select Small Scale Mixing Demonstration Tests*. PNNL-22737. Pacific Northwest National Laboratory, Richland, WA.

Wells BE, DE Kurath, LA Mahoney, Y Onishi, JL Huckaby, SK Cooley, CA Burns, EC Buck, JM Tingey, RC Daniel, and KK Anderson. 2011. *Hanford Waste Physical and Rheological Properties: Data and Gaps*. PNNL-20646. Pacific Northwest National Laboratory, Richland, WA.

Wells BE and JJ Ressler. 2009. *Estimate of the Distribution of Solids within Mixed Hanford Double-Shell Tank AZ-101: Implications for AY-102*. PNNL-18327. Pacific Northwest National Laboratory, Richland, WA.

# 13 Integrated Waste Treatment Flowsheet and Interface Management Strategy

*Jennifer A. Kadinger and Courtney L. H. Bottenus*

This chapter explores tank waste treatment at the Hanford Site, as mentioned earlier in this book, through the lens of interface management and integrated flowsheet strategy development. Interface management is the coordination of interactions among various teams, systems, or processes to ensure they work together effectively within a larger operation or project. An integrated flowsheet is a comprehensive diagram and tool that maps out the sequence and interaction of different processes and material flows in a facility or system of facilities, ensuring the complete system is appropriately defined for analysis, optimization, and/or communication purposes. The Hanford Site provides an excellent case study for this element of environmental remediation because of the site's extensive size and its technical and programmatic complexity.

## 13.1 INTRODUCTION

The U.S. Department of Energy (DOE) Office of River Protection (ORP) manages the River Protection Project (RPP) at the Hanford Site. The mission of the RPP is to safeguard nuclear waste stored in 177 underground storage tanks, arranged in groups known as "Tank Farms," and to manage the waste safely and responsibly until it can be treated at the Hanford Tank Waste Treatment and Immobilization Plant (WTP) prior to final disposition. Therefore, the DOE-ORP is responsible for the storage, retrieval, treatment, and disposal of the Hanford tank waste.

The RPP is comprised of a complex, fully integrated system of waste storage, treatment, and disposal facilities in various stages of design, construction, operation, or future planning at the Hanford Site (Bernards et al., 2020). This integrated system includes a network of subsystems and equipment managed by multiple contracts and contractors. In order to meet and sustain the necessary operations to accomplish the RPP mission, including an increase in the operations tempo as waste treatment begins, closely coordinated management of the interfaces between the RPP's components is essential. For this discussion, the massive scope of the RPP mission will be focused onto what is called the "direct-feed" low-activity waste (LAW) program as the specific example of integrated flowsheet and interface strategy development and management. The term "direct-feed" is used to differentiate this flowsheet configuration from others, but for the purposes of this discussion, the "direct-feed" term will be omitted, simplifying the name to "the LAW program."

For a complex system like the example of the Hanford Site, the definition, management, and control of interfaces are critical to the success of the system's programs and projects. In addition to the above definition, interface management is a process to assist in controlling process or product development when efforts are divided among parties (e.g., government, contractors, geographically diverse technical teams) and to define and maintain compliance among the processes that intersect (Hirshorn et al., 2017). The sections below describe what a flowsheet is, and the pivotal role flowsheets play as a tool in interface management, using Hanford as a demonstrative example.

DOI: 10.1201/9781003329213-17

### 13.1.1 Process Flowsheets

A key component to process interface management common to any manufacturing or production process (e.g., chemical processing plants, refineries) is a flowsheet that facilitates definition and analysis of the connections and interdependencies throughout a system. A complete process flowsheet thoroughly describes the details of a process and is comprised of three primary elements: (i) process flow diagrams (PFDs), (ii) mass and energy balances, and (iii) process control strategies (Cree et al., 2020).

A PFD is often the first thing that comes to mind when one thinks of a flowsheet. This is because it is a key communication tool for visualizing, discussing, and describing the interconnections of the facilities, systems, equipment, and components that comprise a process, and the flow (i.e., material or mass and energy) through these elements. PFDs can vary in level of detail to suit different discussions, allowing communication of high-level interface information, detailed stream-by-stream data, and everything in between. Figure 13.1 shows a simplified PFD for an air conditioner.

To understand the flow between the Hanford facilities, some of which are in various stages of construction from conceptual design to near completion, the integrated flowsheet was developed. As Tank Farms and the WTP prepared for the transition from waste storage and construction, respectively, to waste processing operations, more detailed process flowsheets were needed to support facility operations. In this way, flowsheets are multi-level, progressively becoming more detailed as the perspective moves from the RPP mission level to individual facilities, and then to process operations within a facility, as shown in Figure 13.2.

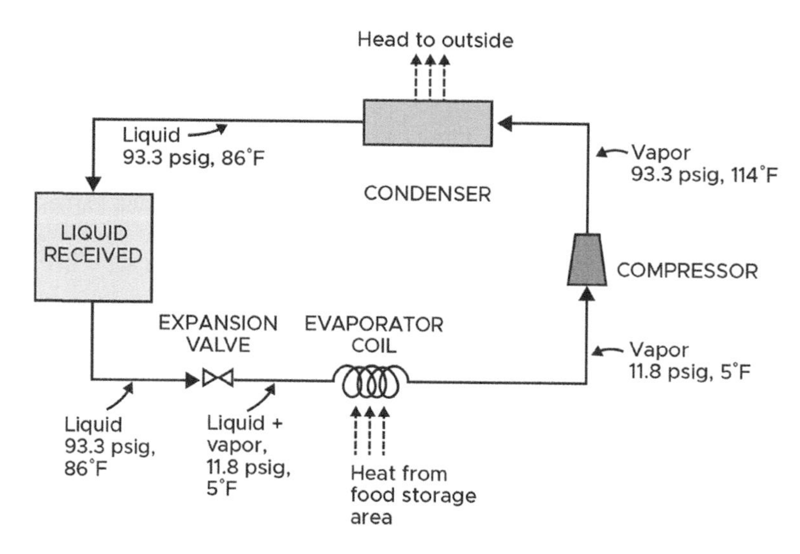


**FIGURE 13.1** Simplified process flow diagram for an air conditioner.

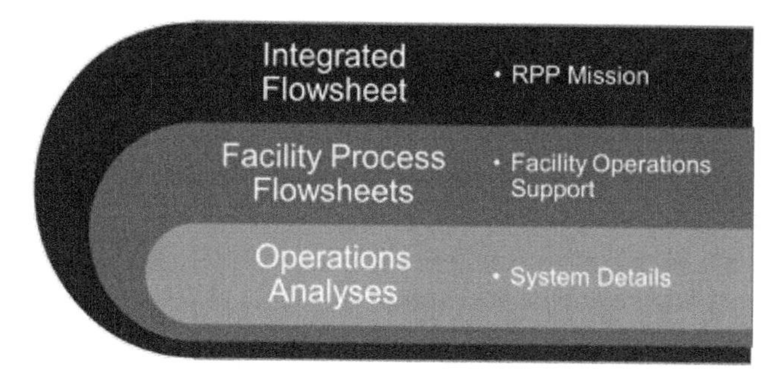


**FIGURE 13.2** Hierarchy of flowsheet detail.

Mass and energy balances account for material and energy (often in the form of heat) flow through the elements of a process. PFDs apply the laws of conservation of mass and energy to identify all process streams, and to evaluate whether material and energy are flowing where intended and whether stream requirements like waste acceptance criteria, design requirements, and process control points are satisfied. Mass and energy balances are often paired with a PFD to illustrate the changes through a process as feed materials are converted to desired products.

With the details of the process outlined by PFDs and associated mass and energy balances, process control strategies describe how the systems need to be operated and controlled to meet desired process performance and product quality. These strategies also include the information necessary to control primary product streams, secondary wastes, recycles, and other non-primary streams to ensure they do not limit the performance of the overall process. While traditionally a PFD is limited to the scope of a single facility or system within a facility, a broader scope can be applied to the same tool. This allows for integrated facilities and systems to be represented in a single PFD that merges multiple flowsheets.

### 13.1.2 The Importance of the Hanford Integrated Flowsheet

The RPP mission is part of the largest and most complex environmental remediation project in the United States of America. The LAW program at the Hanford Site alone comprises multiple prime contractors, two DOE offices (ORP and the Richland Operations Office [RL]), and several smaller contractors and subcontractors. Figure 13.3 shows a very simplified flow diagram of the LAW program, delineated by prime contract. A prime contract encompasses a large scope of work and is characterized by the fact that the requirements of that scope are partially fulfilled by awarding subcontracts. The programmatic complexity depicted in Figure 13.3 paired with the variety of technical challenges and sheer magnitude of the scope of the Hanford Site make the integrated flowsheet an essential tool to ensuring successful integration. The Hanford integrated flowsheet is comprised of process flowsheets that provide key information for integration efforts including the following:

- Definition of facility interfaces,
- Projected material flows and their characteristics,
- Assumptions and their technical bases,
- Identification of opportunities and risks (where risks are potential events or situations that could cause harm or loss and opportunities are those that could lead to positive outcomes or benefits), and
- Data to identify candidate process control parameters, and more.

Although several facilities and processes are constructed and even in operation within the Hanford integrated flowsheet, and although these designs were comprehensively evaluated, including testing and implementation of industry best practices, there remain uncertainties and assumptions carried in each design that introduce potential risks and opportunities. Additionally, the integration of legacy equipment and facilities into a network with newly constructed systems introduces further uncertainties that warrant analysis and evaluation for risk management and process improvement. Technical evaluations, laboratory studies, and analyses assess and mitigate process risk while also maturating opportunities for improvement identified in the process flowsheets. Implementation uses the matured process flowsheets and identified process control parameters to determine the process control strategy and to define the necessary infrastructure, projects, operations strategy, schedule, and plans. Through each step of flowsheet development and maturation, communications and awareness across all contractors and other parties are essential to ensure clarity of roles in the integrated flowsheet and how changes in one part of the flowsheet may impact others.

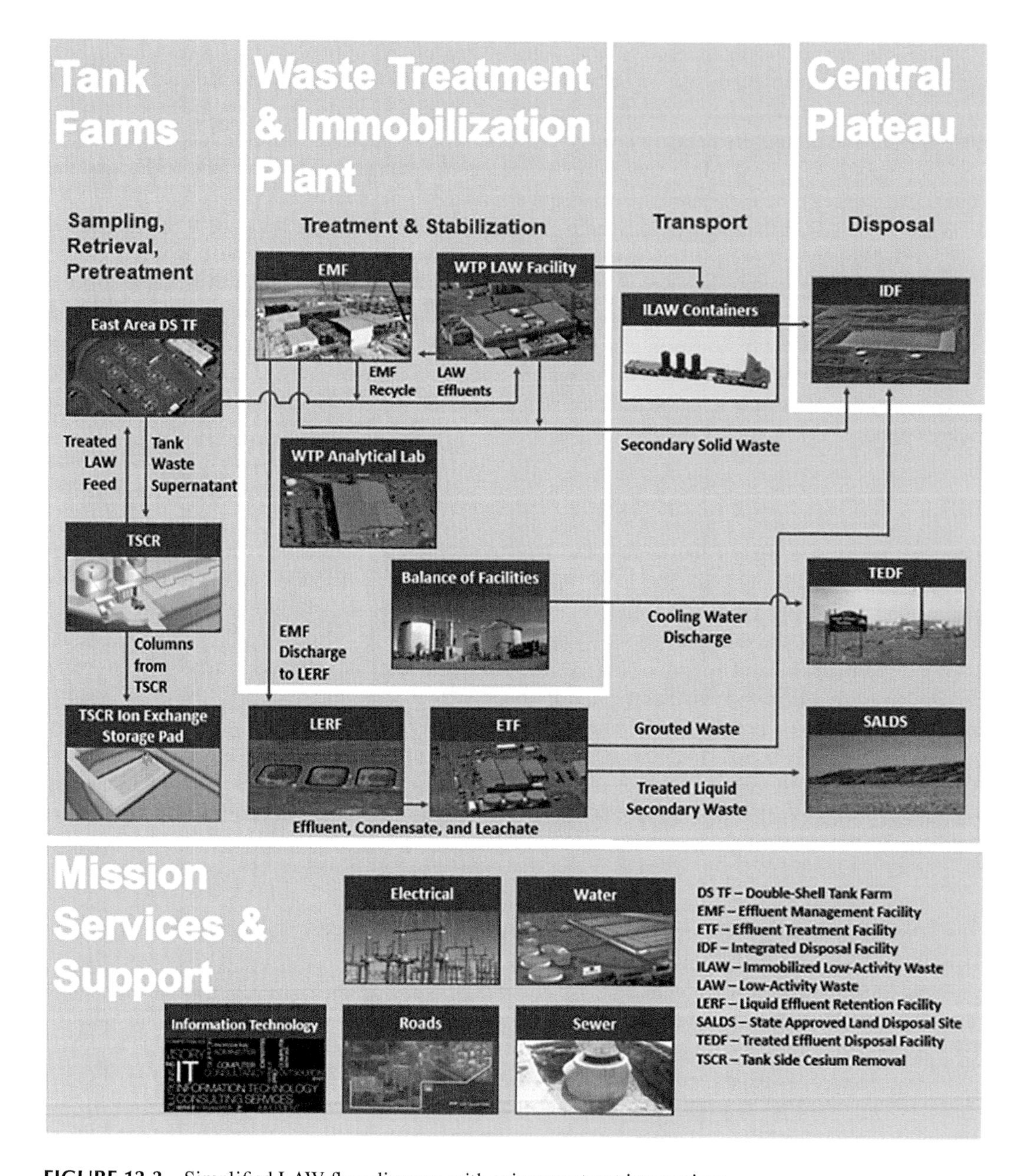


**FIGURE 13.3** Simplified LAW flow diagram with prime contract interactions.

The following sections will highlight this process as they outline the Hanford LAW integrated flowsheet and the details of integrated flowsheet management and interface management for the LAW program.

## 13.2 HANFORD INTEGRATED LOW-ACTIVITY WASTE FLOWSHEET

The LAW flowsheet can be broken into four main functions:

1. **Storage:** The storage and retrieval of waste in the single-shell (SST) and the staging of waste in the double-shell tank (DST) farms

2. **Treatment:** The processing of the tank waste to bring its contents into the state and conditions required for disposal, and the handling of secondary waste streams resulting from tank waste treatment
3. **Disposal:** The transport and waste acceptance into the final disposition site
4. **Analysis:** Characterization and analysis used to facilitate and trigger decision-based events across the other three functions in the flowsheet

The discussion in this chapter covers the path of the liquid tank waste from its storage, retrieval, staging, and waste feed preparation, which enters the front-most system of the tank waste treatment train. Additional details of tank waste storage, retrieval, staging, and feed preparation are discussed in earlier chapters. Figure 13.4 shows a simplified depiction of the LAW Integrated Flowsheet broken into the four main functions. Additionally, Figure 13.4 shows the same process depicted in Figure 13.3 with a focus on highlighting the four main functions of the LAW mission (storage, treatment, disposal, and analysis) and further subdividing the treatment function into three subfunctions:

- Tank Farms treatment, which, in the depicted LAW flowsheet, consists of Tank-Side Cesium Removal (TSCR) operation,
- WTP treatment, which consists of the WTP LAW Facility and the Effluent Management Facility (EMF), and
- Secondary treatment, consisting of the Liquid Effluent Retention Facility (LERF) and Effluent Treatment Facility (ETF), which each plays a crucial role in the follow-on treatment of the secondary waste streams to ensure regulatory requirements associated with the final disposal sites are met.

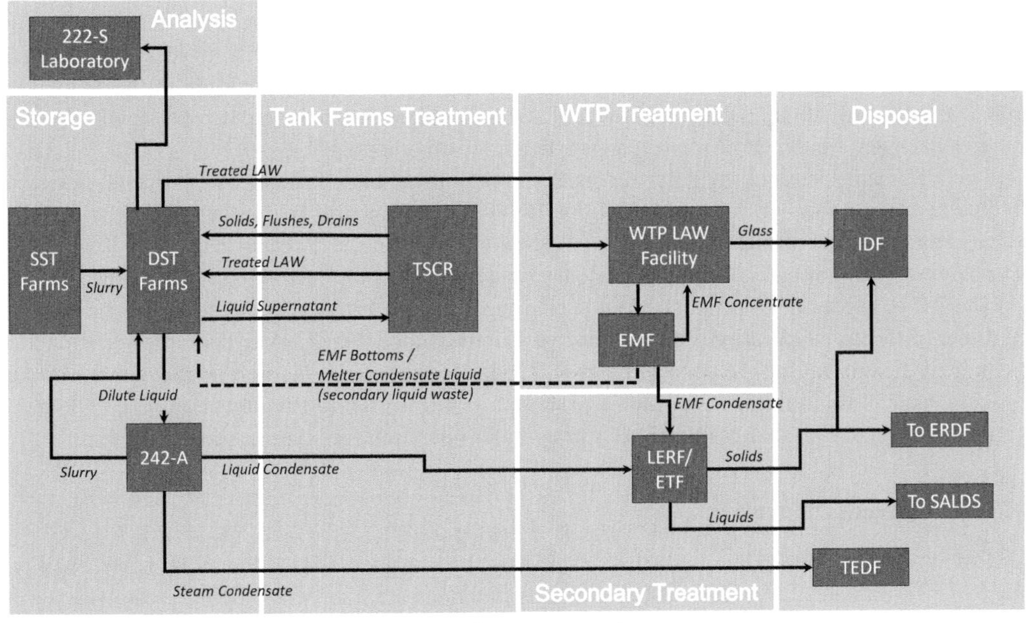


**FIGURE 13.4** Simplified LAW integrated flowsheet diagram. Dashed flow lines represent off-normal operations.

The preparation of tank waste such that it meets the feed acceptance criteria of the WTP and the delivery of that prepared waste is known as "waste feed delivery." Waste feed delivery will be implemented through programs that coordinate and integrate across multiple Hanford Site contractor work scopes. As contract structure at the site evolves over time, specific responsibilities may shift. Among the responsibilities for the waste feed delivery program is the development and maturation of the Integrated Waste Feed Delivery Plan (IWFDP), which describes how the Hanford tank waste will be retrieved, staged (i.e., prepared and qualified), and delivered to the WTP.

With these functions in mind, an understanding of the high-level overview of the flowsheet (shown in Figure 13.5) and the facilities that comprise it is helpful in seeing the role the integrated flowsheet plays in interface management. At the overview level, Hanford Site waste treatment begins with LAW, where liquid supernatant tank waste is staged in DSTs and delivered to TSCR (or successor systems sometimes called "tank farms pretreatment [TFPT] systems"), where most of the cesium is removed using ion exchange and solids are removed via filtration. The pretreated supernatant is then sent to the WTP LAW Facility and immobilized via vitrification. The melter offgas condensate from the WTP LAW Facility is sent to the WTP EMF to be concentrated and recycled through the WTP LAW Facility. The process condensate from the EMF evaporator is sent to LERF for treatment at ETF. Treated liquid from ETF is then disposed of at the State-Approved Land Disposal Site (SALDS). The following subsections briefly describe the facilities involved in the LAW integrated flowsheet to provide additional context to the role of the integrated flowsheet.

### 13.2.1 Tank Farm Facilities

The Hanford Site tank farms are comprised of 149 SSTs, 28 DSTs, and the 242-A Evaporator. Constructed between 1943 and 1964, the majority of the SSTs contain wastes. However, nearly all of the drainable interstitial liquids (where "interstitial liquids" refers to liquids that are contained in the spaces within and between solid materials) have been removed by the SST Interim Stabilization Program (Swaney, 2005). The inventories of the SSTs primarily consist of sludge and crystallized salts with only small amounts of free liquids. The waste remaining in the SSTs will either be retrieved into the DST system where it will be staged for treatment or directly retrieved to other specialized treatment processes, as needed based on composition.

The DSTs are identified by their secondary containment liner and were constructed between 1968 and 1986. As shown in Figure 13.4, the DSTs play an integral role in the RPP mission providing storage capacity and supporting key operations including SST retrievals, 242-A Evaporator campaigns, staging waste, receiving secondary wastes, and staging feed for delivery to the WTP.

The 242-A Evaporator's primary mission is to support tank farm storage by reducing the volume of dilute waste via evaporation, as DST space is limited. Since the 242-A Evaporator began operations in 1977, it has removed approximately 80 Mgal of water from Hanford waste. The resulting process condensate from the evaporation process is sent to LERF for treatment at the ETF. Steam condensate and cooling water from 242-A Evaporator operations are discharged to TEDF.

### 13.2.2 Treatment Facilities

The treatment facilities in the LAW integrated flowsheet include TSCR, the WTP LAW Facility, the WTP EMF, and LERF/ETF. TSCR pretreats liquid supernatant waste by removing solids via filtration and removing most of the cesium via ion exchange. The waste is staged in a DST before being transferred to the TSCR system, where the waste will first pass through a pair of filters to remove solids. Then, the filtered waste is processed through a series of three ion exchange columns to remove cesium. The pretreated waste is passed through a delay tank and gamma detectors for process monitoring before being directed to the WTP feed staging DST. The solids removed via filtration are returned to a separate DST, and the spent ion exchange columns are moved to an interim storage pad for future disposal. The TSCR system enclosures and storage pad are shown

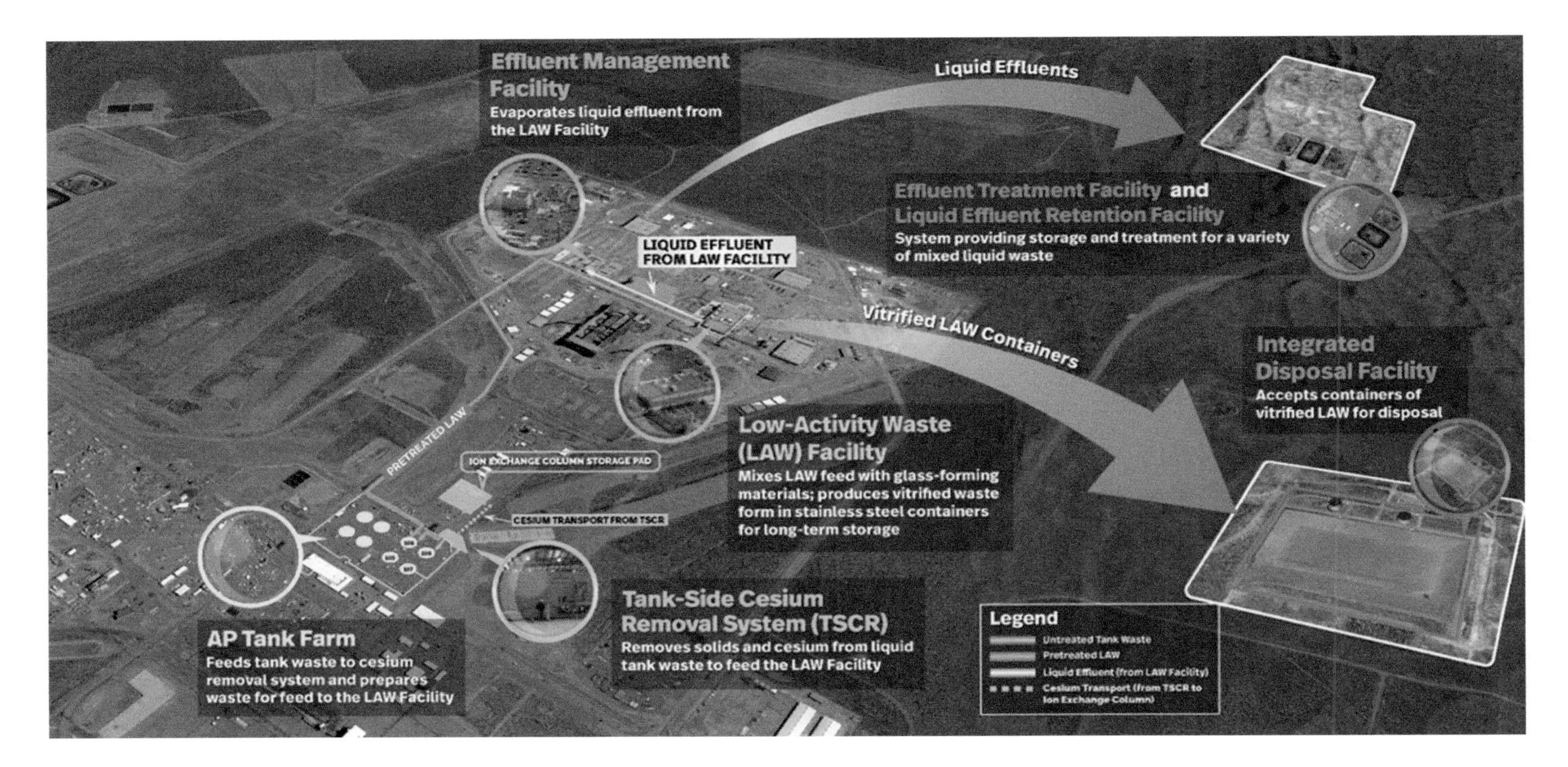


**FIGURE 13.5** Low-activity waste (LAW) integrated flowsheet overview (DOE, 2024a).

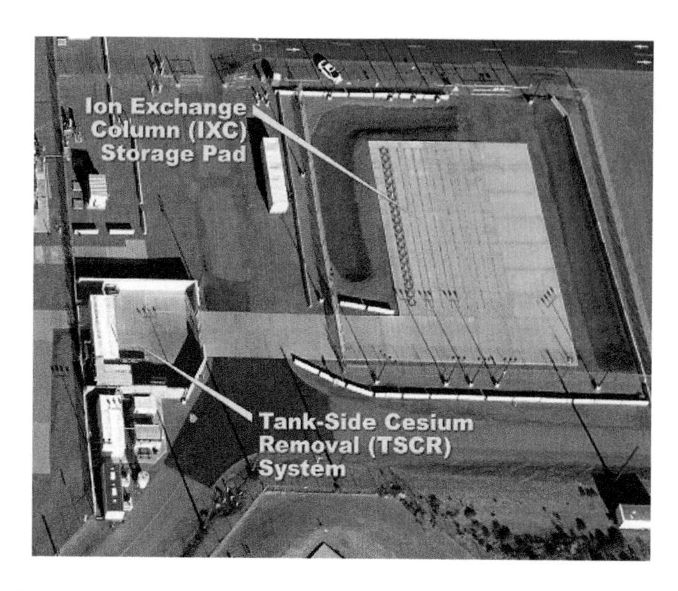


**FIGURE 13.6** Aerial view of the tank-side cesium removal system and storage pad (DOE, 2024b).

in Figure 13.6. The TSCR system is planned to be replaced and/or supplemented by other TFPT systems, the designs of which are yet to be fully defined.

The WTP LAW Facility consists of two melter systems operated in parallel. Each system has a dedicated set of feed preparation vessels, a Joule-heated ceramic-lined melter, and a primary offgas system. The LAW Facility also has a secondary offgas system, which is shared by both melter systems and vessel vents. Pretreated waste is received into receipt vessels and batched for transfer into the melter feed preparation vessels. From these preparation vessels, the waste- and glass-forming chemicals are continuously fed into the melters. Bubblers agitate the melter contents, and an airlift system pours the liquid glass into stainless-steel immobilized LAW (ILAW) containers where the glass cools and solidifies, locking radionuclides into the glass structure. The ILAW containers are then closed and decontaminated before transport to the Integrated Disposal Facility (IDF) for onsite disposal.

The WTP EMF is included as part of the WTP Balance of Facilities (BOF) and provides a means of handling LAW vitrification offgas during the LAW mission and receives secondary effluents from the LAW offgas treatment system, transfer line flushes and drains, and radioactive effluents from the WTP Analytical Laboratory. The EMF concentrates the blended effluent to reduce the total volume, and the concentrate is then recycled through the LAW vitrification process. Other process effluents are sent to LERF/ETF for treatment and disposal. Figure 13.7 shows the WTP site and facilities. Note that the Pretreatment and High-Level Waste Facilities are not associated with the LAW program. The Balance of Facilities and Building 87 support the LAW program (in providing chemical reagents, steam, and electricity, for example) but are not discussed here.

The LERF consists of four lined and covered surface reservoirs that receive and store low-activity, potentially hazardous, aqueous waste generated on the Hanford Site. This aqueous waste is then sent to the ETF for treatment. The ETF consists of a series of wastewater processing units that remove or destroy dangerous organic and radioactive constituents from the aqueous waste. The treated liquid effluent is directed to verification tanks where it is sampled and analyzed to ensure it is below permitted concentration limits before it is discharged to the SALDS. The residue from the treatment processes is concentrated into a brine that can either be loaded into totes for treatment and disposal offsite or be dried into a powder, which is packed into 55-gal drums for disposal at the IDF. Figure 13.8 shows an aerial view of the LERF and the ETF.

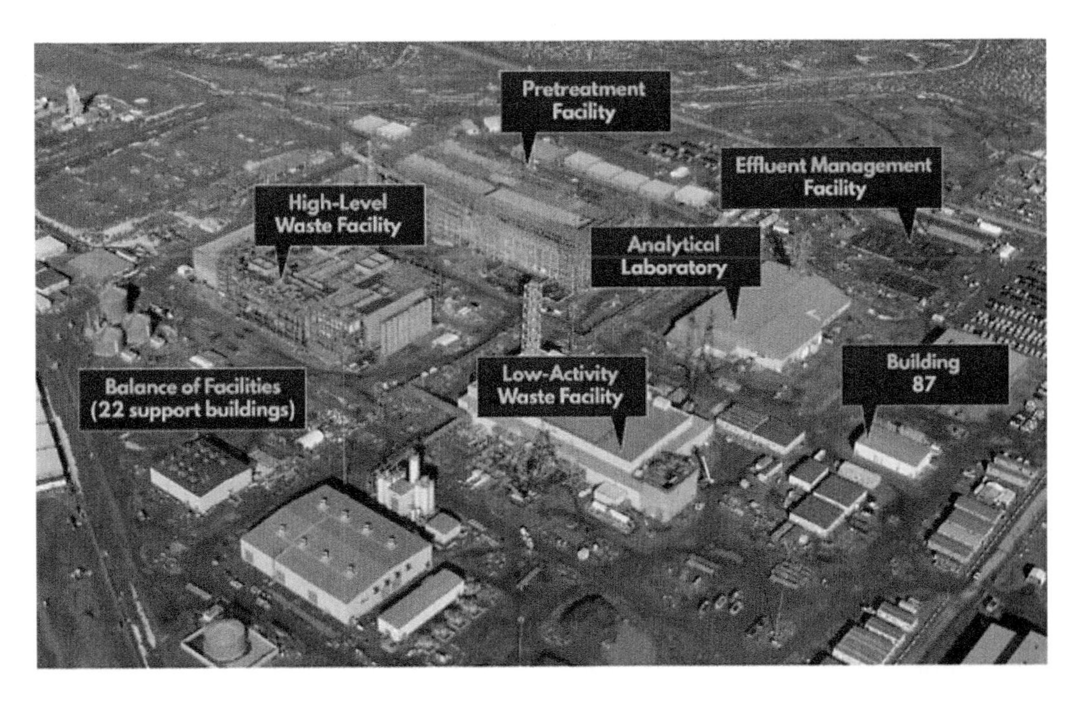


**FIGURE 13.7** Aerial view of the Hanford Waste Treatment and Immobilization Plant site (Goel et al. 2019).

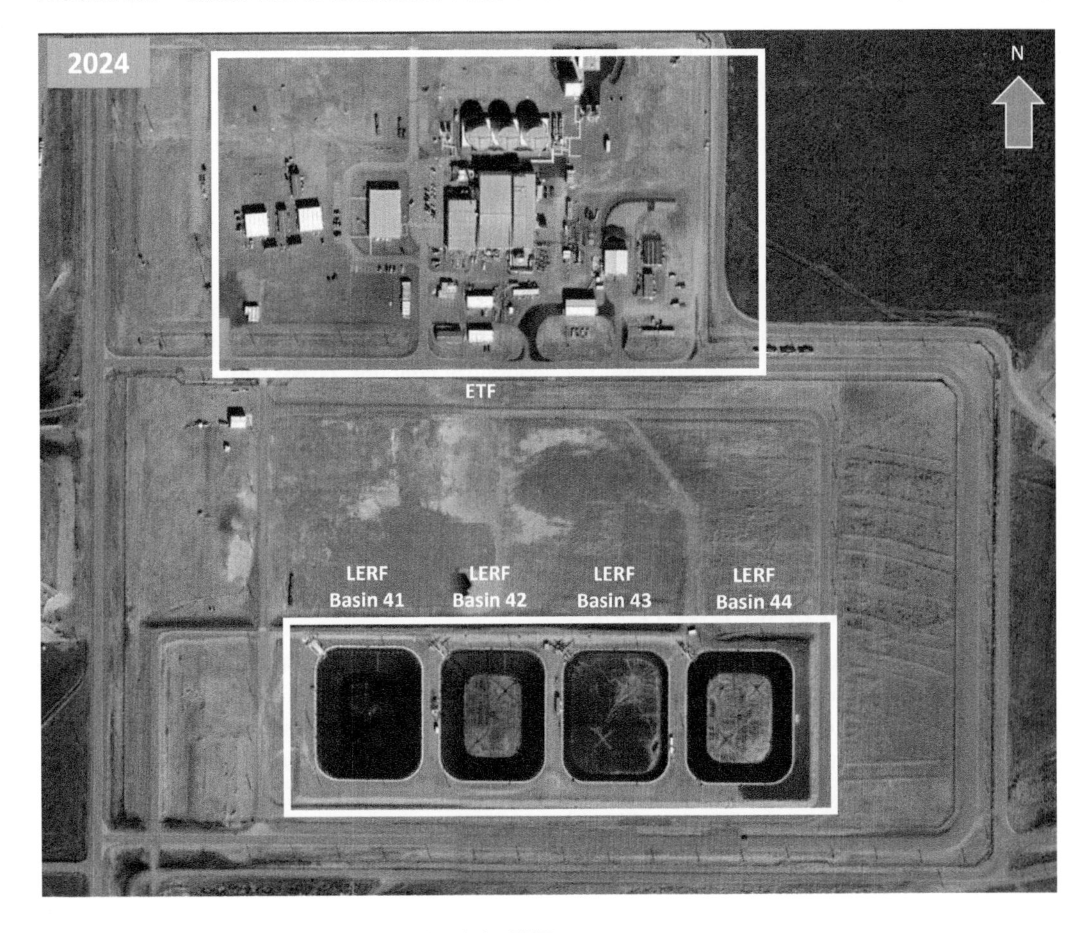


**FIGURE 13.8** Aerial view of the LERF and the ETF.

### 13.2.3 Disposal Facilities

The disposal facilities included in the LAW integrated flowsheet include IDF, TEDF, SALDS, and ERDF. The IDF was constructed in 2006 and consists of two large double-lined disposal cells with a drainage system that collects potentially contaminated water from rain and dust-suppression activities, which has come into contact with the waste packages. Overall, the IDF is approximately 1,500 ft wide, 765 ft long, and 45 ft deep and was designed to be expanded as needed. The IDF provides onsite disposal of low-level wastes and mixed low-level wastes from tank waste treatment operations, waste generated from WTP and ETF operations, onsite non-CERCLA (Comprehensive Environmental Response, Compensation, and Liability Act) sources, Fast Flux Test Facility decommissioning waste, and others. Currently, the dangerous waste permit for IDF within the LAW program only allows mixed low-level waste from IDF operations and ILAW in glass form from the WTP LAW Facility. Other mixed low-level wastes will require a permit modification to be approved by the Washington State Department of Ecology.

The ERDF is a large, engineered landfill located in the center of the Hanford Site that is used for the disposal of low-level radioactive, hazardous, and mixed wastes generated from demolition, remediation, and other cleanup activities across the site. The facility began operations in 1996 and is regulated by the U.S. Environmental Protection Agency under CERCLA. The ERDF also consists of double-lined disposal cells with a drainage system, and the facility currently covers approximately 107 acres and can be expanded further as needed.

The TEDF is a collection, transfer, and disposal system for treated effluents from various 200 Area Hanford Site facilities. These facilities include the 238-W Water Treatment Plant, T-Plant, the 222-S Laboratories, the Waste Encapsulation and Storage Facility, the 242-A Evaporator, and others. The TEDF consists of three pump stations, the TEDF Sampling and Monitoring Building, two infiltration basins, and piping to connect the various facilities in 200 East Area and the 200 West Area to the TEDF system. Liquid effluents not requiring treatment (non-radioactive, non-dangerous liquid effluents) are discharged to the TEDF. As described earlier, the SALDS receives the liquid wastes that were first treated at the ETF.

### 13.2.4 The 222-S Laboratory

The 222-S Laboratory, located in Hanford's 200 West Area, is a 70,000-square-foot analytical facility that began operating in 1951. It is the primary onsite laboratory for analysis of highly radioactive samples in support of all Hanford projects. The laboratory contains 11 hot cells, which enables it to remotely handle highly radioactive samples of tank waste while minimizing radiation dose to workers. The laboratory also contains over 100 pieces of analytical equipment, and 156 fume hoods. Analyses are performed on a wide variety of gaseous, liquid, soil, sludge, and biological samples. During LAW operations, 222-S Laboratory staff will characterize tank waste, as part of feed qualification activities, to ensure it is suitable to be treated at the WTP LAW Facility.

### 13.2.5 The History of the LAW Program

The original version of the RPP integrated flowsheet identified the inter-facility process streams for the flow of the tank waste as it resides in the Hanford Tank Farms through vitrification at the WTP, including final disposition of all secondary waste streams. Since its first iteration, the RPP integrated flowsheet (originally named the Tank Waste Disposition integrated flowsheet and later renamed [Arm et al., 2014]) has undergone three revisions as the flowsheet has matured. The latest iteration of the RPP integrated flowsheet identified 58 LAW process streams, 21 of which will cross current contractual boundaries (Anderson et al., 2014).

Through implementation of continuous improvement frameworks like Lean Management and close coordination with the design process for the LAW program, the RPP integrated flowsheet continued to develop (Cree et al., 2020). A major revision to the RPP integrated flowsheet resulted in an overhaul of the associated flowsheet document, primarily focusing on the LAW phase and incorporation of new sections to describe facility interfaces. These updates included a table of interface flow parameters (IFPs) and their technical bases and a waste acceptance criteria (WAC) analysis.

With the decision to pursue the LAW program came the need for a framework to ensure effective integration and clarity of the functional responsibilities for the program. This framework is built upon the foundations of interface management and the LAW integrated flowsheet.

## 13.3 INTEGRATED FLOWSHEET AND INTERFACE MANAGEMENT

As described in Section 1.2, the complexity of the Hanford Site and its numerous functions, facilities, and contractors and subcontractors necessitates robust interface management. The overall Hanford Site governance model, depicted in Figure 13.9, is designed as a collaborative, transparent decision-making and communication process (Simpson et al., 2021).

Within this governance model is interface management, the role of which is to ensure there is a well-defined set of controlled interfaces to enable inter-contractor work within contractual and operational boundaries. Interface management is made up of people, processes, and tools that are leveraged to effect contractor alignment. These include project liaisons, interface agreements such as interface control documents, service delivery documents, and detailed, documented processes for change control and issue resolution. A fundamental aspect of interface management is communication, and the Hanford Site governance model is structured to enable information to flow quickly between DOE and the site contractors.

To that end, the Hanford Site governance structure includes comprehensive opportunities for collaboration facilitated by working groups and committees such as:

- Bi-weekly Hanford Prime Contractors Interface Management Meetings
- Fire Marshall's Monthly Forum
- Fire Systems Maintenance Monthly Interface Meeting
- Site Space Utilization Committee Meeting
- Information Technology Governance Board Monthly Meeting
- Weekly Resource Allocation Meeting for Craft Services
- Chief Financial Officer Monthly Meeting
- LAW Mission Integration Bi-Weekly Meeting
- LAW Integrated Operations Readiness Committee Monthly Interface Meeting

The charters for these interface working groups flow through the greater Contractor Interface Board (CIB) as outlined in Figure 13.9. In addition to these interface working groups, interface-dependent programs also have charters to ensure effective integration and clarity of responsibilities (Olds, 2019). The LAW program's charter does just that, defining the framework for implementation including details like:

- The leadership model
- Key roles and responsibilities of federal and contractor staff
- Communications management
- Technical interface management and issue resolution
- Performance expectations for individual project stakeholders
- Decision-making processes
- Program management tempo of standing meetings and communications

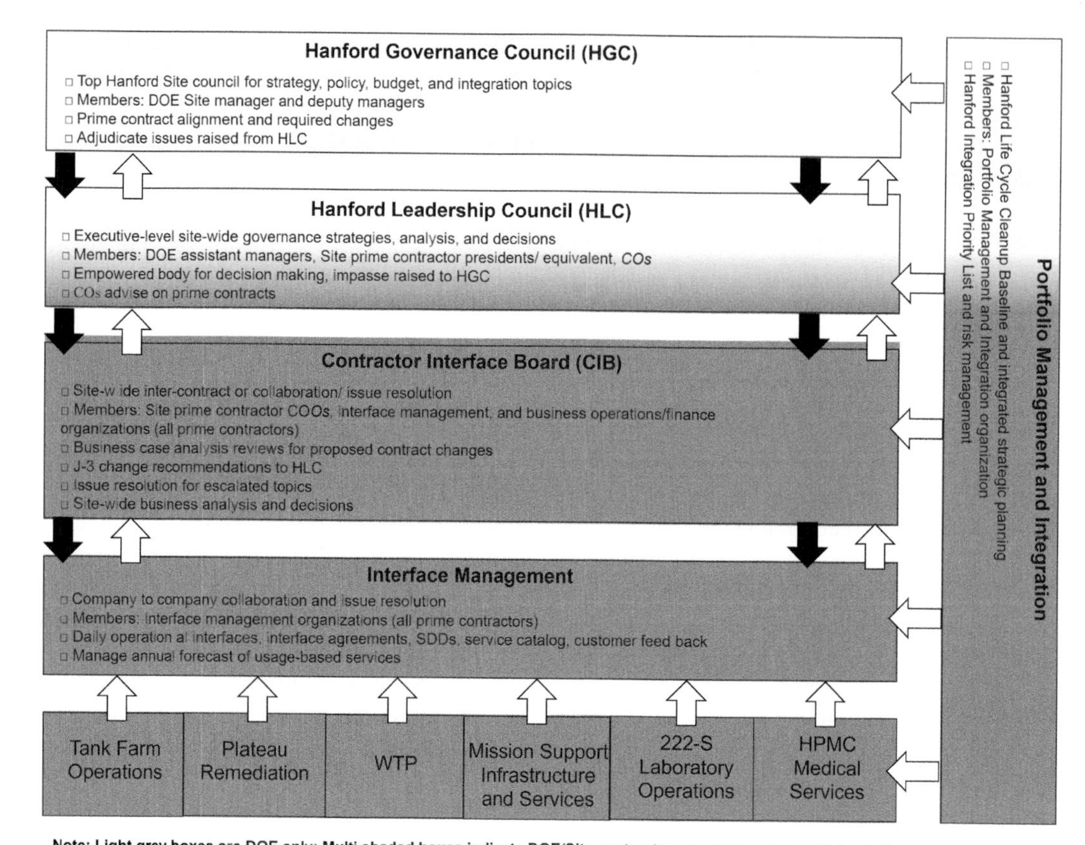


**Note: Light grey boxes are DOE only; Multi shaded boxes indicate DOE/Site contractor management councils/ activities; Dark grey boxes indicate contractor-only interface processes.**

**CIB** = Contractor Interface Board
**CO** = contracting officer
**COO** = chief operating officer
**DOE** = U.S. Department of Energy
**HGC** = Hanford Governance Council
**HLC** = Hanford Leadership Council
**SDD** = service delivery documents
**WTP** = Waste Treatment and Immobilization Plant

**FIGURE 13.9** Hanford Site governance model (HMIS, 2021).

To bolster these interface meetings and discussions, interface agreements document the exchange of services and funds between the Hanford contractors. For the scope of the LAW program, interface control documents (ICDs) are the critical interface agreements. Overall, at Hanford there are two types of interface control documents, Hanford Site ICDs and WTP ICDs. Both types document the physical interfaces and operations demarcation points that are the responsibility of each party to maintain. The WTP ICDs have additional requirements, discussed further below. In fact, because WTP relies on several technical services and operations from other Hanford contractors and will continue doing so into full process operations, the facility has its own active interface management program (Reinemann et al., 2018) in addition to its own set of ICDs.

### 13.3.1 Interface Control Documents

The WTP ICDs identify WTP interface partner contract baseline requirements for shared responsibilities associated with WTP interfaces. Put another way, the external interfaces related to energy, data, or materials transferred between the WTP Contractor and other interface organizations are defined and documented in ICDs. There are several WTP ICDs but for the case study of the LAW program, ICD 30 – Interface Control Document for LAW Feed (Reinemann and Pell, 2021), or simply ICD-30, is of particular importance.

ICD-30 describes the physical and administrative interactions that allow for the transfer of Hanford Tank Farms-treated LAW feed to the Hanford WTP LAW Facility, as shown in Figure 13.10.

The ICD includes definition of each of the LAW feed interface functions and the responsibilities associated with each function for the WTP Contractor and the Hanford Tank Farm Contractor.

The ICD also outlines the interactions and dependencies across the interface relating to nuclear safety and each side's Documented Safety Analysis (DSA). Because each side of the interface relies on the other for interfacing portions of the system, it is imperative that shared nuclear safety responsibilities are clearly understood to ensure compliance. This may include actions or input to technical safety requirements, specific administrative controls, and other safety analysis requirements. The ICD identifies these impacts and provides a mechanism by which the DSAs can be reviewed and amended as necessary to ensure these interfacing requirements are systematically captured.

In conjunction with the ICD, to ensure compliance with applicable safety, permitting, and technical bases of Tank Farms and the WTP LAW Facility, an integrated feed qualification program was jointly developed by the Hanford Tank Farms Contractor and the WTP Contractor. Implementation of this program ensures the feed acceptance criteria and qualification requirements are met for the authorized transfer of waste feed to TSCR and then of treated waste to the WTP LAW Facility (Colby et al., 2021). Figure 13.11 outlines the feed qualification program and highlights the information flow between interfacing parties in addition to the flow of samples and waste material.

These integrated and embedded tools, the governance model, ICD, feed qualification program, and feed acceptance criteria are implemented as the foundation of interface management for the LAW program. From this understanding of the overall structure of these programs and tools, the critical role of an integrated flowsheet becomes clear. Development, maturation, analysis, and communication of the integrated flowsheet ensure continued understanding of the impact that changes or decisions may have within systems, across interfaces, and throughout the whole flowsheet. These technical data bolster integrated strategic decision-making and interface communication.

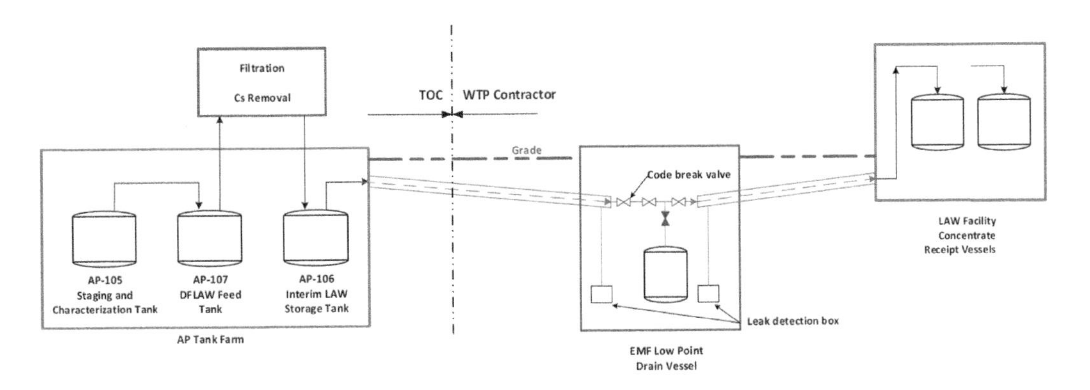


**FIGURE 13.10** LAW feed transfer interface (Bechtel, 2021).

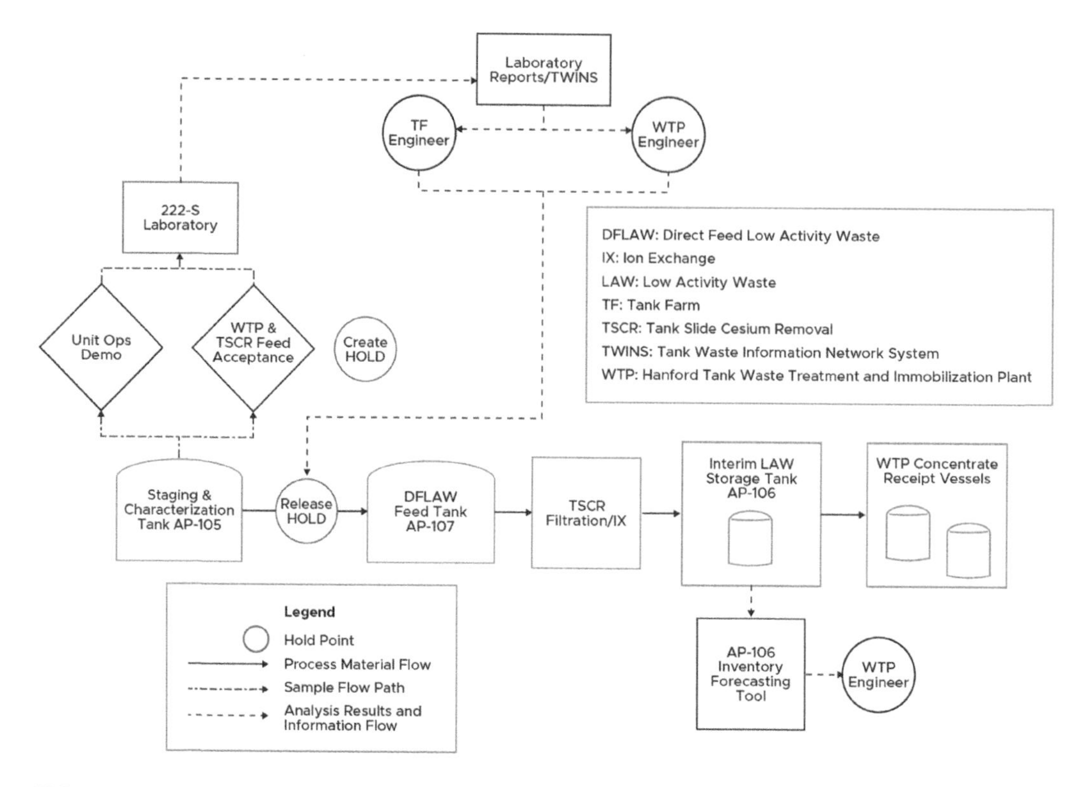


**FIGURE 13.11** Simplified integrated LAW feed qualification program.

## 13.4 KEY TAKEAWAYS FOR INTERFACE MANAGEMENT PROGRAMS

Due to the Hanford Site's size and technical and programmatic complexity, interface management is critical to the success of the RPP mission. This exploration of the case study of the LAW program has highlighted a number of key tools and characteristics that have built a robust interface management program, mostly notable of which is the integrated flowsheet. This foundational tool grows and matures with the process design and programmatic structure, outlining paths forward at the outset of design and evolving with greater and greater detail to provide decision-makers the information and data necessary to understand the push-and-pull impacts of changes throughout the greater system.

Armed with an understanding and knowledge of the integrated flowsheet, the other tools and mechanisms of the overall interface management program described above can be implemented with added clarity and rigor. This is certainly evident from a technical perspective as several of these tools focus on the technical data and information that describe the systems and interfaces, but it is also true from a communication perspective, which is equally as important to successful interface management. Each of these tools and processes works to not only determine the necessary information for robust interfaces but also to provide tools, pathways, routines, and triggers for comprehensive, complete, and timely communication and documentation of that information. These characteristics, tools, and processes, as demonstrated by the Hanford Site LAW program, are key for implementing and maintaining a complex and successful interface management program driven by data, technical bases, and clear communication.

## REFERENCES

Anderson, K.A., M.D. Britton, J.M. Colby, L.H. Cree, M.S. Fountain, D.W. Nelson, V.C. Nguyen, and M.E. Stone, 2014, *One System River Protection Project Integrated Flowsheet*, RPP-RPT-57991, Rev. 2, Washington River Protection Solutions, LLC for the U.S. Department of Energy Office of River Protection, Richland, WA.

Arm, S.T., J.M. Colby, M.S. Fountain, V.C. Nguyen, R.M. Russell, L.M. Sasaki, and M.E. Stone, 2014, *One System: Tank Waste Disposition Integrated Flowsheet - River Protection Project Reference Integrated Flowsheet*, RPP-RPT-57991, Rev. 0, Washington River Protection Solutions, LLC for the U.S. Department of Energy Office of River Protection, Richland, WA.

Bechtel. 2021. *ICD 30 - Interface control document for DFLAW Feed.* 24590-WTP-ICD-MG-01-030, Rev 1. Bechtel, Richland, WA.

Bernards, J.K., G.A. Hersi, T.M. Hohl, R.T. Jasper, P.D. Mahoney, N.K. Pak, S.D. Reaksecker, A.J. Schubick, E.B. West, L.M. Bergmann, G.R. Golcar, A.N. Praga, S.N. Tilanus, and T.W. Crawford, 2020, *River Protection Project System Plan: Safely, Effectively, and Efficiently Treat Tank Waste and Close Hanford Tanks*, ORP-11242, Rev. 9, Washington River Protection Solutions, LLC for the U.S. Department of Energy Office of River Protection, Richland, WA.

Colby, J.M., D.W. Nelson, and A.J. Kolb, 2021, *Integrated DFLAW Feed Qualification Program Description*, RPP-RPT-59314, Rev. 2, U.S. Department of Energy Office of River Protection, Richland, WA.

Cree, L.H., C.R. Kimura, V.C. Nguyen, S.N. Randall, and B.M. Tardiff, 2019, *River Protection Project Integrated Flowsheet*, RPP-RPT-57991, Rev. 3, Washington River Protection Solutions, LLC for the U.S. Department of Energy Office of River Protection, Richland, WA.

Cree, L.H., T.J. Wagnon, and S.T. Arm, 2020, "Refining the Approach to Process Flowsheet Development and Maturation at Hanford - 20049," *Proceedings of Waste Management Symposia 2020*, Phoenix, AZ.

DOE. 2024a. "Direct-Feed Low-Activity Waste." website accessed 6/26/2024. https://www.hanford.gov/page.cfm/DFLAW

DOE. 2024b. "Hanford marks 2022 priority, treats first batch of tank waste." website accessed 6/26/2024. https://www.energy.gov/em/articles/hanford-marks-2022-priority-treats-first-batch-tank-waste

Goel, A., McCloy, J.S., Pokorny, R., and Kruger, A.A. 2019. Challenges with vitrification of Hanford High-Level Waste (HLW) to borosilicate glass - An overview. *Journal of Non-Crystalline solids: X*, vol. 4. https://doi.org/10.1016/j.nocx.2019.100033

HMIS. 2021. *Hanford Site Interface Management Plan and Governance Process*. Hanford Mission Integration Solutions. HMIS-2102886. Richland, WA.

Hirshorn, S.R., L.D. Voss, and L.K. Bromley, 2017, *NASA Systems Engineering Handbook*, NASA/SP-6105, Rev. 2, National Aeronautics and Space Administration, Washington, DC.

Olds, T.E., 2019, *Hanford Site Operations Direct-Feed Low-Activity Waste Program Charter*, U.S. Department of Energy Office of River Protection, Richland, WA.

Reinemann, D., L. Gonzales, and A. Harshfield, 2018, *Interface Management Plan*, 24590-WTP-PL-MG-01-001, Rev. 11, Bechtel National, Inc., Richland, WA.

Reinemann, D., and M. Pell, 2021, *ICD 30 - Interface Control Document for DFLAW Feed*, 24590-WTP-ICD-MG-01-030, Rev. 1, Bechtel National, Inc., Richland, WA.

Simpson, C.A., B.H. Von Bargen, and J.A. Reno, 2021, *Hanford Site Interface Management Plan and Interface Governance Process*, HMIS-2102886 Rev. 0, Hanford Mission Integration Solutions, LLC, Richland, WA.

Swaney, S.L., 2005, *Single-Shell Tank Interim Stabilization Record*, Rev. 9, HNF-SD-RE-TI-178, Babcock Services, Inc., Richland, WA.

# 14 Tank Waste Disposal

*R. Matthew Asmussen*

## 14.1 CONCEPT OF WASTE DISPOSAL

Consideration of all the back-end products and wastes of any chemical or radiological process operation is crucial to ensure environmental protection and human health. A principal goal of waste disposal, whether nuclear, industrial, or municipal, is to isolate the hazardous components from the public and biosphere. Landfill locations are selected where public exposure would be limited from direct exposure, water transport, or food chain migration, and this approach is similar worldwide. The main difference for radioactive waste disposal compared to industrial or municipal wastes is the long time for which isolation is required. Any radioactive waste disposal operator must work with governing regulators and local stakeholders to ensure that requirements are met and confidence exists in the selected disposal approach. Effective siting of a disposal location can provide significant advantages, such as a lack of a pathway to potable water and stable geology – see, for example, the Clive Utah site (USA) or Waste Control Specialists in TX (USA). Geographic isolation, however, is only a single prong of the approach used in the disposal of radioactive wastes. The engineering design of waste forms and waste packages (both covered in Chapter 8 using glass as the example), disposal facilities, and final closure structures such as caps all play a crucial role in providing protection to the environment and human health. The selection, design, and development of a disposal location for the various wastes in the Hanford mission, covered in this chapter, are no different. This chapter covers the history of the identification of waste management approaches at Hanford, then discusses the three main facilities in use: the Integrated Disposal Facility (IDF), which is planned to receive immobilized waste forms principally from tank waste treatment; the Environmental Restoration and Disposal Facility (ERDF), which was crucial in facilitating River Corridor cleanup and facility decommissioning and is still in use today; and the State-Approved Land Disposal Site (SALDS) to highlight the management of all liquids generated in a waste treatment flowsheet.

## 14.2 HANFORD DISPOSAL PLAN HISTORY

The novelty of the plutonium production processes performed at Hanford meant a significant waste handling challenge was presented. When production processes were developed and operations initiated, there was limited knowledge on the compositions of the wastes being produced and beyond storage of the spent fuel no known processes for handling the high volumes of waste and contaminated materials being generated beyond discharge of perceived low-level liquids to cribs ponds and trenches. However, the importance of plutonium production to national defense left no other viable options. Contaminated liquid and solid wastes were concentrated and stored in the large underground tanks described in Chapter 6, while final disposal was deferred. The current costs of that decision and the added environmental risks serve as a constant reminder to build back-end handling of all waste streams early into any chemical or nuclear process flowsheet or developmental technology.

In general, the Hanford Site geology and weather lend itself to shallow disposal of radioactive low-level wastes. Limited rainfall and higher temperatures limit soil moisture required for transport of contaminants in the subsurface. Geographically, the presence of the central plateau provides separation from both the water table (an unconfined aquifer below the site) and the Columbia River providing geological isolation (USDOE 2018). During the early operational period at Hanford, transuranic (TRU) waste (radioactive waste containing more than 100 nanocuries of alpha-emitting transuranic


DOI: 10.1201/9781003329213-18

isotopes per gram of waste, with half-lives greater than 20 years), non-TRU and industrial wastes were disposed in shallow burial grounds (Geiger et al. 1977). The burial grounds were constructed near operational sites. In all, there were 26 sites in the 100 area, 28 sites in the 200 area, and 11 in the 300/600 area. Beginning in the 1970s, the TRU waste management was fully transitioned to engineered structures such as caissons, cribs, culverts, buried vaults, and tunnels. These locations have remained fairly undisturbed with the exception of in situ vitrification of the 116-B crib (Luey et al. 1992) and stabilization of the PUREX facility storage tunnels; an ~20′ portion of Tunnel 1 suffered a localized collapse in 2017 (Farabee and LaBonty 2020). The tunnel collapse exposed rail cars of contaminated solids to the open atmosphere. Tunnel 1 was backfilled and stabilized using an engineered grout following the collapse in 2017, and Tunnel 2 was stabilized in 2019 as a preventive measure (Farabee and LaBonty 2020). Deep wells and mined spaces within the basalt layers below the Hanford Site were also evaluated for the disposal of the tank wastes in the 1960s and 1970s, but abandoned based on practicality and high economic burdens (Isaacson 1969).

There have been ongoing efforts to retrieve, repackage, characterize, and dispose of TRU wastes present at Hanford (Bannister 2016). Known as the TPA M-091 Milestone series, these efforts pursue complete retrieval and removal of the backlog of Hanford Site mixed low-level waste (MLLW) and transuranic mixed (TRUM) waste. These wastes include cleanup of the Plutonium Finishing Plant (see Chapter 3), K Basins, U Plant, 618 burial grounds, and the 200-PW operable units and in the future may also include the BC cribs and PFP cooling water ditches. The TRU waste will be transferred to the Waste Isolation Pilot Plant (WIPP), while any wastes recovered in this effort meeting low-level waste (LLW) or MLLW requirements will be sent for trench disposal, lessening the burden on the tank waste treatment mission.

The disposal mission covering the largest environmental risk at Hanford has been, and remains, the immobilization of the wastes currently present in the 177 underground storage tanks, including 149 single-shell tanks (SSTs) and 28 double-shell tanks (DSTs). With the chemical and radiological complexity of the Hanford wastes, unique management strategies are required to treat, immobilize, and dispose of the waste. Two distinctions on the waste will be used in the forthcoming discussions. The highly radioactive tank wastes, including supernate, saltcake, and sludge wastes that originated from fuel reprocessing were historically managed during storage as high-level waste (HLW) prior to formal waste determination and disposition. Retrieved wastes that have been processed to remove solids and a portion of the radionuclides (i.e., pretreated) produce a low-activity waste (LAW) feed that can be subsequently immobilized for disposal as MLLW. The original disposal plan at Hanford focused on the DST wastes to be immobilized as a grout in large vault monoliths (van Beek and Wodrich 1990). During the 1980s, both the Hanford Site and the Savannah River site (SRS) developed grout-based approaches for the immobilization of low-level salt wastes (Wilhite 1987, Serne et al. 1992). The liquid wastes would be mixed with a formulation of dry reagents, then pumped as a slurry into large concrete vaults. SRS has implemented this approach with the generation of large saltstone grout disposal units since 1990. Saltstone is the grout waste form produced at SRS from mixing low-level salt waste with a formulation of blast furnace slag, fly ash, and Portland cement. This use of grout vaults at Hanford is a history that is far more complex.

The grout at Hanford would have been produced at the Grout Treatment Facility (GTF; Figure 14.1), which was comprised of four components: a dry reagent silo facility, a 3800-m$^3$ feed tank, the grout processing plant called the Transportable Grout Equipment, and the disposal vaults. All of these were constructed in the 1980s and planned to be operated for several decades.

At the time the GTF was evaluated, Resource Conservation and Recovery Act (RCRA) regulations required a single-liner system for hazardous waste landfills and surface impoundments, while the RCRA Minimum Technology Guidance (MTG) recommended a double-liner system with a leachate collection system between the liners. With stringent rules in place for the long-term performance required of the waste form and the nature of the grouted waste, the disposal system design considered for the Hanford GTF provided more protection than the MTG. The grout vaults would first comply with the requirements of a surface impoundment to receive the liquid waste and then

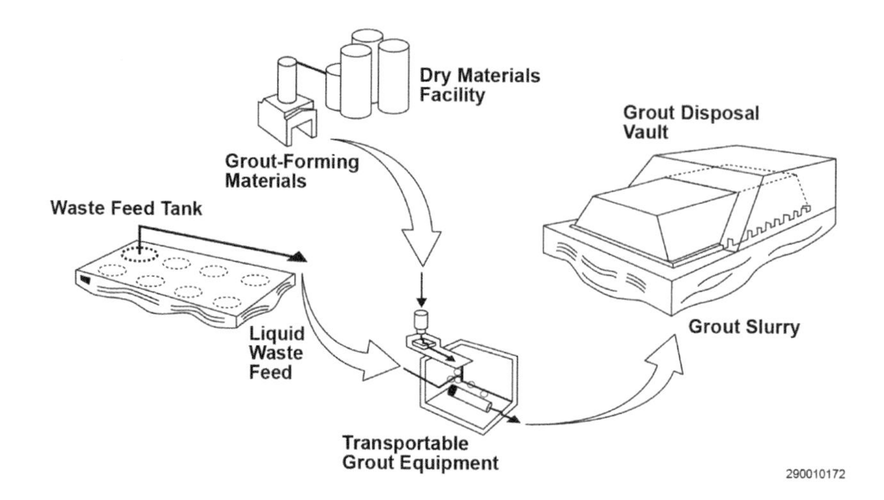


**FIGURE 14.1** Grout treatment facility concept. Reprinted with permission from van Beek and Wodrich (1990).

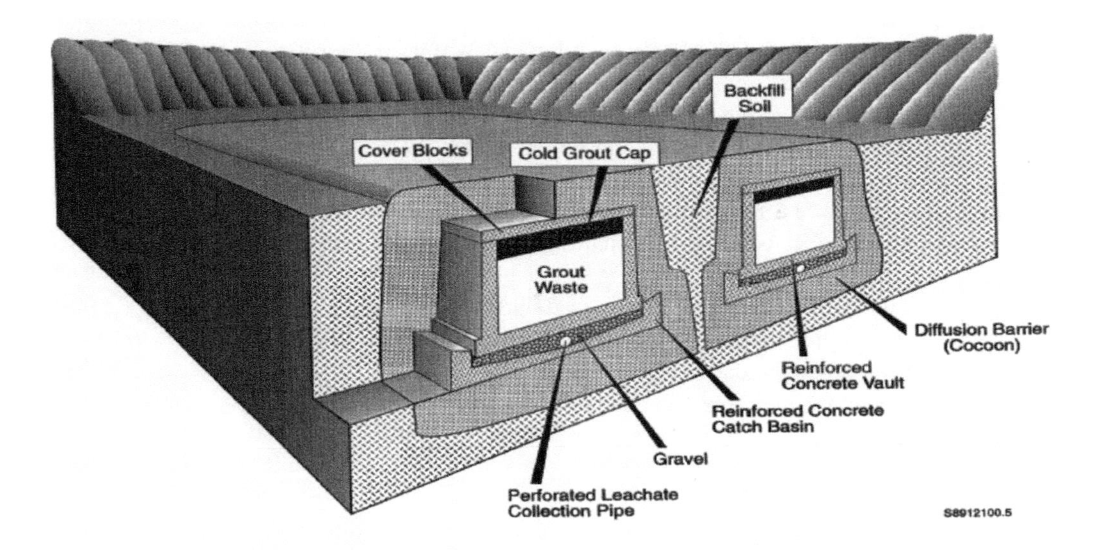


**FIGURE 14.2** Schematic of the grout vaults slated for use at the Hanford Site. Reprinted with permission from USDOE (1992).

comply with the hazardous waste landfill requirements once managing the solidified waste form. Along with the double-liner and leachate collection system, the grout vaults also contained a solidified waste form, an asphalt-covered gravel diffusion barrier, and a multi-layer cover system (van Beek and Wodrich 1990). The extra barriers were implemented to meet DOE long-term performance criteria. The grouted waste would be poured into the center of the disposal system within a 37.6 m × 15.4 m × 10.4 m high reinforced concrete vault. The concrete vault had an internal liner of asphalt to limit water ingress. A drainage path made of high-density polyethylene (HDPE) surrounded the vault to move any liquids to a leachate catch basin. Finally, the closed vault would have a cover system placed on top. The disposal vault concept is shown in Figure 14.2.

A 1 Mgal demonstration vault was prepared that was 38 m long, 15 m wide, and 10 m deep (Figure 14.3). The vault system began construction in 1985 and was completed in 1988, and the

**FIGURE 14.3** PSW demonstration vault prior to addition of the grouted waste.

grout filling campaign ran from August 1988 through July 1989 (Cline et al. 1990). The liquid immobilized in the demonstration vault was low-level spent decontamination solution phosphate/sulfate waste (PSW) from the Hanford N-reactor. A mixture of 41 wt% ordinary Portland cement, 40 wt% fly ash, 11 wt% attapulgite clay, and 8 wt% potters' clay with tributyl phosphate was used as an air de-entraining agent at 0.03 vol% (relative to the grout slurry). The final waste form was non-hazardous under RCRA and is still buried on the Hanford Site.

The source of waste to be immobilized in the grout vaults was the low-level liquid waste present in the 28 DSTs. These wastes were hazardous waste under U.S. Environmental Protection Agency's (EPA) federal regulations, and dangerous waste under the Washington-specific regulations, and at the time of disposal would be considered mixed waste (van Beek and Wodrich 1990). The radioactivity of the liquid was projected to be 0.3 Ci/L, with 99% of that dose coming from Cs-137 (it should be noted that current waste processing at Hanford will remove the Cs and other radionuclides from the retrieved supernate/saltcake waste using the Tank Side Cesium Removal system to significantly lower levels; Barker et al. 2019). Tc-99 and I-129 were considered the radionuclides of primary concern for long-term performance (and still are to this day). To account for the long-term performance and chemical nature of the wastes, the grout formulation was altered from the PSW vault to a 40 wt% limestone, 28 wt% blast furnace slag, 28 wt% fly ash, and 4 wt% Portland cement formulation since the PSW vault did not contain radionuclides (Serne et al. 1992). To immobilize all the DST waste, a 100,000-$m^2$ area in the Hanford 200-E area was designated to house 43 of the mixed waste disposal vaults below the surface. Each vault interior would measure 37.6 m long, 15.4 m wide, and 10.4 m high. The first four vaults planned to receive DST wastes were constructed in the late 1980s (Figure 14.4). While most of the DST waste was considered LLW or MLLW, there was an acknowledgment at the time that some DST waste would require pretreatment for solid separation to generate a HLW fraction and a LLW fraction. Prior to 1993, the LLW was to be grouted and dispositioned on the Hanford Site and the HLW would be vitrified and disposed of at a federal repository. The vitrification of HLW followed by federal repository disposal remains the baseline plan today.

Development of the waste treatment and immobilization pathways at the Hanford Site ran in parallel with development at the SRS, where grouting of LLW and placement in large vaults coupled with vitrification of HLW have been used (Wilhite et al. 1988). The SRS has been successfully executing this combination of approaches for over 30 years producing Saltstone Disposal Units

**FIGURE 14.4** Location of the grout vault disposal system in the Hanford 200-E Area looking west (left) and the construction of the first four grout vaults planned to receive DST waste (right).

(SDU) for the saltstone grout and glass produced at the Defense Waste Processing Facility (DWPF) being stored on-site pending final disposal at a planned federal geologic repository. However, the planning for the immobilization of LLW/LAW at Hanford changed drastically in 1994. In 1989, a Tri-Party Agreement (TPA) was signed as a comprehensive cleanup and compliance agreement (Hanford Federal Facility Agreement and Consent Order) between DOE, Washington Department of Ecology, and EPA that defined key milestones to be met in the Hanford cleanup mission.[1] The TPA states that tank waste is managed under RCRA requirements with regard to tank waste treatment and disposal. The original TPA agreed to 14 grout vaults to be completed by 1994 (milestone M-01-00 of the TPA). In the early 1990s, a series of evaluations were performed based on the expected performance of the grout vaults and life cycle cost projections (Kincaid et al. 1995). A revision to the TPA was made in 1994 to change from grout vaults as the immobilization pathway for the LLW/LAW to vitrification. Multiple reasons drove this change including stakeholder concern for long-term performance of the vaults, immobilization of the waste from the SSTs that would require expansion up to 200 vaults, sophistication of the barrier system of each vault, non-retrievability of the grout vault in the event of setting of the grout being limited due to radiation levels or organics presence, and cost estimates at the time showing equal costs for vitrification. It should be noted that many of these challenges have since been alleviated in updated assessments, including those commissioned by Congress and supported by a National Academies of Sciences, Engineering, and Medicine Review of a Federally Funded Research and Development team report (2020). With the TPA revision, a separate waste treatment program was initiated to produce glass waste forms from the LAW, and thus brings us to today (1997–2023) with the startup of the Hanford Waste Treatment and Immobilization Plant (WTP) and the on-site disposal facility known as the Integrated Disposal Facility (IDF).

## 14.3 INTEGRATED DISPOSAL FACILITY

### 14.3.1 History and Development

With the conceptualization of WTP in the 1990s through various efforts including the Hanford Waste Vitrification Project (HWVP; Westsik et al. 1988) and the Tank Waste Remediation System (TWRS; Keeney and von Winterfeldt 1996), the disposal location of the immobilized LAW glass (ILAW, also referred to as vitrified low-activity waste, VLAW) had to be defined. HLW glass disposal has always been destined for a federal off-site deep geologic repository. The first plan for

ILAW was to modify the existing four former grout vaults to receive the packages of ILAW; these four vaults have enough space to receive 10% of the overall projected LAW volume (Kincaid et al. 1995, Mann 1999). Following the filling of the four vaults with ILAW glass packages, additional concrete vaults could be constructed to receive the remaining ILAW inventory. Other options considered at the time were the construction of a RCRA-compliant large concrete vault, a landfill trench with or without RCRA liners dependent on the RCRA classification of the waste, the use of existing concrete buildings on site like the reprocessing canyons (cost prohibitive to modify and not enough internal capacity), or placement into the SSTs (cost prohibitive) (Mann et al. 1998, Mann 1999). In 1999, the decision was made to use a trench-like design for a RCRA-compliant landfill (double-lined trench with a leachate collection system), similar to the radioactive mixed burial trenches already at Hanford (Taylor 1999, Mann et al. 2001). The facility would receive the ILAW glass only, as at the time, no plan was made for the secondary wastes (which are contaminated liquids or solids that result from vitrification of tank waste). With continued design, this facility was given the name Integrated Disposal Facility (IDF) in the early 2000s (Mann 2003). The IDF will receive the ILAW glass, IDF operational wastes, and likely the solidified secondary waste forms (e.g., grout).

### 14.3.2 Design and Construction

The IDF is a ~13-m-deep trench excavated in the 200-E area of the Hanford Site southwest of WTP and is projected to be 422 m × 330 m in size. Its location was selected to be near existing tank farms, on unused land and inside the 200 area (Shord 1995). Individual cells will be constructed within the IDF. The facility will have a RCRA-compliant dual-liner system and leachate collection system. Specifically, the liners are a geosynthetic clay and an admix layer made mostly of soil and bentonite to manage water transport. The bottom of the facility is sloped to direct leachate to sumps that will move it to the leachate collection system. The first two cells of the IDF were constructed in 2006 (Figure 14.5).

Waste form packages will be placed into the IDF as individual lifts. Following each lift, the packages will be backfilled with the native sand originally excavated from the site; first as a low-density backfill around the packages to prevent soil arching that can lead to voids around the packages, then as a high-density packed backfill. Once full, the IDF trench will have a surface barrier cap emplaced to minimize physical intrusion and limit infiltrating water recharge. While the specific cap has not been defined to date, it is anticipated to be a RCRA Subtitle C–compliant barrier made up of topsoil, sand, aggregate, and asphalt. The presence of the cap will also ensure that the packages remain at least 5 m below the surface to provide shielding from radioactive material and deter intrusion. The IDF in

**FIGURE 14.5** Aerial view of the IDF and the first two constructed cells (looking south).

its current design (circa 2022) is 82,000 m$^3$, and a proposed disposal capacity is 505,000 m$^3$ that is currently under review. A potential expansion capacity of up to 2,260,000 m$^3$ has also been discussed (Bates 2022). In the 2018 iteration of the IDF PA, the four largest radionuclide inventories for the IDF were Sr-90 (258,000 Ci), Tc-99 (26,400 Ci), and Cs-137 (14,500 Ci; Table 3–29 of USDOE 2018).

The geology of the Hanford Site Central Plateau (where the Hanford 200 Area and IDF are located) is ~150 m of unconsolidated or semi-consolidated sediments that lays on basalt bedrock. The IDF sits 90–100 m above an unconfined aquifer (~70 m from the bottom liner). Between the IDF and the aquifer are poorly sorted pebble-to-cobble gravel and fine-to-coarse grained sand (Hanford formation) and fluvial gravel layers (Ringold Formation; Reidel and Fecht 2005). For performance compliance of the IDF, the point of determination is a 100-m well that is downgradient of the IDF.

### 14.3.3 Regulatory Requirements for the IDF

The disposal of any waste form at the IDF must meet the requirements of DOE Order 435.1 Radioactive Waste Management for wastes for near-surface disposal. The long-term performance requirements for disposal are addressed through a performance assessment (PA) that evaluates the long-term impact of near-surface disposal and provides the quantitative demonstration of compliance with the performance objectives for the long-term protection of the public and the environment. The PA assesses many different scenarios for exposure to the disposed inventory of radiological wastes including waste form release, inadvertent intrusion, and agricultural activities. Several PAs of the IDF have been performed to date with the most recent iteration published in 2018 (Mann et al. 1998, Mann 1999, Mann et al. 2001, USDOE 2018).

The IDF is also permitted as a RCRA facility, and waste acceptance criteria (WAC) are in place (Borlaug 2019). Any immobilized wastes to be disposed must comply with 40 CFR 268 Land Disposal Restrictions. Compliance with LDR requirements can be demonstrated through approved treatment or conducting an acidic extraction test (EPA Method 1311, Toxicity Characteristic Leaching Procedure, TCLP) and ensuring content of hazardous components is below listed levels. Any container must be 90% full, and free liquids must be below 1 vol% of the package. Under the current permit, the IDF can only receive wastes into Cell 1 including WTP ILAW and IDF operational wastes such as immobilized leachates. The approval process is ongoing to expand disposal into Cell 2, to allow for the disposal of secondary wastes from WTP operations such as a grouted waste form and to add a waste storage and waste treatment pad for grout micro- and macroencapsulation.

## 14.4 ENVIRONMENTAL RESTORATION DISPOSAL FACILITY (ERDF)

The IDF is not the only disposal site in operation on the Hanford Site. Wastes from Hanford Site environmental remediation and decommissioning and demolition cleanup activities are disposed at the Environmental Restoration Disposal Facility (ERDF; Figure 14.6). ERDF is an EPA- and Washington State Department of Ecology–authorized disposal facility that is regulated under CERCLA (Benson et al. 2007, Mehta et al. 2013). Construction of ERDF finished in 1996 between the 200-E and 200-W Areas of the Hanford Site. ERDF was critical for facilitating the cleanup of the Hanford River Corridor in the 100 Area and 300 Area (Hawkins 2008). When CERCLA cleanup activities initiated at Hanford in 1992, the contaminated materials recovered were staged at their production location. Once ERDF opened, the wastes were disposed of at ERDF, which continues today. Without access to ERDF, the>7 million tons of contaminated soil demolition debris generated during the River Corridor cleanup effort would have to have been shipped to an off-site disposal facility, which would have increased the duration and complexity of the cleanup mission (Hawkins 2008).

A series of 152 m wide × 152 m long and 21 m deep cells comprise ERDF and have been made in a modular fashion to allow expansion up to 4.1 km$^2$ when required. ERDF has a double-liner system made with a sand operational layer, geotextile separators, gravel drainage layers (at bottom), HDPE geomembranes, and admix layers. This barrier is designed to direct percolating water to a leachate collection system.

**FIGURE 14.6** Environmental Restoration Disposal Facility (ERDF) on the Hanford Site.

The wastes disposed of at ERDF are generally high volumes of slightly contaminated soils from CERCLA activities on the Hanford Site, facility rubble, solids from the Effluent Treatment Facility (ETF) resulting from evaporating the processed brine and metals from reactor parts. Soils are dispersed within the disposal cell and compacted. Under RCRA and the Washington Administrative Code LDR treatment requirements, solid waste that meets the definition of debris can be treated using macroencapsulation with grout and disposed of at ERDF. Macroencapsulation is used to fill void space and capture mobile contaminants or radionuclides. The macroencapsulation approach for ERDF is flood grouting of the trench. Solid items to be macroencapsulated are driven into the disposal trench and placed on concrete blocks or pads to allow the grout to flow around and cover the waste items. After curing, the grouted debris is covered with soil. The resultant waste complies with CFR 268.45, "Treatment Standards for Hazardous Debris." Like the IDF, the characteristics of ERDF that assist with isolation of the waste are its location on the central plateau well above the confined aquifer (~100 m) and away from the Columbia River, the engineered barriers, and the nature of the waste. Following closure, a RCRA-compliant closure cap will be placed onto ERDF. Current projections are that the largest radionuclide inventories in ERDF at closure will be Cs-137 (~400,000 Ci), Sr-90 (~200,000 Ci), H-3 (23,000 Ci), and 60 MT of U (Mehta et al. 2013).

## 14.5 STATE-APPROVED LAND DISPOSAL SITES

For the disposal of decontaminated liquids as a result of waste processing on the Hanford Site, a pair of land disposal facilities exist: the SALDS and the Treated Effluent Disposal Facility (TEDF) located in the 600 Area. SALDS is located north of the 200-W Area, and TEDF is east of the 200-E Area. The ETF at Hanford receives dangerous, radioactive liquid effluents from a variety of site operations and will eventually receive liquid effluents from WTP. After liquid effluents are treated at the ETF, they are staged in verification tanks. The liquid is sampled, and if concentrations meet both the Washington State Water Discharge Permit limits and the delisting petition limits, the liquid is transferred via pipe to SALDS. The liquids are discharged as a non-dangerous, delisted waste. SALDS is a drain field where water is discharged onto the surface. It is projected that SALDS will receive over 600 Mgal of treated effluent in the duration of the Hanford mission, with discharges of up to 600,000 gal/day (not a constant delivery rate; Bernards et al. 2020). If the liquids do not meet the discharge limits, they can be returned for additional treatment at ETF.

Any liquids collected from site operations that do not require treatment to meet disposal limits (i.e., are non-radioactive and not dangerous) are sent to TEDF (Bernards et al. 2020). TEDF is comprised of two man-made disposal ponds that are 2 ha in size. The water will percolate into the subsurface from the disposal ponds (Borgstrom 2004). Waste liquids sent to TEDF include cooling system condensate from buildings, rainwater from paved areas, potable water, boiler blowdowns, and building drains with controlled usage limits (Cree et al. 2019).

## 14.6 SUMMARY

In all, Hanford stands as an example of the needs to define waste handling and managing practices during the development and deployment of new technologies. The significant contribution of Hanford to national security cannot be understated, but the significant cleanup mission left in its wake stands as a constant reminder of the importance of managing waste disposal in any reprocessing effort. When final disposition paths are not put in place prior to process initiation, the associated costs, environmental risks, and complexity associated with the waste disposal can be inflated over time. The need for safe waste disposal from any process must be met upfront while working with appropriate regulators and stakeholders. Despite the complexity and scope of the Hanford waste treatment mission, the disposal pathways are becoming more complete and clearer as the site approaches waste treatment operations.

The disposal sites located on the Hanford Site each meet unique regulatory disposal requirements, heavily driven by the presence of an unconfined aquifer below the site and the proximity to the Columbia River. Waste disposal must ensure the continued safety of these two features. Recent evaluations have considered alternate disposal locations for Hanford wastes, including commercially operated facilities outside of the state (Bates 2022). The immobilized Hanford LAW waste forms are likely to meet the WAC of these sites. Most importantly, many of these alternate sites do not have a pathway to potable water for the disposed contaminant and radionuclide inventory. Placing some Hanford wastes at these sites would meet the overall goal of any waste disposal in isolating the hazards from the public and biosphere.

## NOTE

1 https://www.hanford.gov/page.cfm/TriParty.

## REFERENCES


Bannister R. 2016. *M-091 Transuranic Mixed/Mixed Low-Level Waste Project Management Plan.* HNF-19169 Rev. 18. CH2MHill, Richland, WA.

Barker T, K Ard and B Chamberlain. 2019. *Hanford Tank Side Cesium Removal System: Process Description and Performance Metrics-19613.* WM Symposia, Inc., Tempe, AZ.

Bates WF. 2022. *Follow-on Report of Analysis of Approaches to Supplemental Treatment of Low?Activity Waste at the Hanford Nuclear Reservation. Vol. I.* Savannah River National Laboratory, Aiken, SC .

Benson CH, WH Albright and DP Ray. 2007. *Evaluating Operational Issues at the Environmental Restoration Disposal Facility at Hanford.* US Department of Energy, Office of Environmental Management, Hanford Operations. CRESP: Consortium for Risk Evaluation with Stakeholder Participation

Bernards JK, GA Hersi, TM Hohl, RT Jasper, PD Mahoney, NK Pak, SD Reaksecker, AJ Schubick, EB West, LM Bergmann, GR Golcar, AN Praga, SN Tilanus and TW Crawford. 2020. *River Protection System Plan.* ORP-11242, Rev. 9. Washington River Protection Solutions, Richland, WA.

Borgstrom, MSCCM. 2004. *Final Hanford Site Solid (Radioactive and Hazardous) Waste Program Environmental Impact Statement Richland,* Washington. US Department of Energy Richland Operations Office, Richland, WA.

Borlaug W. 2019. *Waste Acceptance Criteria for the Integrated Disposal Facility.* IDF-00002. CH2MHill, Richland, WA.

Cline MW, AR Tedeschi and AK Yoakum. 1990. *Phosphate/Sulfate Waste Grout Campaign Report.* WHC-SA-0829. Westinghouse Hanford Company, Richland, WA.

Cree LH, CR Kimura, VC Nguyen, SN Randall and BM Tardiff. 2019. *River Protection Project Integrated Flowsheet.* RPP-RPT-58991, Rev. 3. Washington River Protection Solutions, Richland, WA.

Farabee A and A LaBonty. 2020. Lessons learned from Hanford's Purex Tunnel 2 structural stabilization project - 20509. *Conference: WM2020: 46. Annual Waste Management Conference*, Phoenix, AZ, 8–12 March 2020; Other Information: Country of input: France; Available online at: https://www.xcdsystem.com/wmsym/2020/index.html. United States: Medium: X; Size: 33 pages.

Geiger JF, DJ Brown and RE Isaacson. 1977. *Assessment of Hanford Burial Grouns and Interim TRU Storage.* RHO-CD-78. Rockwell Hanford Operations, Richland, WA.

Hawkins AR. 2008. *Regulation at Hanford - A Case Study. Challenges in Radiation Protection and Nuclear Safety Regulation of the Nuclear Legacy*. Springer Netherlands, Dordrecht.

Isaacson RE. 1969. *The Hanford Exploratory Deep Well*. ARH-SA-47. Atlantic Richfiel Hanford Company, Richland, WA.

Keeney RL and D von Winterfeldt. 1996. *Value-Based Performance Measures for Hanford Tank Waste Remedition System (TWRS) Program*. PNNL-10946. Pacific Northwest National Laboratory, Richland, WA.

Kincaid C, J Voogd, J Shade, JH Westsik Jr, G Whyatt, M Freshley, MG Piepho, K Blanchard, K Rhoads and B Lauzon. 1995. *Performance Assessment of Grouted Double Shell Tank Waste Disposal at Hanford*. WHC-SD-WM-EE-004, Rev.1. Westinghouse Hanford Company, Richland, WA.

Luey J, SS Koegler, WL Kuhn, PS Lowery and RG Winkelman. 1992. PNL-8281). Pacific Northwest National Lab (PNNL), Richland, WA.

Mann FM. 1999. *Performance Objectives for the Hanford Immobilized Low-Activity Waste (ILAW) Performance Assessment*. Hanford Site (HNF), Richland, WA.

Mann F. 2003. *Risk Assessment Supporting the Decision on the Initial Selection of Supplemental ILAW Technologies*. RPP-RPT-17675. CH2M Hill, Richland, WA.

Mann FM, RJ Puigh, S Finfrock, EJ Freeman, R Khaleel, DH Bacon, MP Bergeron, BP McGrail, SK Wurstner, K Burgard, WR Root and P LaMont. 2001. *Hanford Immobilized Low-Activity Tank Waste Performance Assessment: 2001 Version*. DOE/ORP-2000-24Rev.0. US Department of Energy Office of River Protection, Richland, WA

Mann FM, RJ Puigh, PD Rittmann, NW Kline, J Voogd, Y Chen, C Eiholzer, C Kincaid, BP McGrail, AH Lu, G Williamson, N Brown and L LaMont. 1998. *Hanford Immobilzied Low-Activity Tank Waste Performance Assessment*. DOE/RL-97-69. U.S. Department of Energy, Richland Operations Office, Richland, WA.

Mehta S, WJ McMahon, R Khaleel, SM Baker, B Sun, C Courbet and N Hassan. 2013. *Performance Assessment for the Environmental Restoration Disposal Facility, Hanford Site, Washington*. WCH-520-Rev 1. Washington Closure Hanford, Richland, WA.

National Academies of Sciences, Engineering and Medicine. 2020. *Final Review of the Study on Supplemental Treatment Approaches of Low-Activity Waste at the Hanford Nuclear Reservation: Review# 4*. National Academies of Sciences, Engineering and Medicine, Washington, DC.

Reidel SP and KR Fecht. 2005. *Geology of the Integrated Disposal Facility Trench*. Pacific Northwest National Laboratory, Richland, WA.

Serne R, R Lokken and L Criscenti. 1992. Characterization of grouted low-level waste to support performance assessment. *Waste Management* 12(2–3): 271–287.

Shord A. 1995. *Tank Waste Remediation System Complex Site Evaluation Report*. WHC-SD-WM-SE-021, Rev. 0. Westinghouse Hanford Company, Richland, WA.

Taylor WJ. 1999. *Contract No. DE-ACO6-99RL14047 - Decision to Change the Immobilized Low-Activity Waste (ILA w) Disposal Baseline to Proceed with the Remote- Handled Trench Alternative*. Letter 99-DPD-066 Department of Energy Office of River Protection, Richland, WA.

USDOE. 1992. *Grout Treatment Facility Dangerous Waste Permit Application*. DOE/RL 88-27 Rev 2 - Vol 1. US DOE Richland Field Office, Richland, WA.

USDOE. 2018. *Performance Assessment for the Integrated Disposal Facility, Hanford Site*. RPP-RPT-59958, Rev. 1. Washington River Protection Solutions, LLC, Richland, WA.

van Beek JE and DD Wodrich. 1990. *Grout Disposal System for Hanford Site Mixed Waste*. WHC-SA-00694. Westinghouse Hanford Company, Richland, WA.

Westsik JH, WL Kuhn, BA Pulsipher and JL Nelson. 1988. Evaluation of strategies for controlling HWVP (Hanford Waste Vitrification Plant) glass by process simulation. *Conference: Spectrum '88: International Topical Meeting on Nuclear and Hazardous Waste Management*, Pasco, WA, 11 September 1988; Other Information: Portions of this document are illegible in microfiche products. United States: Medium: ED; Size: Pages: 7.

Wilhite E. 1987. *Concept Development for Saltstone and Low Level Waste Disposal*. DP-MS-86-110. Du Pont de Nemours (EI) and Co., Savannah River Lab., Aiken, SC.

Wilhite EL, CA Langton, HF Sturm, RL Hooker and ES Occhipinti. 1988. Saltstone processing startup at the Savannah River Plant. *Conference: 10. Annual DOE Low-Level Waste Management Conference*, Denver, CO, 30 August 1988; Other Information: Portions of this document are illegible in microfiche products. United States: Medium: ED; Size: Pages: 18.

# Part V

## Long-Term Stewardship and Future Land Use

Vicky L. Freedman and Nicolas J. Huerta

The challenge of complex site remediation requires answers to important questions on future land use: What are the potential future uses of the site? What level of cleanup is required and achievable to enable those potential uses? Do site owners and stakeholders agree on future uses? These questions have complex social, technological, and regulatory dimensions, but they are crucial to defining an end state for a site where human health and the environment are adequately protected. More concretely, remediation efforts to achieve the targeted end-state site condition must be based on an *anticipated future land use*, as well as a *comprehensive land use plan*. The plan for land use must meet applicable environmental regulations and the needs of a diverse set of stakeholders.

The future use of a site can be a major factor in determining the eventual cleanup costs. Future land use options may vary from conditions requiring resource-intensive physical and institutional controls, like limiting site entry, to a "greenfield" condition with no land use restrictions at all. The expected land use is important in the remediation decision-making process because risk assessment helps determine the extent of cleanup needed to be protective of human health and the environment. At Hanford, for example, stakeholders include the site owner, Tribal Nations, and state and local governments and their regulatory agencies as described in Chapter 2. Additional costs may still be incurred once a site reaches its end-state land-use. Although a site may be remediated to the selected land-use, there may still be a need for long-term waste management and monitoring.

In addition to a site's next intended use, an end-state vision is influenced by practical constraints associated with technology limitations, resource availability, costs, timing, and unintended consequences of remedial actions. Therefore, environmental remediation goals often need to balance risks and costs of contaminant removal versus leaving contamination in place. For example, in regions where contaminant concentrations are high but immobile, the intended site use should be evaluated in conjunction with the risk of remobilization with an aggressive remedy. In contrast, the

DOI: 10.1201/9781003329213-19

technological challenges, cost, and time to reach remedial objectives of cleaning up low levels of contamination distributed over large areas in the subsurface may require different considerations for a site's next intended use.

At complex sites, different land uses may apply to different areas due to a site's proximity to waterways, the nature of contaminants, the best available technologies for treatment, and the extent of surrounding natural habitat. The goal of land use planning is to determine how the land can be used in the future by balancing competing interests (e.g., industrial, agricultural, ecological preservation, cultural, and residential). Decisions on future use are often based on socioeconomic factors. For highly complex contaminated sites such as Hanford, decision-making also needs to consider the extent of contamination, the technologies available for cleanup, and the risks of leaving contamination in place versus the risks associated with a remedial approach. Hence, end-state determination at Hanford has identified the risk of exposure to radiological contaminants and chemicals (DOE/RL 2005) as an important component, with continued federal control being a key component of successfully achieving an end state. Chapter 15 describes the history of identifying end states at Hanford, and key considerations that may serve as lessons learned for both Hanford and other sites.

## REFERENCE

DOE/RL. 2005. *Hanford Site End State Vision.* DOE/RL-2005-57. U.S. Department of Energy, Richland Operations Office, Richland, WA.

# 15 End State
## *Vision for Future Land Use at Hanford*

*Vicky L. Freedman and Nicolas J. Huerta*

### 15.1 HANFORD COMPREHENSIVE LAND-USE PLAN

DOE has long recognized that the determination of future land use is a shared responsibility, with community input being crucial to the process (DOE 1996). To this end, four years after weapons-era plutonium production ceased and the Hanford Site mission shifted to environmental cleanup, DOE published a notice of intent to write the Hanford Remedial Action Environmental Impact Statement (HRA-EIS) to evaluate alternative land uses that were compliant with the Tri-Party Agreement (TPA).

In 1992, a citizens' Working Group, Hanford Future Site Use Work Group, was convened to identify potential future land uses and associated remediation requirements (e.g., unrestricted future access would require greater decontamination than restricted future access) for the Hanford Site (Hanford Future Site Uses Work Group 1992; Abbotts and Weems 2010). The Working Group included representatives from the agricultural industry, environmental groups, and local government. Tribal Nations participated directly in government-to-government deliberations with the DOE and state agencies. The Working Group identified a range of potential future land uses and remediation alternatives (Hanford Future Site Uses Working Group 1992), and TPA agencies committed to using the Working Group's findings as part of the Hanford remediation and land-use decision process.

Discussions on future land use continued over the next six years. General agreement was reached with an industrial end state for the Central Plateau where waste management would occur for the foreseeable future. However, there was widespread disagreement among different Tribal Nations and stakeholder groups for other areas at Hanford. This resulted in groups working independently of one another and identifying a variety of visions for future land uses at Hanford. For example, agricultural interests favored farming, whereas environmental interests favored conservation land uses. Local governments also claimed rights to future land-use planning and favored returning Hanford land to agriculture and residential use (Mercer 2002). The challenge with these disparate land-use plans is that the impact on required cleanup levels and related remedial approaches is vastly different in time frame, cost, and technology development and deployment depending on the type of land use selected.

While DOE continued to solicit stakeholder and Tribal Nation input, DOE was guided by an engineering approach to identify Hanford end states, with a preferred alternative that would result in an "optimal use of site land" based on the "principles of ecosystem management and sustainable development" (DOE/RL 1999). During this time, DOE gathered environmental data (e.g., vegetation, geomorphology, topography, springs), soil and groundwater contaminant data, waste disposal site history, cultural resource site locations, and information on endangered species and their habitats to assist in the objective-based assessment of future land use. Because the primary focus was the protection of a sensitive 51-mile stretch of the free-flowing Columbia River called the Hanford Reach, the U.S. Congress also became involved. Although there was initial disagreement in the

DOI: 10.1201/9781003329213-20

approach to protect the Hanford Reach, Congress eventually established a local commission to manage the Hanford Reach and surrounding lands in 1996. Ultimately, in 2000, President Clinton proclaimed the Hanford Reach a national monument, designating the U.S. Fish and Wildlife Service to manage it.

DOE released the HRA-EIS in 1996 (U.S. Department of Energy 1996), with an emphasis on preservation and light recreational land uses. However, local community groups and Tribal Nations denounced the HRA-EIS. One disagreement was with respect to land ownership, because local governments wanted to control land use for areas of the Hanford Site that DOE no longer needed for its mission. There were also several disagreements with respect to land-use type. Agricultural stakeholders wanted some of the land dedicated to its historical agricultural land use. City planners wanted more residential and industrial land-use designations. Preservationists wanted more restrictions on livestock grazing. Also, Tribal Nations and other stakeholders criticized the study, because it implied incomplete cleanup.

DOE revised the HRA-EIS based on site data and input from the community. DOE released the Final Comprehensive Land-Use Plan Environmental Impact Statement (CLUP) in 1999 (U.S. Department of Energy, Richland Operations Office 1999), with a future land-use plan based on continued federal ownership of the site, input from stakeholders and Tribal Nations, and public comments received during scoping and drafting of the document (U.S. Department of Energy, Richland Operations Office 1999). The principal goals of the CLUP were to describe environmental impacts from remedial actions, identify future land uses, and develop a comprehensive land-use plan to support decision-making over the next 50 years. This was achieved by striking a balance with stakeholder and Tribal Nation interests identified in the development of the HRA-EIS, including (i) continued DOE land-use needs, (ii) preservation of the ecology and cultural sites, and (iii) continued economic development in the local community. The CLUP identified five distinct geographic areas: the Wahluke Slope, the Columbia River Corridor, the Central Plateau, the Fitzner-Eberhardt Arid Lands Ecology Reserve, and all other areas of the site. The CLUP also identified six alternative land-use scenarios across the site. Although stakeholder groups and Tribal Nations had different end-state visions and expectations, shared values resulted in common land uses and protections (Burger et al. 2020), including the protection of the Columbia River and the designation of the Central Plateau as an industrial waste management area.

The DOE issued the corresponding Record of Decision (ROD) in November 1999, selecting the preferred alternative described in the Final EIS, because it provided the best balance among DOE mission needs, local economic development, and the need to protect environmental resources. Remediation of the Hanford Site makes land available for alternative uses, including preservation, conservation, and recreation. The Central Plateau was designated as an industrial use area that would transition to a long-term waste management area after appropriate remedial activities are completed (which would also require other activities such as long-term monitoring, institutional controls). Although the Tribal Nations and DOE agreed upon the treaty-reserved rights for fishing along the Hanford Reach, they disagreed over the rights to hunt, gather plants, and pasture livestock on the Hanford Site (Tribe and Baptiste 2005). This disagreement has not yet been resolved.

## 15.2 RISK-BASED CONSIDERATIONS

Environmental remediation decisions are guided by the reduction of potential exposures to human health and the environment. Site remediation is generally guided by conservative policies that are not always practically achievable within reasonable time frames. End-state determinations that factor in risks provide an approach for prioritizing cleanup actions and cleanup levels.

A risk-based approach can be used to identify interim actions that result in *interim* end states that advance remediation toward a *final* end state. To this end, a phased remedy approach replaces a final solution when the latter goal results in limited progress toward a site's end state. A phased approach, especially for complex site remediation that does not readily offer immediate final solutions, can accelerate the environmental cleanup process. This approach reduces risk incrementally, even if it is initially a short-term risk reduction. For example, groundwater pump-and-treat may be deployed to contain a contaminant plume and prevent further plume expansion. However, it is often unable to remediate to required cleanup levels and is often used as an initial, interim measure. As long as final solutions are not jeopardized by interim remedial actions, then interim end states can be achieved in a stepwise fashion until technology challenges are overcome and a final end state can be achieved.

The amount of time needed to achieve an end state is also an important consideration in its selection. Although prolonged cleanup times are not desirable, additional time allows for taking advantage of natural attenuation processes that reduce risk. For example, at Hanford and other DOE sites, tritium, a short-lived radionuclide (half-life of ~12.5 years), is present in groundwater above regulatory standards. Given that there is currently no technology for remediating tritium, allowing the tritium to decay naturally over several decades to levels meeting remedial action outcomes, with containment and monitoring in place as needed, is a viable approach. Consequently, continued federal control of contaminated sites is a key factor in limiting exposure pathways until a site can reach its next intended use, whether that is an interim or final end state.

## 15.3 RISK-BASED END STATES AT HANFORD

In 2003, DOE headquarters issued a "Use of Risk-Based End States" policy to conduct cleanup at Hanford and other DOE sites that would achieve clearly defined, risk-based end states (RBES) (DOE 2003). Each site within the DOE complex was required to comply with applicable regulatory requirements, with an end-state vision formulated in cooperation with affected governments, Tribal Nations, and stakeholders (U.S. Department of Energy 2003). This policy deviated from historical practices, allowing for potentially different regulatory requirements and remediation decisions at different DOE sites based on the affected governments, Tribal Nations, and stakeholders.

Hanford issued its own risk-based end-state vision (U.S. Department of Energy 2003), identifying remediation targets consistent with the planned future use of the site and using the land-use plans described in the CLUP (1999) to formulate the risk scenarios. Because the 2003 RBES vision document stated that the policy was not a "license to do less," the proposed variances (i.e., deviation from the historical practice) for cleanup implied less cleanup would be performed at Hanford. Moreover, the TPA regulators, Tribal Nations, and stakeholder groups did not concur with the Hanford risk-based end-state vision, citing that risk was only one of nine criteria in the CERCLA process used for remediation decision-making (Abbotts and Weems 2010). Because the draft document was issued without stakeholder involvement, workshops were subsequently held to gather more external contributions to the 2003 RBES vision.

A revised RBES vision document including input from stakeholders and DOE headquarters was released as a final end-state document in 2005 (DOE/RL 2005). The final 2005 vision document not only dropped the word "risk" from its title, but also explicitly stated that risk was not the sole basis for remediation decision-making. The document reinforced that existing RODs would not be changed on the sole basis of risk and could only be changed through the existing regulatory processes that also include extensive community participation. It reaffirmed the TPA commitment to prioritizing and advancing remediation. The primary technical basis for the variances was risk scenarios based on land use (e.g., the use of a preservation land use instead of a more conservative, rural residential scenario). A second technical basis for variances was to leave waste in place, which increased the public perception of a reduced commitment to cleanup. The RBES vision clearly stated that variances would be pursued through existing regulatory processes. Stakeholder input on end states was also more effectively included in the document, providing a stepwise progression toward a final end state. For example, along the River Corridor, preservation and Tribal activities (e.g., fishing) were identified for the next 50 years (i.e., interim end state), with unlimited land use after 50 years (i.e., final end state).

DOE continues to manage Hanford Site land use according to the CLUP (DOE 2009, 2015) and is responsible for working with stakeholder groups and Tribal Nations to provide access and use that is consistent with the CLUP. However, as more data are collected and remediation alternatives identified, land-use designations for areas still under DOE control at the Hanford Site have further narrowed to mostly industrial and conservation uses. This is because any areas within the Hanford Site that have agricultural, research, or recreational designations are to be managed by the U.S. Fish and Wildlife Service – possibly requiring additional decision-making and coordination. Figure 15.1 details the current layout of the Hanford Site; Figure 15.2 shows overlay of the proposed future land use.

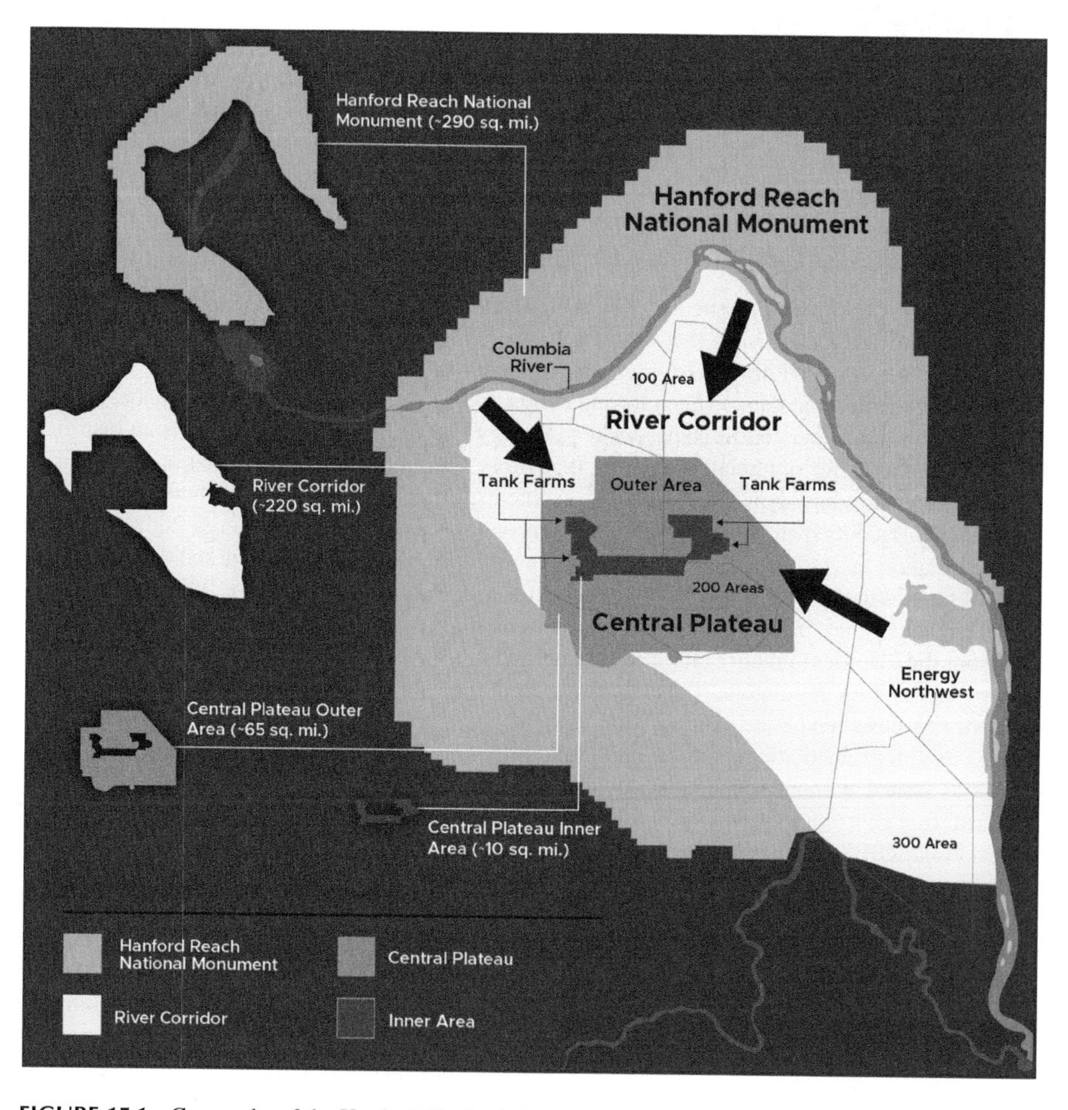


**FIGURE 15.1** Geography of the Hanford Site lends itself to a Cleanup Completion Framework with a geographic approach considering active cleanup footprint reduction moving from the outer areas and toward the Central Plateau consisting of River Corridor remediation, then pivoting to accelerate Central Plateau remediation, and establishing a long-term waste storage mission in the inner area.

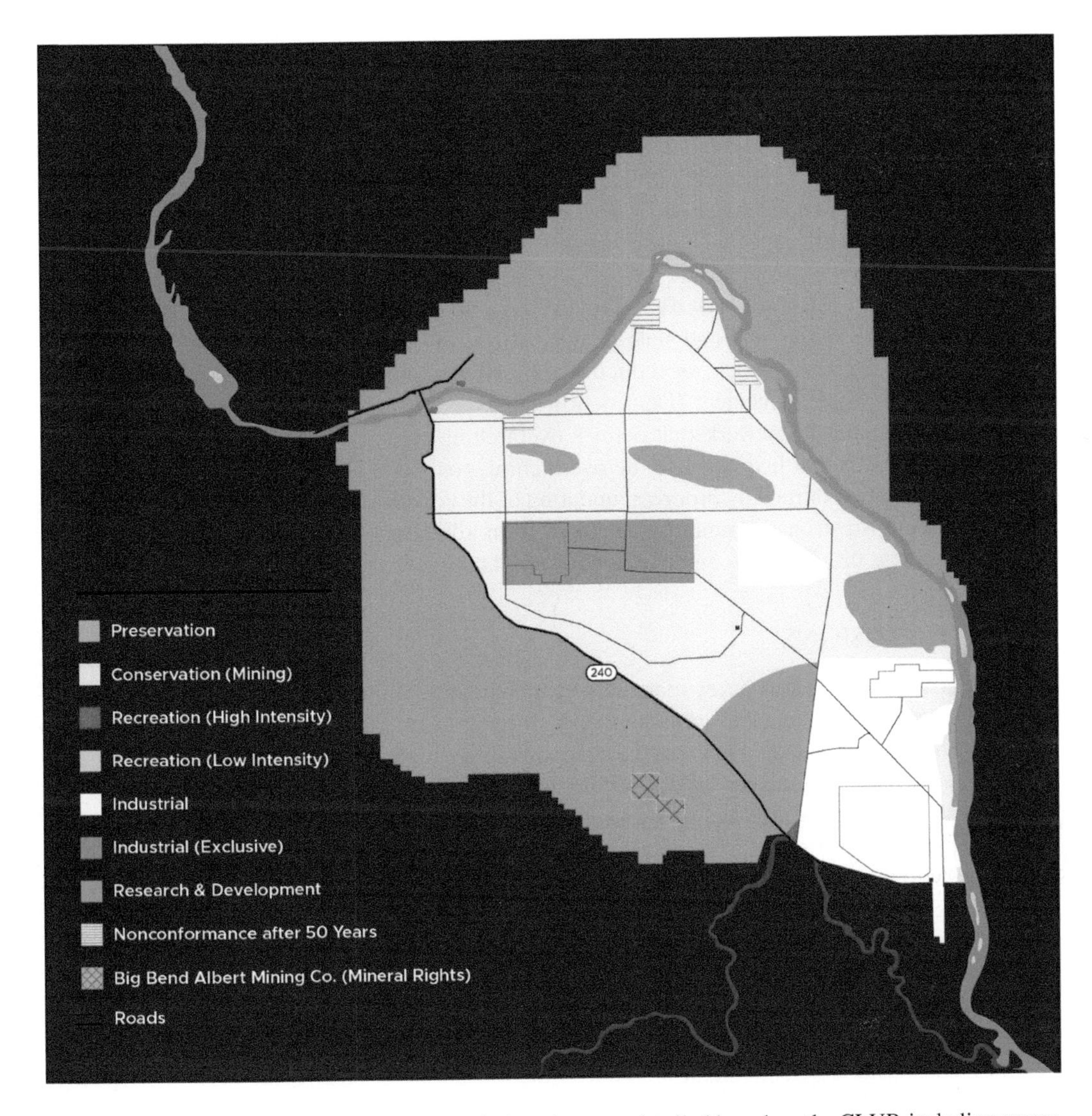


**FIGURE 15.2** Hanford Site future land-use designations are detailed based on the CLUP, including preservation, conservation, recreation, industrial, and research and development.

The land-use designations are defined in the CLUP and summarized here:

- **Preservation:** An area managed for preservation of archeological, cultural, ecological, and natural resources with no new consumptive uses (e.g., mining) and limited public access
- **Conservation (Mining):** An area for protection and management of ecological, geological, archeological, and cultural resources with limited public access, potentially allowing limited mining activities with an approved permit
- **Recreation:** An area for visitor-serving activities including high-intensity (e.g., golf courses or recreational vehicle parks) and low-intensity resources (e.g., trails and permitted campgrounds)
- **Industrial (Exclusive):** An area for treatment, storage, and disposal of hazardous, dangerous, radioactive, and non-radioactive wastes

- **Research and Development:** An area for basic or applied research that requires the use of a large-scale or isolated facility
- **Non-conformance:** An area with an exception based on previous or pre-existing land uses established prior to land-use designation

Future land uses may change over the long time frames anticipated to reach remediation targets (DOE 1999), but uses are expected to be consistent with the CLUP, including land for Tribal Nations use, public recreational use along the river, residential, industrial, and commercial development, and preservation of open spaces (DOE 2009). Although the use of RBES is no longer a DOE policy (DOE 2011), risk-reduction and end-state goals continue to support remediation decision-making at Hanford. A stepwise approach for reducing risk and reaching end-state goals can be achieved through adaptive site management (Demirkanli and Freedman 2021). This approach provides a formal and systematic framework that defines overarching objectives for remediation, supported by interim objectives that help achieve the overarching goals (see Chapter 1). It requires not only continuous evaluation of remedy progress, updating of the conceptual site model as new knowledge is acquired, but also strong community engagement in all phases of the planning, execution, and evaluation process that move site remediation toward its targeted end state.

## 15.4 CONCLUSIONS

Stakeholder engagement was a key element of end-state determination at Hanford. However, the existence of multiple stakeholder groups made it difficult to meet all stakeholder expectations for land use. The Central Plateau Area was the exception to the disputes associated with other areas of the site, reaching concurrence on an industrial land-use end state, though some Tribal Nations still publicly voice disagreement with any action that limits use of even the inner area of the Central Plateau. This agreement stems from a common understanding of the current technology limitations associated with the Central Plateau, the long time frames needed for cleanup, and the distance to potential receptors at the Columbia River. Containment of vadose zone and groundwater contaminants to the Central Plateau Area (to the extent possible) will be a key component of effectively implementing the industrial land-use policy.

Although DOE demonstrated a preference for using risk to inform achievable cleanup standards, stakeholder groups interpreted the use of risk as a justification for cleaning up less. Apart from the Central Plateau, risk was not a factor in determining the site's end state. Given current technology limitations and budget constraints; however, risk *can* inform decision-making by prioritizing site cleanup and *will* impact remedial options available and timelines for cleanup. Certainly, those sites with the highest risk of dose and exposure to potential receptors should be cleaned up first, even if remediation technologies are not available to clean up to current regulatory standards. Interim actions can mitigate risk while advancing land use toward a final end state. This approach is consistent with a phased remedy approach associated with adaptive site management. The overarching objective is protection of human health and the environment, while an interim objective that leads to an interim end state may advance that goal through a targeted concentration reduction or containment within a defined area.

Based on the Hanford experience, DOE has initiated a more systematic, technical basis for establishing risk and end states as demonstrated through land-use identification in the CLUP and ongoing characterization activities that contribute to remedy management. Similar to a phased remedy/adaptive site management approach, land-use identification and risk evaluations that drive cleanup priorities are iterative and often non-linear. Although it can be challenging to maintain clear and open communication in an iterative decision-making framework, transparency is a key tenet of end-state determination.

## REFERENCES

Abbots, J. and C. Weems. 2010. Remediation, land use, and risk at Hanford. *Remediation*, 20(3): 133–149.

Burger, J., M. Gochfeld, D. S. Kosson, K. G. Brown, J. Salisbury, and C. Jeitner. 2020. A paradigm for protecting ecological resources following remediation as a function of future land use designations: a case study for the Department of Energy's Hanford Site. *Environmental Monitoring and Assessment* 192: 1–29.

Demirkanli, D. I., and V. L. Freedman. 2021. *Adaptive Site Management Strategies for the Hanford Central Plateau Groundwater*. No. PNNL-32055; DVZ-RPT-076. Pacific Northwest National Lab (PNNL), Richland, WA (United States).

DOE. 1996. *Draft Hanford Remedial Action Environmental Impact Statement and Comprehensive Land Use Plan*. DOE/EIS-0222D. U.S. Department of Energy, Washington, D.C

DOE. 1999. *Final Hanford Comprehensive Land-use Plan Environmental Impact Statement*. DOE/EIS-0222-F. U.S. Department of Energy, Richland, WA. https://digital.library.unt.edu/ark:/67531/metadc715109/

DOE. 2003. Policy DOE P 455.1. Use of risk-based end states. U.S. Department of Energy, Washington, D.C. https://www.directives.doe.gov/directives-documents/400-series/0455.1-APolicy/@@images/file.

DOE. 2009. *Hanford Site Cleanup Completion Framework*. DOE/RL-2009-10, Rev. 1. U.S. Department of Energy, Richland, WA.

DOE. 2011. DOE DOE N 251.106. U.S. Department of Energy, Washington, D.C. https://www.directives.doe.gov/directives-documents/200-series/0251.106-CNotice/@@images/file.

DOE/RL. 2005. *Hanford Site End State Vision*. DOE/RL-2005-57. U.S. Department of Energy, Richland Operations Office, Richland, WA.

Hanford Future Site Uses Work Group. 1992. *The Future for Hanford: Uses and Cleanup*. Department of Energy, Richland Operations Field Office, Richland, WA.

Mercer, D. 2002. Future-histories of Hanford: The material and semiotic production of a landscape. *Cultural Geographies*, 9(1): 35–67.

Tribe, N. P., and K. L. Baptiste. 2005. Hanford Tribal Stewardship. APA. https://commons.clarku.edu/cgi/viewcontent.cgi?article=1001&context=nez

DOE. 2015. *Supplement Analysis of the Hanford Comprehensive Land-Use Plan Environmental Impact Statement*. DOE/EIS-0222-SA-02 Rev 0. Richland, WA. https://www.energy.gov/nepa/eis-0222-hanford-comprehensive-land-use-plan

# 16 Afterword

*Nikolla P. Qafoku*

The idea to write a new book about legacy complex waste site remediation illustrated with a set of case studies focused on the Hanford Site circulated for the first time right after the establishment of the Center for the Remediation of Complex Sites (RemPlex) in late 2019. This idea was immediately supported by research scientists and the management of Pacific Northwest National Laboratory's Earth Systems Science and Nuclear Science Divisions and the Environmental Management Market Sector.

Over the last four to five decades, there have been many efforts – characterization studies, laboratory and field research, and technology development and deployment – related to environmental remediation at the Hanford Site. These efforts have intensified during the last three decades, with significant progress in cleanup. Moreover, major national research programs sponsored by the U.S. Department of Energy (DOE) have focused on the Hanford Site among other sites. Because the Hanford Site is one of the most complex environmental challenges in the world, the knowledge and experiences gained during decades of site-related research and cleanup, which are summarized in this book, should be helpful to professionals remediating complex sites worldwide.

This is not the first book about Hanford Site remediation, and likely not the last, given the decadal timeframe envisioned for remediation. For example, *Hanford: A Conversation about Nuclear Waste and Cleanup*, by Roy E. Gephart (2003), is a great resource for understanding the history and environmental challenges of the Hanford Site. In this work, we strive to update and expand on Gephart's book and others to communicate the general challenges of remediation and cleanup of a complex site via specific case studies from the ongoing cleanup of the Hanford Site.

The Hanford Site is indeed complex mainly because of its association with the production of nuclear materials during the Manhattan Project and the Cold War. Some background is summarized, especially in the specific case studies representing different phases of the environmental remediation process with references to previous work and with additional details added for the most enthusiastic readers.

Some key aspects of the Hanford Site that contribute to its complexity are as follows:

- **Nuclear Legacy:** The Hanford Site played a pivotal role in the production of the atomic bomb during World War II, and it was one of the three main sites (together with Los Alamos, New Mexico, and Oak Ridge, Tennessee) involved in the Manhattan Project. The site's historical significance and its contributions to the nuclear arms race, specifically as the first industrial-scale plutonium production facility, make it unique.
- **Scale of Production:** The scale of production was unprecedented, with nine nuclear reactors and their associated chemical processing facilities, making it the largest plutonium production complex in the U.S.A. The Hanford Site was also geographically massive, currently encompassing approximately 586 square miles (1,518 $km^2$) of land that could potentially be chemically and radiologically contaminated. The size of the Site, the scale of production, and the sheer number of facilities involved set Hanford apart from other nuclear sites.
- **Environmental Contamination:** The wartime political and project imperatives and the drive to produce plutonium for nuclear weapons at all costs led to significant contamination of buildings, soil, and the subsurface. Radioactive and chemically hazardous waste fluids of different and complex compositions were stored for future treatment or intentionally or unintentionally released into the soil, contaminating the sediments of the thick vadose zone and


DOI: 10.1201/9781003329213-21

even the deeper groundwater. This contamination poses long-term environmental and health risks, making the site unique in terms of the magnitude and complexity of the cleanup efforts required.
- **Scientific Research:** Despite the contamination, the Hanford Site serves as a unique research and development environment. As demonstrated in different chapters of this book, researchers have been able to study the mobility of contaminants, the effects of long-term radioactive and chemical exposure on the local environment, and the effectiveness of waste treatment, decommissioning, and remediation methods. This research contributes to our understanding of the impacts of nuclear activities and helps develop methods for remediation and future risk mitigation.

The cleanup of the Hanford Site is an ongoing and highly complex process. The contaminated material and infrastructure at the site include radioactive waste liquids and solids, spent nuclear fuel, equipment, buildings, soil, and groundwater. Further, the contaminated subsurface spans a depth of more than 200 ft in some locations, and there is a significant seasonal exchange of groundwater and river water. The scale and complexity of the cleanup efforts call for a broad spectrum of advanced technological and remediation solutions, which distinguish it from many other contaminated sites.

The cleanup of the Hanford Site is complex and unique, and some of the challenges underpinning the cleanup effort that were highlighted in the chapters of this book include the following:

- **Radioactive and Hazardous Materials:** The site contains vast amounts of radioactive and hazardous materials resulting from decades of nuclear weapons production. This includes radioactive waste, spent nuclear fuel, and various chemical contaminants. Handling and disposing of these materials safely and effectively require specialized expertise and advanced technologies.
- **Complex Infrastructure:** The Hanford Site was originally designed for production, not for waste management or cleanup. It comprises numerous chemical processing facilities, nuclear reactors, underground storage tanks, and transfer pipelines spread across approximately 586 square miles. Some of the infrastructure has deteriorated or is outdated, which makes accessing and removing the waste challenging.
- **Storage Tanks:** Hanford has 177 underground storage tanks, most of which were constructed between the 1940s and 1960s. These tanks were used to store millions of gallons of radioactive and chemical waste. Over time, some of the tanks have leaked, leading to further contamination of the environment. Tank waste retrieval is complex, and specialized equipment and techniques are employed to safely remove the waste for treatment and proper disposal. Treating and immobilizing the waste stored in these tanks are also complex and delicate processes and are crucial activities at the Hanford Site.
- **Scale:** The large size of the Hanford Site and the large volume of waste that needs to be managed make the cleanup a massive undertaking.
- **Cost:** It is estimated that the cleanup will take many more decades to complete and cost hundreds of billions of dollars. Securing funding for such a long-term and expensive project is a continuous challenge.
- **Stakeholder Involvement:** The Hanford Site cleanup involves multiple stakeholders including city, state, tribal, and federal governments and their agencies, as well as local community and special interest groups. Coordinating and incorporating the diverse interests and perspectives of these stakeholders is complex. However, those leading the cleanup effort must be transparent with the myriad of stakeholders to be successful.
- **Environmental and Health Risks:** The contamination at Hanford poses significant environmental and health risks. Cleanup activities must be conducted in a way that minimizes further releases of hazardous materials and protects the surrounding ecosystem (particularly the Columbia River) and communities. Ensuring worker safety and public health during the cleanup process adds another layer of complexity.

The cleanup of the Hanford Site requires leveraging technologies developed and implemented at other sites (using lessons learned) and developing new technologies and scientific advancements. The waste is often highly radioactive and chemically complex, requiring advanced methods for containment, treatment, and disposal. Developing and implementing these technologies, as well as understanding the long-term impacts of the contamination, present ongoing challenges. Despite these challenges, continuous efforts are being made to address the legacy of the site and mitigate the environmental and health risks associated with its contamination.

Each section of this book covered a step of the environmental remediation process, including a discussion of some of the remediation techniques applied at the Hanford Site to immobilize, degrade, or remove contaminants such as organics, metals, and radionuclides. The Hanford Site has been undergoing extensive remediation efforts to address the environmental impacts and clean up the legacy of nuclear weapons production. Various remediation methods have been employed to mitigate contamination and achieve the end-state vision, such as pump-and-treat to contain and remediate groundwater plumes; soil excavation, removal, and transport to a designated disposal facility; soil flushing to accelerate extraction and transport of contaminants from deep soils to the pump-and-treat system; in situ treatment of contaminants in locations too deep to excavate; and waste treatment and immobilization of highly radioactive wastes – a major cost and complexity – and Hanford due to the magnitude of the facilities needed to process and immobilize this waste to enable long-term, safe disposal.

Each remediation technique has its advantages and limitations with regard to specific site conditions. Often, a combination of techniques, that is, a remediation strategy, is employed to address different aspects of contamination and achieve the desired end states. Site characterization, conceptual site models, feasibility assessments, active remediation, and monitoring play crucial roles in selecting and implementing the most suitable remediation techniques. The overall goal of remediation is to clean up the site safely and effectively, protect human health and the environment, and mitigate the long-term risks associated with the historical nuclear activities at Hanford.

This book's Introduction highlights that a new generation of scientists and engineers has taken on the responsibility of developing and implementing environmental remediation techniques and strategies, carrying forward the mission of cleanup. To reflect this generational shift, many of the Hanford chapters in this book were intentionally authored by early-career engineers and scientists actively engaged in laboratory and field-scale environmental cleanup projects. They are being mentored by professionals in the later stages of their careers. This approach aligns with the assertion that remediating complex waste sites is a multigenerational endeavor. Each generation shoulders the duty of preparing and passing on knowledge and experience to the next, emphasizing the importance of continuity in this ongoing effort.

In the end, I would like to emphasize that the Hanford Site has a variety of similarities with many other complex sites around the globe. For this reason, this book serves as a reference and an information source to many professionals who are working to solve similar complex characterization and remediation problems. I believe this book will be useful to both readers new to environmental remediation and to professionals with an interest in learning about the specific case studies covered for the Hanford Site.

I would like to commend the outstanding work of the book editors, Drs. Stuart T. Arm and Hilary P. Emerson, and the authors who have so excellently described the challenges and successes of Hanford Site cleanup to inform and inspire those who carry on the crucial work of environmental remediation to safeguard human health and the environment around the world.

# Index

Note: **Bold** page numbers refer to tables and *italic* page numbers refer to figures.

For Product Safety Concerns and Information please contact our EU representative GPSR@taylorandfrancis.com Taylor & Francis Verlag GmbH, Kaufingerstraße 24, 80331 München, Germany